OXFORD MEDICAL PUBLICATIONS

Paediatrics

Oxford Core Texts

CLINICAL DERMATOLOGY
ENDOCRINOLOGY
PAEDIATRICS
NEUROLOGY

PAEDIATRICS
Understanding Child Health

Edited by

Tony Waterston
Consultant Paediatrician,
Newcastle City Health Trust

Martin Ward Platt
Consultant Paediatrician,
Royal Victoria Infirmary,
Newcastle Upon Tyne

and

Peter Helms
Professor of Child Health,
University of Aberdeen

Oxford   New York   Tokyo

Oxford University Press

1997

Oxford University Press, Great Clarendon Street, Oxford OX2 6DP

Oxford   New York

Athens   Auckland   Bangkok   Bogota   Bombay   Buenos Aires
Calcutta   Cape Town   Dar es Salaam   Delhi   Florence   Hong Kong
Istanbul   Karachi   Kuala Lumpur   Madras   Madrid   Melbourne
Mexico City   Nairobi   Paris   Singapore   Taipei   Tokyo   Toronto
and associated companies in
Berlin   Ibadan

Oxford is a trade mark of Oxford University Press

Published in the United States
by Oxford University Press Inc., New York

© T. Waterston, M.P. Ward Platt and P. Helms, 1997

A catalogue record for this book is available from the British Library

Library of Congress Cataloging in Publication Data
(Data applied for)

ISBN 0 19 262564 0 (h/b)
0 19 262563 2 (p/b)

Typeset by
EXPO Holdings, Malaysia

Printed in Hong Kong

Preface

Why a new textbook of paediatrics? Because the emphasis in the curriculum is changing, and source texts must reflect this change if they are to be of use to students. Many medical schools have already moved to a new emphasis in medical teaching, and the rest are following. In this new curriculum, the scope of learning is reduced to a 'core', and the style to a model of directed self-learning rather than a passive absorption of facts. The focus is on a problem-based clinical approach; on primary care; on epidemiology and public health, ethics, communication skills, and a recognition of the emotional aspects of illness—all of central importance to child health.

Paediatrics aims to be the source book for the new curriculum. It is presented in a problem based format brought together under broad system headings. Each section commences with a problem and follows through to the diseases or conditions which could cause it. Sources of further information for those students undertaking special study modules or elective/project work are given at the end of each section.

We believe that students need to tackle real situations and learn how to evaluate them in a scientific fashion, rather than trying to learn details about obscure diseases which they may never see. We have given due weight to epidemiology, social paediatrics, prevention, and the increasingly important problems of child abuse and disability. We hope that students will see emotional factors as a strand running through paediatric illnesses. When dealing with paediatric diseases, we have highlighted summary points in boxes and hope that students will find these to be useful *aides memoire* for revision.

As the style of this book is unique we look forward to receiving comments which could be incorporated into a second edition.

Newcastle and Aberdeen   T.W.
March 1997   M.W.P.
         P.H.

How to use this book

Having bought or borrowed this book you need to ensure the maximum return for your investment of time and/or money. It has been designed to help you acquire important factual knowledge as well as how to approach problems as they present in real life. Although there is no substitute for personal experience in this regard, the authors and contributors have attempted to give you, the reader, opportunities to use your own knowledge and experience to solve the problems presented.

If you know little or nothing about the problem posed follow through the solution—if you think you can tackle the problem try to work it out for yourself before reading on. There is plenty of space for your own notes in the generous margins so we hope you will 'customize' the book to your own requirements.

We have supplied extensive cross-referencing so you can follow through the different facets of a particular problem: paediatrics and child health are by their very nature holistic disciplines with a strong interplay between organic disease, growth and development, family styles and socioeconomic conditions.

Summaries. Restate the key features of the problems presented and should prove useful for revision purposes. View them as the 'skeletons' of your body of knowledge.

Additional information is for completeness and may require further reading or discussion with your clinical tutors or teachers.

Further reading for deeper knowledge and exploration is suggested at the end of each chapter. These together with the original articles and reviews in the key journals listed on page 403 should provide a rich resource for special study modules or project work where greater detail may be required. Learning is an interactive process and we have learnt a great deal in preparing this book. We would wish this process to continue and will be pleased to receive your comments and suggestions for future editions.

Contents

Contributors

Ian Auchterlonie
Consultant Paediatrician, Royal Aberdeen Children's Hospital

Michael Bisset
Consultant Paediatrician and Gastroenterologist, Royal Aberdeen Children's Hospital

Philip Booth
Consultant Paediatrician, Neonatal Unit, Aberdeen Maternity Hospital

William Church
Consultant Ophthalmologist, Aberdeen Royal Infirmary

Gaynor Cole
Consultant Paediatric Neurologist, Royal Aberdeen Children's Hospital

John Dean
Consultant Clinical Geneticist, Aberdeen Royal Hospital

James Ferguson
Consultant in Accident and Emergency, Royal Aberdeen Hospital

Leonora Harding
Clinical Psychologist, Royal Aberdeen Children's Hospital

Peter Helms
Professor of Child Health, University of Aberdeen

Stephen Jarvis
Donalt Court Professor of Community Child Health, University of Newcastle Upon Tyne

David Kindley
Consultant Paediatrician and Director, Raeden Centre, Aberdeen

Derek King
Consultant Haematologist/Oncologist, Royal Aberdeen Hospital

Camille Lazaro
Senior Lecturer in Forensic Paediatrics, University of Newcastle Upon Tyne

David Meikle
Consultant ENT Surgeon, Freeman Hospital, Newcastle Upon Tyne

Jean Robson
Associate Specialist in Orthopaedics, Freeman Hospital, Newcastle Upon Tyne

P.J. Smail
Consultant Paediatrician, Royal Aberdeen Children's Hospital

Violet Smith
Head Orthoptist, Royal Aberdeen Children's Hospital

Ross J. Taylor
Senior Lecturer in General Practice, University of Aberdeen

Martin Ward Platt
Cosultant Paediatrician, Royal Victoria Infirmary, Newcastle Upon Tyne

Tony Waterston
Consultant Paediatrician, Newcastle City Health Trust

Marion White
Consultant Dermatologist, Aberdeen Royal Infirmary

George G. Youngson
Consultant Paediatric Surgeon, Aberdeen Royal Hospital

Section A

1

The child in society

Growing up in the UK in the 1990s

Growing up in the UK towards the end of the 20th century is different from what it was 30 years ago and extraordinarily different from 100 years ago. In many ways it is much better—children in our society have physical security, adequate nutrition, good access to health care, and universal education. These improvements are measurable by the changes in child mortality shown in Fig. 1.1. Not only has there been a dramatic fall in child mortality, but the reasons for death are quite different. We should not look back to a 'golden age' when children had their basic needs met yet were not exposed to the pervasive influence of television, 'junk' foods, and drugs: it never existed, or if so it did so for only a tiny minority of children. Children and young people are better off now compared with the past, and compared with most children around the world they are hugely better off (Fig. 1.2).

Yet we could do better: there are children who sleep on the streets; children who are prostitutes; children on drugs; children who are sexually abused; children who are exposed to unacceptable traffic risks; and above all, children who live in poverty—nearly one in three on current estimates (Fig. 1.3). In terms of income our society is as unequal as it has ever been, with more very rich people but also more families 'on the margin' who have great difficulty obtaining their basic needs. All these factors affect health and it is no surprise that children living in poverty are more prone to infections, to undernutrition, to disabilities, to accidents, and to child abuse than those in better off families. At present, the trend is towards greater rather than less inequality.

This chapter looks at child needs, at the major social changes in recent years, and then describes what it is like growing up in the UK at the turn of the 20th century, from the perspective of children from a variety of backgrounds.

Child needs

The needs of children are easy to describe, but not so easy to meet. Children need adequate nutrition, love and security at home, good housing, a safe environment,

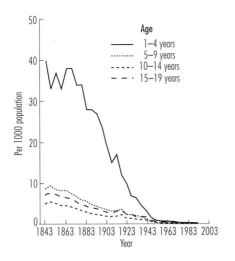

Fig. 1.1. Trend in mortality under 20 years (1841–5 to 1986–90, England and Wales). (From Office of Population Censuses and Surveys.)

Fig. 1.2. Mortality in children under 5 years of age from all causes and from nutritional deficiency and immaturity. (Reproduced with permission from Sanders, D. (1985) *The struggle for health* Macmillan, Basingstoke.)

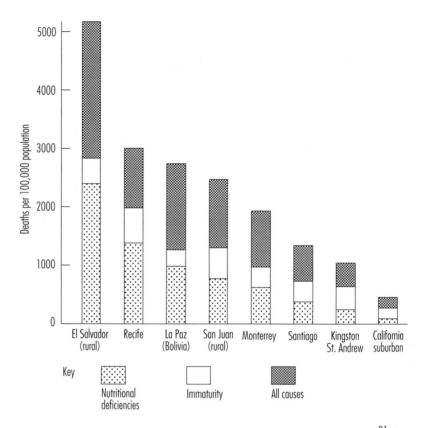

Fig. 1.3. Children and whole population in poverty (living in households below half the average income), (1979–91, Great Britain). (From Oppenheim, C. (1990). *Poverty: the facts.* London, Child Poverty Action Group.)

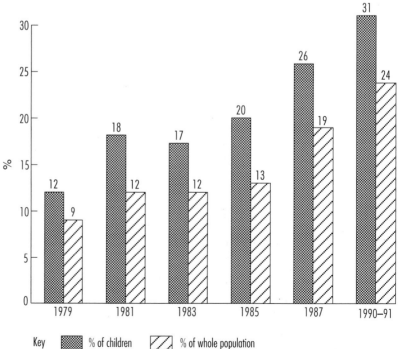

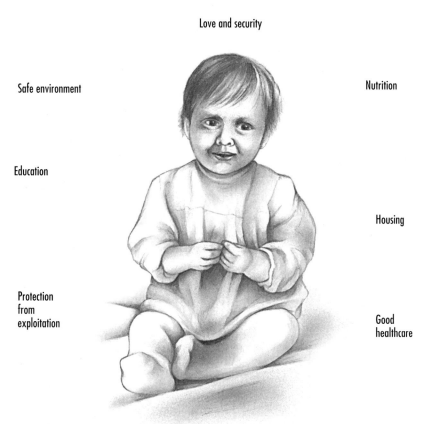

Love and security

Safe environment

Nutrition

Education

Housing

Protection
from
exploitation

Good
healthcare

Fig. 1.4.   A child's needs.

education, access to good health care, and protection from exploitation. The assistance of the state is needed for almost all of these, even the first, as the provision of good nutrition is very difficult for parents on low incomes and without job security. As we have become a richer nation there have been a number of social trends which have made it harder for parents to meet their children's basic needs. These include increasing income inequalities, the growth of the television culture, changes in family structure (particularly more divorce and separation of parents), attitudes to discipline, food marketing, the motor car, pollution of the environment, substance abuse, and racism. Each of these will be dealt with briefly with a mention of their effect on health.

Social trends and effect on health

Income inequalities

Figure 1.3 illustrates the increasing disparities in income in recent years. There are three reasons for these changes: (1) increased unemployment; (2) growth in single parent families; and (3) changes in benefit structure and taxation. There is not space here to deal with all these topics in detail. Suffice it to say that income inequality almost certainly has a marked adverse effect on health for those at the bottom end, even if their overall income is rising. It has been shown that

**Summary
Child needs**

- Adequate nutrition
- Good housing
- Safe environment
- Education
- Good health care
- Protection from exploitation

countries with the greatest social inequalities also have the greatest health inequalities. The reason for this is not fully understood but it has been ascribed to the issue of 'exclusion' from the benefits enjoyed by the better off.

Growing up in a low-income family has many adverse effects on children. Primarily, it affects the family's ability to provide love and security. Family break-up, single parent families, and high stress levels are all commoner in poor families. The parents are less able to provide a healthy diet, the availability of safe place-space is restricted, the opportunities for recreation are reduced, education is likely to be less good (certainly for poor families living in cities), and there will be greater difficulty in accessing health care. In addition, children in poor families will often live in neighbourhoods with a high crime rate, a lack of social cohesion, and a higher risk of exploitation (child labour, child abuse).

Television

The explosion of television culture in the post-war years has had a major impact on child upbringing. Children from all social classes spend many hours in front of the television (1.4 hours per day on average in the UK), and the growth of video games has provided an additional recreation. It is difficult to measure the effects of television and there is much unsupported conjecture. Among the possible adverse effects of increased television watching are:

- increased consumption of fatty and sugary foods due to advertising;
- reduced exercise levels;
- reduced communication within the family;
- increased tendency to violence and increased sexual promiscuity (probably confined to vulnerable children).

However, there are also beneficial effects of television as an educational medium. The task of educators and legislation in the future should be to reduce the adverse effects, while promoting the beneficial effects.

Changes in family structure

As a result of changing gender attitudes and expectations, marital breakdown is now more common than it ever was (40% of first marriages end in divorce) and single-parent families are widespread. Figures and trends are shown in Fig. 1.5. A single-parent family is generally poorer than a two-parent family. Families in conflict are less able to provide the love and security a child needs—emotional stress prevails and emotional, behavioural, and conduct disorders become much commoner in the children.

Attitudes to discipline

It is often suggested that children are out of control because of parents' permissive attitudes, which have been promoted by child-care experts since the war, particularly by the American paediatrician, Dr Benjamin Spock, although subsequent British writers including Jolly, Stanway, and Leach have followed a similar philo-

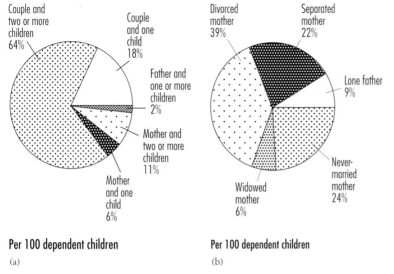

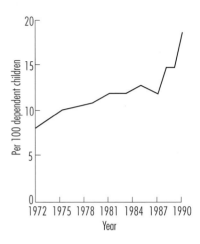

Trend in children in lone parent families 1972–90.

Fig. 1.5. (a) Structure of the family (1990, Great Britain). (b) Children in one-parent families (Great Britain). (From Haskey, J. (1991). OPCS Population Trends 65, 35–43)

sophy. It is true that attitudes to child rearing are altogether different now from what they were pre-war—physical punishment is discouraged, parents are encouraged to listen to their children and ignore minor misdemeanours, the concept of children's rights has been promoted. Much influence has come from television culture and popular magazines, as well as from child-care manuals. However, the use of physical punishment as a form of discipline remains very common. Perhaps some parents feel that without recourse to smacking, they are unable to keep discipline; perhaps parents feel that they need to set wider limits than they themselves were subject to. Many parents do not like saying 'no'. A major factor must be the greater number of broken families with likely inconsistencies in parental control. Penelope Leach, in *Children first*, describes the confusion over what discipline is actually *for*. She calls for parents to give more time to their children—asserting that it is those lacking in attention who are most 'undisciplined'.

Society attitudes to discipline have certainly altered. Whether they are responsible for the present epidemic of behavioural and conduct disorders is uncertain. The increase in poverty, the influence of television, and a consumer-led society may well be just as important. It is essential that parents have adequate information and support at times of stress.

Food marketing

The marketing of food has increased dramatically in recent years and much of this is aimed at children. The reduction in breast-feeding is related to the advertising of infant formula; the massive increase in use of so-called 'junk' foods— soft drinks, sweets, snack bars, and desserts—has followed the wide advertising of these substances on television, hoardings, and in magazines. Cola drinks are

**Additional Information
Prevalence of smacking of
children in the UK (1985)**

- % of mothers who smacked
  their 1-year-old babies   63

- % of mothers who never
  smack their 4-year-old   3

- % of 4-year-olds smacked
  once a day or more   7

- % of mothers of 4-year-olds
  who believed in smacking   83

- % of boys aged 7 who had
  been hit with an implement   26

From Newell, P. (1985). *Children are people too*. Bedford Square Press, London.

Fig. 1.6. (a) A single parent family.

Fig. 1.6. (b) A two-parent family.

Summary
Effects of food marketing

- Increased intake of soft drinks
- Snacking throughout the day
- Increased intake of high-calorie and high-fat foods
- Increased intake of salt (added as a flavour enhancer)
- Increased intake of food additives

now part of youth culture yet are of little value nutritionally, and when given to young children are very harmful to developing dentition. The other foods which are heavily marketed have a high sugar and often animal fat content, frequently contain many additives, and have very limited food value. Many parents also use commercial weaning foods for their infants: these foods are costly and often nutritionally unbalanced, but convenient. Poorly balanced nutrition is common in UK society and is largely the result of the marketing of 'junk' foods. Undernutrition, on the other hand, still occurs in poor families and is largely cost-related. Families on low incomes contribute a higher proportion of their income to food than the better off and therefore food choices are harder to make and economy is all important.

The motor car

The car has had a much greater adverse effect on children's health than is generally realized. Although there are many benefits of car ownership in terms of access to recreational and health facilities and time saved in travelling, the adverse effects are massive—namely, accidents, pollution, and restriction of exercise.

Fig. 1.7. The expansion of food marketing has led to an increased consumption of 'junk' foods, particularly in low-income families.

The major cause of death in children is accidents. The most serious type of accident is a motor vehicle accident and children are extremely vulnerable victims (see Fig. 1.8). Children may be victims as pedestrians, as cyclists, or as passengers, and none of these is easy to prevent without either a restriction of the number of vehicles on the road or a separation of vehicles from pedestrians. Both these solutions should become political imperatives (see Chapter 2).

Cars cause pollution through the emission of carbon monoxide which causes headaches and also cardiovascular effects, and through the emission of nitrogen oxides, which are respiratory irritants, and also contribute to the formation of ozone, a harmful secondary pullutant. Particulates from diesel engines are also implicated in respiratory morbidity. Cars contribute to global energy depletion through their total reliance on a non-renewable energy resource. They are also a major contributor to global warming through carbon dioxide emission. Another immediate effect of pollution on children is lead poisoning, if leaded petrol is used. In 1991, only 40% of UK petrol sales were unleaded, showing that there is still a high potential risk of lead ingestion from this source.

As the numbers of cars on the road increase, so parents are less willing to allow their children to travel independently. This restricts children's mobility and the amount of regular exercise they indulge in. Since 1971, children's independent mobility has decreased considerably. This reduction in regular exercise is likely to have long-lasting effects on cardiovascular function and also contributes to obesity.

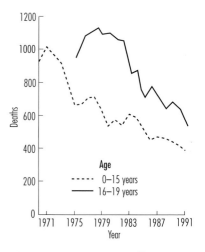

Fig. 1.8. Deaths in road traffic accidents (1970–91, Great Britain). (From Department of Transport; 1992).

Pollution of the environment

Pollution is widely perceived to have increased, but its effects are insidious. With small increases each year it is difficult for an individual to recognize or monitor

Additional Information
Children's independent mobility

% of 7- to 8-year-olds travelling alone to school	
1971	80
1990	9
% of journeys by bicycle (11- to 15-year-old girls)	
UK	4
The Netherlands	60

Summary
Effects of the motor car

- Accidents
- Emissions causing respiratory illness
- Restriction in mobility and exercise
- Global warming

Fig. 1.9. Teenage smoking leads to disease in later life.

the changes. Many of the effects are on plants and the soil cycle rather than directly on humans. Table 1.1 shows the main types of pollution thought to affect children.

Drug and substance abuse
Smoking

Smoking is still very common in children and is not showing the same reduction as in adults. A quarter of 15-year-olds and a third of 18-year-olds are smokers, and smoking is increasing among young women. Only a third of 15-year-olds have never smoked. In addition to the direct effects of tobacco on children who smoke themselves, many are harmed by their parents' smoking. Children of smoking parents are more subject to respiratory complaints and also twice as likely to smoke themselves as children of non-smoking parents. Young women have been specifically targeted by the tobacco industry and the cigarette advertising expenditure in women's magazines in 1988 was £9.7 million.

In order to maintain sales of cigarettes at their present level, the tobacco industry needs to recruit 300 new teenage smokers a day in the UK to replace older customers who have died.

Drinking

In 1989, a tenth of teenage drivers killed on the roads and a fifth of drivers of all ages killed had alcohol levels over 80 mg/100 ml of blood. Heavy drinking by the parents affects the life and health of children: 30% of married men living with a

Table 1.1.
Sources of air pollution

Indoor		Outdoor	
Source	Pollutant	Source	Pollutant
Cigarettes, cigars	Tobacco smoke	Motor vehicles	Lead
Heating & cooking	Carbon monoxide		Nitrogen oxide
appliances	Nitrogen dioxide		Hydrocarbons
			Carbon monoxide
			Carbon dioxide
Building products	Formaldehyde		
	Volatile organic		
	compounds		
	Lead	Coal	Ozone
Subsoil	Radon, methane	Power stations	Sulphur dioxide
		nuclear	radiation
		Agriculture	Pesticides
		Industry	Toxic chemicals (e.g. dioxin)

child under 5 drink heavily (over 21 units/week); 20% of 11- to 15-year-olds had a drink in the last week and 60% of 16- to 17-year-olds; 13% of 16- to 17-year-old young men are said to be heavy drinkers.

Drugs

In a survey of 15-year-olds in England and Wales, 15% of girls and 18% of boys had been offered, at least once, LSD, heroin, Ecstacy, cocaine, or crack ('hard' drugs), while 19% of girls and 20% of boys had been offered either cannabis, amphetamines, or tranquillizers ('soft' drugs).

Substance abuse deaths are rising—60 per year in 1984 and 100 per year in 1990. Substances commonly sniffed or inhaled include glue, aerosols, nail varnish remover, fire extinguishers, cigarette lighter refills, and typewriter correcting fluid.

Risk-taking behaviour in young people is extremely common and perhaps inevitable. There is an onus on schools to ensure that children are adequately supplied with information on the nature of the risk, and on parents to set a good example and provide a supportive but firm environment over the teenage years.

Racism

Racism, meaning the assumption of superiority/power of one ethnic or religious group over another, is pervasive throughout our society and has a significant effect on mental health in the racial groups who are stigmatized. Racism particularly affects those of different skin colour but also those of different religious persuasion from the majority—especially Jews. Racism is present in the health service and is apparent in the disparaging or neglect of culture and values around naming, diet, and childbirth. It is important to understand and respect the traditions of other cultures and to recognize the difficulties faced by those whose first language is not English.

It is apparent that growing up in the UK in the 1990s is by no means easy or necessarily pleasant, particularly for those excluded from the major benefits of our society. The following case studies are based on families known to the author (their main characteristics altered to ensure anonymity). Remember that for most children, growing up is still a wonderful experience—our task is to ensure it becomes so for all.

Summary
Drug and substance abuse

- A quarter of 15-year-olds and a third of 18-year-olds are smokers

- The tobacco industry must recruit 300 new teenage smokers every day to replace those who have died from smoking

- 20% of 11- to 15-year-olds and 60% of 16- to 17-year-olds had a drink in the last week

- 15% of girls and 18% of boys aged 15 years have been offered at, least once, LSD, heroin, Ecstacy, cocaine, or crack

- There were 100 deaths from substance abuse in 1990

Case studies
Low income

James (age 4) is one of four children living in an inner city housing estate in the North of England. He lives with his parents and brothers and sisters—June is 7,

Jacquie 5, and Ron 2. His father used to be a welder but has been unemployed for 6 years and doesn't have much hope of getting a job. (The statistics of the estate where he lives are shown in Table 1.2.) The family's income is below half the national mean household income (the poverty line as defined in the European Union). They live in a council house with three bedrooms—the two younger children and the two older ones share bedrooms. The house is damp and expensive to heat—the family keep only one room heated and the bedrooms have moisture and fungus on the walls. They have no telephone and no car. All the children have health problems. June has asthma, Jacquie has asthma and eczema, James has recurrent coughs and colds and grommets in his ears, and his speech is delayed, Ron has discharging ears and a squint. They have all had to attend hospital on various occasions. June and Jacquie are both on inhalers. James has to attend ear, nose, and throat out patients for his grommets and Ron attends another hospital for his squint which will need an operation.

James is due to start school next year but his behaviour is difficult and his speech is delayed—the words are difficult to understand and his health visitor thinks he won't cope in reception class. He is overactive and his parents find it difficult to control him. He has attended hospital for his behaviour and the parents were told he needs a special diet avoiding additives. They don't think they can afford this. He should have speech therapy but this can only be provided at a centre which is two bus rides away.

Financial factors are paramount for this family. The costs of hospital visits are high and an additive-free diet is more expensive than an ordinary diet, unless the

Table 1.2.
Effects of low income (inner city housing estate)

	Estate	City average
Unemployed	50%	27%
Council houses	81%	40%
Use of car	9%	38%
Access to telephone	55%	81%
Children in one-parent families	35%	20%
Children with height below 3rd centile at school entry	4.5%	2.5%
Immunization uptake at 1 year	70%	90%
Accidents in children	2 × city mean	
Child abuse notifications	7 × city mean	

family is able to shop around, which is difficult if you don't have a car. 'Healthy' food (i.e. a properly balanced diet) is more expensive than a high-sugar, high-fat, low-fibre diet, without much fruit and vegetables. It is important that health professionals recognize these constraints and do not make hasty judgements based on their own life experiences.

The family are referred to a community paediatrician who works mainly outside hospital and has close contact with social services, preschool teachers, and therapy services. The paediatrician recognizes that James is a 'child in need' in terms of the Children Act (1989) and hence eligible for special assistance. (See the additional information boxes at the end of the chapter.)

This family is caught in the 'poverty trap', which leads to low morale and stress as well as a susceptibility to illness both physical and mental. The services can help if there is working together between health, social services, and education: these services need to be mobilized as families often find it hard to make their way through the system. The social services have a lead role in providing assistance but there is a key role for health in identifying 'children in need'.

Social services are consulted so that financial support can be arranged under Section 17 of the Children Act. A community dietician visits the family at home to give advice on healthy eating on a low income, and the whole family agree to modify their diet to include fewer additives. With the assistance of a social worker, a day nursery place is found for James and speech therapy is provided in the nursery. An arrangement is made with the hospital that further follow-up will take place in the GP surgery. Advice is sought on the family's entitlement to benefits, and they are able to obtain help with heating bills. Closer attention is paid to the control of the other children's asthma and it is found that their inhaler technique is poor. Careful instruction is given by the practice nurse.

James' behaviour and speech improve considerably in the nursery and he is able to enter the normal reception class at primary school. His parents feel more able to cope and take on a behavioural management regimen with the support of a psychologist.

The family's problems were eased by a team approach with co-operation between different agencies. However, there are structural problems of unemployment, low income, and poor housing which must be tackled to prevent such situations recurring.

Additional Information
Benefits for children and families

- *Non-contributory*
  Child benefit
  Disability living allowance
- *Means-tested*
  Income support
  Housing benefit
  Family credit
  Social fund
  Community charge benefit

Ethnic minority

Zafir (age 6) is the second child of three and has an elder brother of 8 and a sister of 3. His father came to England 10 years ago from Bangladesh and his wife followed after his brother's birth 8 years ago. They now live in Yorkshire and Zafir's father works in a shop. His mother helps there too when she can. The father speaks good English but the mother has no English and feels very isolated. The father works long hours, but does not wish her to mix with local women. The family has been subject to racial abuse recently and hence the mother is further discouraged from going out.

Zafir is having difficulties at school. He is slow at learning English and does not speak much. He is not yet able to write the alphabet and does not seem interested. His teachers feel he is tired. He is expected to help in the shop after school. An appointment was made to see the school nurse but the parents did not attend—it was during shop hours, and Zafir's mother was reluctant to attend on her own.

It is important to exclude organic problems. Possible causes of school failure include deafness, poor vision, anaemia, migraine, and bullying. It is important that an interpreter is used and this should not be a family member as confidentiality is then removed.

The nurse talks to Zafir after obtaining the parents' permission, using a Bengali interpreter. She suspects that he has poor hearing, and is being bullied.

Further testing reveals that Zafir's hearing is poor due to glue ears (see Ch 12, p. 226). He admits that he has been bullied by two boys in the same class and this makes him very anxious. Also, he is tired because he works late each night in the shop. The nurse talks to the head, who agrees to take action on the bullying and develop an anti-bullying policy in the school.

Communication with Asian families is not easy for health workers who do not understand the language and culture of the parents. A translation service is the first priority: siblings or relatives should not be used as translators. Facilities for women to learn English and develop skills should be provided in a neutral environment, recognizing men's sensitivities as well as women's rights. Racial-awareness teaching in the schools and community will help in combatting the widespread racism in British society.

Summary
Health care in ethnic minority families

- Use an interpreter who is not a family member when English is poor
- Understand gender roles in Muslim families
- Families are frequently exposed to racist harrassment
- Respect local culture and traditions
- Understand naming systems used in different ethnic groups

The school nurse explains to the parents, on a home visit with an interpreter, about the nature of Zafir's problems and his need for a good night's sleep. It is agreed that he will work in the shop for an hour only after school and at weekends. Zafir's mother is encouraged to learn English and join a local group of other Bangladeshi mothers.

Disability

Mary (age 4) is a younger child of two who was found to be poorly sighted at the age of 3 months. She has a congenital disorder which is non-progressive and which allows her to distinguish light from dark but she cannot make out shapes. She is otherwise healthy. She attends the ophthalmologist and has a special teacher.

Her father is unemployed and the family live in a council house in an inner city housing estate in London. She obtained a place at a special nursery for poorly sighted children when she was 3 but has been unable to take it up because of transport difficulties. She is on a morning-only place at nursery finishing at 12 noon (the staff feel she could not cope yet with a full-time place), but the local authority will only provide transport at 8.30 a.m. and 3.30 p.m. She is becoming difficult to handle at home as she is an active girl; her mother does not find it easy to help her to learn. The family cannot afford the daily taxi fare.

Mary is under the statutory school age and there is no obligation for the education authority to provide transport. A school bus runs at the beginning and end of the day but funding is not available for transport at other times. Under the Children Act (1989), Mary is classed as a 'child in need' and hence is eligible for section 17 funding.

The GP appeals to the education authority and also requests the social services department to provide financial assistance. The former say they have no funding and social services say it is an educational responsibility. While negotiations proceed, the family went to the press and their story appeared on the front page of the local newspaper. As a result, a taxi firm offered to provide transport free. After the first term, social services took over the funding of transport until Mary was of statutory school age.

Sometimes, a health worker takes on the role of 'advocate' for a child or family when they seem to be stuck in the system. This should happen only rarely as it is time-consuming and not always equitable. However, it is important that doctors working with children do stand up for their interests, and help to make the services more child-orientated.

Malnutrition

Joseph (age 3) is the fourth of five children and lives with his parents in a small village in southern Uganda (see Table 1.3 for statistics in Uganda). His father is a farmer who has recently obtained a job as a gardener in the city 20 miles away to obtain extra income, and only returns home at weekends. His mother now works in the fields as well as looking after the children. Joseph suffered measles recently—he was not immunized because his mother had no time to take him to the monthly clinic as she was working in the fields. Since then he won't eat the staple food, the bulky maize porridge, and his mother can't afford the oil and groundnuts that are recommended. He is becoming thin and his legs are puffy. She now has to pay to attend the local clinic because of 'structural adjustment policies' as the Ugandan government has introduced user charges in the health service. She has made the 20 mile journey to the nearest hospital where children are still seen free, and after a two hour wait sees the outpatient doctor.

Malnutrition in children is widespread around the world and is a major cause of mortality and morbidity. Its causes are multiple. Socioeconomic factors are just as important as physical ones. Infectious disease if often a precipitant of acute malnutrition, which may present as *Kwashiorkor*: the form of malnutrition in which oedema is a main feature. 'Kwashiorkor' is a Ghanaian word meaning the illness of the displaced child—displaced off the breast by a new baby. The differences

Additional Information
Structural adjustment

As a condition of giving loans to poor countries, the World Bank and International Monetary Fund requires that certain policies are implemented by the country concerned. These are called *Structural Adjustment Policies* (SAPs) and require:

- Reduced public funding on services (including health & education)
- The introduction of user charges (i.e. people should pay for health care)
- Privatization
- Restriction of imports

SAPs have been criticized by many institutions, including UNICEF, because of their adverse effects on the poor.

Additional Information
Main causes of malnutrition

- Poverty and food costs
- Infectious disease (measles, diarrhoeal disease)
- Staple foods are very bulky (e.g. maize meal)
- Food shortages (especially during drought)
- Urbanization (i.e. entry into the cash economy)
- Lack of education of parents

Table 1.3.
Statistics: Uganda and the UK (1992)

	Uganda	UK
Population (million)	18.7	57.7
Population growth rate	2.9%	0.2%
Mortality (under age 5/1000)	185	9
Rural population with access to safe water	30%	–
Adult literacy	Male 62%	
	Female 35%	

between kwashiorkor and *marasmus* (malnutrition characterized by wasting) are shown in the additional information box. In Uganda, poverty and hence malnutrition are heightened by the debt crisis. Uganda is repaying more to rich northern countries in interest repayments than it receives in the form of aid. Uganda has had to seek loans from the World Bank to repay its debt. A condition of these loans is the *Structural Adjustment Policy*, which can have an adverse effect on child health.

The doctor makes a diagnosis of early kwashiorkor together with a respiratory infection which could be tuberculosis. He asks for the child to be admitted but the mother is not keen because of her other children at home. The hospital has recently commenced an outreach nutrition programme and the mother agrees to join this. Investigations are carried out for tuberculosis which are found to be negative.

Treatment of malnutrition is relatively straightforward and relies on the control of infection and the provision of a high-calorie diet which may need to be given by nasogastric tube as the child is often unwilling to eat. Complications during the recovery phase are, however, quite common, in the form of diarrhoea or metabolic problems. The major difficulty in rehabilitation is relapse on returning home, because the circumstances which lead to the malnutrition in the first place have not changed. Hence, it is important to develop outreach or community-based programmes so that the child is treated in the home environment. Food supplements are provided and help is given to the mother to enable her to become self-sufficient in the home setting. This is a question not just of education but also of politics, as the women may have no say in what food is grown or who owns the land: vital issues in the prevention of malnutrition.

Joseph improves slowly on a diet of maize porridge, but supplemented with oil and pounded groundnuts. His mother joins a group of local women who have started a food production co-operative. They are growing food suitable for the local community to eat and are using this as a source of income. They plan to request the local administration to set up a health clinic in the village, staffed by trained village health workers, which will provide basic treatment and immunization free.

In many poor developing countries there are many difficulties that face a local community trying to make life healthier for their children. Much depends on policies in the rich northern countries, particularly in trade and aid. Unless policies on loans are made more equitable, and overconsumption in northern countries is reduced, disparities in mortality will continue to increase.

Additional Information

Kwashiorkor
Semi-acute onset
Oedema present
Child apathetic, poor appetite
Metabolic upset (hypoglycaemia, hypocalcaemia, hypothermia)

Marasmus
Chronic onset
Marked wasting, no oedema
Child alert, hungry
Normal metabolism

There may be very little difference in the diets of children suffering from kwashiorkor and marasmus

Services for children

In the UK, services for children are the responsibility of the Local Authority and the Health Authority. The *Local Authority* is under the democratic control of local councillors elected each year at the local elections. Local authority departments employ professional administrators who are accountable to a committee whose chairperson is a councillor (e.g. the education committee or the social services committee). Attempts are being made to improve liaison between the different departments that provide services for children and between the local and health authorities, but these are imperfect as yet.

The social services

Several services for children and their families are provided by social services departments. However, these have been cut back considerably in recent years owing to rate-capping. Welfare benefits are not provided by social services but by social security, which is administered locally but is not part of the Local Authority.

Many parents need assistance to improve their uptake of benefit as many families are not fully aware of their entitlements.

Education department

The Local Authority education department is responsible for the provision of education. Since the introduction of local management of schools, decision-making has been devolved to the headteacher and governing body of the individual school. The main function of schools is to deliver the national curriculum. However, the education department provides additional *support services*.

The main points of contact between the health and education departments are in relation to health education, to the care of children at school with illnesses, and to the management of children with disabilities.

The health authority

The Health Authority differs from the Local Authority in its absence of democratic control. The chief executive and the chairperson of the health authority are both appointed by the government. The Health Authority is the *purchaser* of services *provided* by hospital or community-based NHS Trusts, which employ professional and support staff who are in direct contact with patients (Table 1.4). The function of the purchaser is to assess the health needs of the population, determine priorities in service provision, and set contracts with provider trusts to ensure that appropriate services are delivered. This system is not yet working effectively as it is difficult to determine priorities between cure and prevention and between hospital and community services.

The 'trust' system was instituted with the 1990 National Health Service reforms. A trust has a high level of financial control: it can set its own terms of service for

Additional Information
Social services: provision for children and families

- Adoption and fostering
- Day care
- Residential care
- Home helps
- Child protection
- Welfare rights advice
- Individual social worker advice support
- Respite care

Additional Information
Education department support services

- Careers service
- Curriculum support service
- Education welfare service
- Educational psychology service
- Special educational needs teaching

Table 1.4.
Structure of the local health system

Purchaser/Commissioner	Provider
Health Authority ←Contract→ GP fund-holding practices	Units or trusts (e.g. hospital trust, community trust, mental health trust)
Purchases health services	Provides health services
Determines health needs of population	Employs doctors, nurses, therapists, laboratory staff, etc.
Funded by Department of Health according to size of population	
Public health departments are normally sited at the Health Authority—all other doctors are normally in provider units. GP practices are normally providers but fundholding practices are also purchasers.	

employed staff; and it can 'compete' with other trusts for the provision of health care at a more 'economic rate'.

Staff in the health service working with children

The following groups of staff work entirely or mainly with children:

Medical/Dental	*Nursing*	*Others*
Paediatricians	Paediatric nurses	Clinical psychologists
Child psychiatrists	Health visitors	Speech therapists
Paediatric surgeons	School nurses	Physiotherapists
Children's dentists	Dental nurses	Occupational therapists

Child health surveillance

The main service for children in the community provided by the Health Authority is child health surveillance. This is a programme offered both by GPs and (usually) community trusts and is the programme of preventive activities provided by doctors and health visitors (see Chapter 4 for a detailed discussion).

Legislation for children

The main legislation concerning children and health in the UK is contained in the Children Act (England and Wales 1989, Scotland 1995), the Education Act (1993), and the UN Convention on the Rights of the Child (1989, ratified by the UK 1991). The essential components of these Acts are summarized in the additional information boxes.

Additional Information
Health needs assessment requirements

- Epidemiological data (e.g. mortality, morbidity rates)
- Service data (e.g. immunization uptake, child admissions rate)
- Socioeconomic data (e.g. deprivation index)
- Behaviour and related data (e.g. alcohol intake, smoking)
- Competency and cultural data (e.g. language, self-esteem)
- Community perceptions (e.g. people's views of health problems)

Additional Information
Children Act (1989)

- Defines parental responsibilities
- Gives paramount consideration to welfare of child
- States that the child's wishes and feelings be respected
- Defines responsibilities for 'children in need'
- Prescribes register for children with disabilities
- Describes court orders in relation to custody and access, to education, to child protection

Conclusions

Many of the health problems faced by children have social origins. Doctors need to be aware of social factors in disease as well as how they can influence them, through data collection, working with staff from other disciplines, or advocacy. These aspects are discussed further in Chapter 4.

Further reading

Leach, Penelope (1994). *Children first*. Michael Joseph, London.

Spencer, N. (1996). *Poverty and child health*. Radcliffe Medical Press, Oxford.

UNICEF. *State of the world's children*. Oxford University Press. (Published annually).

Woodroffe C., Glickman M., Barker M., and Power C. (1993). *Children, teenagers and health—the key data*. Open University Press, Milton Keynes.

2

Setting priorities: epidemiology in child health

Setting priorities in health care for populations is analogous to the process of making a diagnosis and prescribing medical treatments for individuals. Problems concerning the health of populations present in an undifferentiated way and require investigation and refinement (diagnosis). Epidemiology is the use of medical statistics to study the distribution and determinants of health in populations. It is the principal diagnostic tool for the planning and evaluation of health services. In the practice of population medicine it is also essential to have access to a formulary of effective 'treatments' which can be prescribed once the diagnosis has been made. These health care interventions for populations may well correspond to aggregates of the same medical or surgical treatments offered by clinical services to individuals. However, the 'prescribing' of health services for populations has several extra dimensions.

1. It is often 'marginal' in the sense that *unmet* need for effective health services is the principal object of diagnosis and treatment.
2. It is 'comprehensive', in that an attempt is usually made to examine the appropriate package of services from primary care through to recovery/rehabilitation (and including prevention of the original event if possible).
3. It is 'balanced' in so far as some judgement may have to be made between the competing demands for the same resources.
4. The treatment for a population health problem may involve health care programmes managed at a population level (e.g. immunization, screening, contact tracing) or delivered through environmental changes (either in the physical environment—sanitation, road safety—or in the social environment—health promotion campaigns, community development, welfare rights).

Summary
Health services for populations

- *Marginal*: unmet need is the principle objective

- *Comprehensive*: cover primary, secondary, tertiary care, and prevention as well as cure

- *Balanced*: judgement must be made between competing demands for resources

- *Solutions* come from outside as well as from within the health services

Definitions

An epidemiological study requires the collecting of data about the health status of groups of people (population). This involves material from those who clearly have the condition/diagnosis of interest (the *numerator*) together with an account of the whole group of people from whom these 'cases' are drawn (the *denominator*). The numerator and denominator are combined in a *rate* (e.g. the mortality rate for UK children aged 10–14 years was about 0.2 per 1000 in 1990, based on 680 deaths from among 3.5 million children alive at that age).

This rate is an '*incidence*' rate as it describes the occurrences of health-related events during a fixed period of time. However, for some types of health data where the condition lasts for a considerable period, the more usual measure is a '*prevalence*' rate. For instance, a survey of one thousand 10 to 14-year-olds might reveal that 20% of them claim to have smoked at least one cigarette during the previous week (period prevalence) and about 15% of them have elevated salivary cotinine levels, indicating very recent tobacco smoke exposure (point prevalence).

You might insist that smoking is not a health-related 'event' in the same sense as death. Smoking does, however, illustrate an important category of health data,

Summary
Definitions

Numerator
Total number of cases

Denominator
Total population from which the cases are drawn

Incidence
Risk of acquiring a disease or a characteristic thereof

Prevalence
Risk of having the disease or a characteristic at any given point or within a period of time

Risk factor
Factor that confers an increased risk of a disease or condition (e.g. smoking is a risk factor for lung cancer)

Morbidity
Population rate of a non-fatal condition

that is, 'risk factors' for future disease. If the strength of causal evidence is such that this property of the individual confers a definite increased probability of future ill health, then it is of considerable interest. For instance, it may be used as a measure of 'outcome' of health service interventions in just the same way as reduction in mortality or morbidity.

'Morbidity', in an epidemiological sense, is usually measured as the population rate of a defined non-fatal condition, either acute or chronic. The boundary between acute and chronic morbidity is poorly defined, but is most easily captured by saying that acute morbidity is measured by incidence rates, whereas chronic morbidity is measured as prevalence.

For the purposes of illustration, the following four case studies are drawn from across the spectrum of health problems for children. First, there is health promotion (i.e. a service aimed at the primary prevention of disease by altering risk factors). The second concerns screening (a service directed at detection of disease in its early, preclinical stages—sometimes called secondary prevention). The third is an example where the issue is acute morbidity (in this case, accidental injury—one of the commonest causes of hospital admission in childhood). Finally, there is an example of how mortality data can be analysed to specify priorities for health care in the perinatal period.

Case studies
Health promotion

Anne Henry (the practice-attached health visitor) has just returned from a routine home visit to a mother with an only child, aged $2\frac{1}{2}$ years. Like many on this large estate, the family live in a terraced house with no garden and with few toys or books. The parents are teenagers, both heavy smokers, and the child seems to eat little more than sweets, chips, and sugary cakes. Anne (who has a caseload of about 60 births per annum and numerous other tasks) can only spend about an hour on these visits, and is discussing with others in the primary health care team what they can usefully do to promote such children's health.

First, deciding on priorities requires an insight into the important influences at this age on children's present and future health. Second, there is the question of which of these influences are specific problems locally. Third, it is important to act on these influences in the most effective way.

At this age, children need at least adequate nutrition for growth, a secure and stimulating environment for normal psychosocial and language development, and protection from potentially dangerous aspects of their home environment (including air pollution).

(a)   (b)

Fig. 2.1.   (a) and (b) Accidents waiting to happen.

The practice look at their *consultation records* and decide that there appears to be particular problems with asthma, infant wheezing, and middle ear disease (problems known to be associated with passive smoking) among preschool children from the estate. A request to the Director of Public Health produces a special analysis of material from a district-wide survey of smoking among postnatal mothers two years ago, showing that although half the mothers (and some fathers) on this estate gave up during the pregnancies, the prevalence of exposure within the household to smoking at the age of one month postnatally was back up to 80%. They decide to examine the literature and to consult local experts for guidance concerning the most effective way to change this pattern. There are no evaluated *interventions trials* among families with children of this age, but a consensus is arrived at that a three-pronged campaign may produce results.

First, an attempt is made to extend the successful midwife-mediated antenatal campaign messages, both beyond delivery and also to fathers. Second, there is a particular effort to influence smoking parents during '*opportunistic*' general practice contacts when their children are presenting respiratory symptoms, and last, a smoking

Summary
Primary care consultation records

All general practices keep individual clinical records. Many have also developed computerized databases with at least selected health event and risk data recorded

Summary
Intervention trials

- Randomized trials are difficult in clinical practice, especially in primary care
- Reviews from the literature are an essential resource
- Controlled trials should be used whenever possible

Summary
'Opportunistic' contacts

- About 90% of families with preschool children will consult their GP at least once a year
- These represent opportunities to offer advice in a setting where further resources and professional explanation are readily available

Summary
The 'focus group'

- A qualitative research method
- Provides valuable information on health-related behaviour
- Provides more in-depth information than a quantitative survey
- Consists of a taped group discussion using standard open-ended questions
- Tape is analysed for recurring and key themes

cessation programme is widely advertised with nicotine patching and an accessible weekly support group.

The practice group also decide that part of Anne's time for a month should be spent in introducing a simple postal questionnaire survey among practice families before home visits, to monitor progress of the $2\frac{1}{2}$-year-olds. Alongside this, the practice manager performs an audit on the practice computer records to ascertain the proportion of contacts with parents of children under 5 years, where household passive smoking status is established.

It should be noted here that the 'outcome' of a successful smoking cessation programme, in common with many forms of primary prevention, cannot be judged by changes in health status (e.g. bronchitis, lung cancer) because of the small number and long time scale of pathological events. Health risk prevalence, however (e.g. smoking), can be used as an intermediate outcome, and because of the causal evidence, can be reasonably assumed to represent true changes in health status. When conducting such surveys, always look for previously validated methods (and larger partners, e.g. the district department of public health medicine).

Over the next year, Anne's questionnaire for $2\frac{1}{2}$-year-olds is extended to the two other health visitors working with the practice and they accumulate data on over 150 families. This allows them to see how their problem is distributed within the practice area and concentrate their resources. In addition, they get help from the Health Promotion department in conducting '*focus group*' discussions with groups of local mothers and with the subsequent design of an appropriate health promotion strategy. From these local group sessions and their survey data, it becomes clear that smoking is heavily concentrated among a group of young single mothers, who have very limited incomes. For the large part, they are confined indoors with their children and have little social support from extended families or their neighbours, who live in equal poverty. Many of the mothers are acutely conscious of the potential harm to their children's health and have already made unsuccessful attempts to cut down on their smoking. A specific issue which arises from these discussions is the need expressed by many mothers for a regular meeting place with play facilities as a way of sharing their problems and of sustaining their efforts to avoid smoking.

Didactic health education will have little effect on culturally integrated behaviour. The evidence is that most people are already aware of what is more or less healthy in their 'lifestyle' but factors such as poverty, poor local environment, and low self-esteem prevent them from changing their behaviour. A sensitive approach, tailored to meet locally expressed needs, is essential. Great care must also be taken not to induce unrelieved guilt and anxiety in severely stressed families. (See also Chapter 1, p. 14.)

The health visitors take the problem to the tenants' association on the estate. With the support of the evidence from the primary health care team, the tenants' association put a proposal to the social services department of the local authority that space in an empty house on the estate should be made available for a mother and toddlers group. Although this is eventually successful, there continue to be difficulties in financing the recurring staff costs of a part-time play co-ordinator to maintain the rota of volunteers and to supervise the security of the premises.

In this scenario, the priorities for action have been derived from a sequence of epidemiological studies (e.g. GP data, surveys of smoking behaviour), together with more qualitative work to illuminate the local context. The action itself involves the advocacy by health professionals in concert with local activists to tap into both statutory and voluntary resources.

Screening

Dr Peter Brooke, one of the consultants in community child health for a northern English city, is concerned about the efficiency with which local children with undescended testes (UDT) are diagnosed and offered surgical correction (orchidopexy). This is stimulated by a recent report confirming that early treatment improves fertility and reduces the propensity of such testes to develop malignancy. There is also literature to suggest that the frequency of undescended testes has doubled to 3% in the last 40 years.

For many years, a screening service for UDT has been included in the routine medical examination of all local babies at 6 weeks and again at 12 months, on the premise that otherwise the problem will not be reliably noticed by their parents until much later.

Undescended testes appear to satisfy many of the requirements for a useful screening programme—namely, a relatively common problem with potentially serious consequences which is amenable to reliable detection by a simple test and where there is evidence that an acceptable form of treatment early in the natural history will significantly improve prognosis. (See also Chapter 23, p. 390 for surgical treatment of UDT.)

Additional Information
Surveillance visits

- Every child should be offered a sequence of screening examinations during the preschool period.
- The schedule is based on that proposed in *Health for all children* (Hall 1996)
- Screening contacts should result in a centrally returned record attached to the local register of all children living in the district

Summary
Sources of health/illness data on individual children

- Coded data held on computer concerning admissions
- Hospital medical records
- Primary care records: GP and health visitor
- Computerized population child register for a district

To obtain an overview of what is happening locally, Dr Brooke asks his information manager for a print-out of all local children under the age of 15 who have had orchidopexy operations in the last 10 years. From the hospital number on the print-out, he traces the children's original records, both hospital and community, in an attempt to establish how they came to be referred for operation. In addition, he looks to see whether they are actually screened for UDT by physical examination during surveillance visits at 6 weeks and 18 months, as recorded on the central register of all local children maintained in his office. (See also Chapter 4, p. 75.)

There are several sources of routinely maintained data on individual children which may be used to examine children's progress through a set of related health care episodes. First, there is coded data held centrally on computer concerning each individual hospital admission, which include, among other items the age, place of residence, diagnosis, and nature of any operative procedure. Second, there are the more comprehensive written medical records which are kept in the hospital for each patient. Third, there are consolidated clinical records kept by the children's GP based on contacts with the primary health care team and including correspondence following clinical care by other agencies. Health visitors also keep their own records which may contain additional information (e.g. weights). Fourth, there is the child population register maintained across the whole city to allow centralized management of the community-based screening service with records of individual screening tests, such as the 6-week UDT check.

Dr Brooke first investigates if there is any evidence to suggest that orchidopexy operations are performed at increasingly young ages, as one might expect if *screening were effective*. He groups the children according to their years of birth (i.e. in successive cohorts) and draws graphs to show how the cumulative rate of orchidopexy increases with age. (See Fig. 2.2.)

Analysis is made by birth cohort, rather than year of operation as the issue, is to see whether screening at a fixed age has changed the secular pattern of diagnosis, referral, and treatment. As can be seen in Table 2.1, the experience of the more recently born children is incomplete as some of them have not reached the age at which orchidopexy might still be performed.

The expected number of children with undescended testes from any birth cohort is largely determined by the total number of live births. This denominator for the rate is readily available for any district. Although the prevalence of undescended testis is higher amongst babies of low birthweight, these tend to represent a very stable fraction of all births in a given community (about 7%).

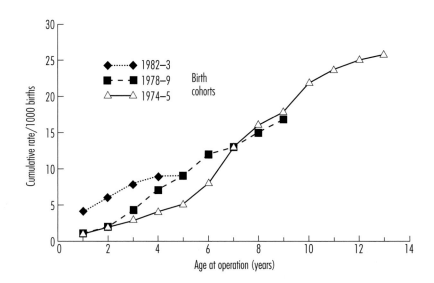

Fig. 2.2. Cumulative rates of orchidopexy for three birth cohorts. (From Tamhne, R., Jarvis, S., and Waterston, A. 1990. Auditing community screening for undescended testes. *Archives of Disease in Childhood*, 65, 888–90.)

Table 2.1.
Postneonatal screening for undescended testes

Cohort	Age 6 weeks			Age 1 year		
	No. of records seen*	No. who had check done	No. detected (No. who had no action taken)	No. of records seen	No. who had check done	No. detected (No. who had action taken)
1980	13	8	2 (2)	13	8	3 (1)
1981	10	8	2 (2)	10	4	3 (1)
1982	5	4	2 (1)	3	3	2 (2)
1983	4	2	1 (1)	4	4	3 (1)
1984	4	4	4 (3)	1	1	1 (1)
1985	3	3	2 (1)			

* Records were traced for 90% of children who had orchidopexies and who were eligible for screening (i.e. unreferred by this age).

Additional Information
Screening effectiveness

- The 'target condition' of the screening (in this case UDT) should receive improved treatment

- Demonstrating the beneficial longer-term outcome of such earlier treatment (e.g. prevented cancer, improved fertility) is often a research exercise

- A frequently omitted component of the 'effect' of screening is the anxieties that are raised among false positives

- Many of these potential 'dis-benefits' which undermine the effectiveness of screening can be monitored and avoided using local data collected during the operation of a screening programme

This analysis seems to reinforce the notion that the more recent cohorts of children have earlier operations and also reveals a total cumulative rate in the older cohorts, with more complete data, which approximates to that reported in other parts of the UK (about 2.5%). However, Dr Brooke still does not know how these more recently operated children came to be diagnosed—were they referred from community screening or is there some other explanation for this trend to earlier treatment?

Examination of the individual clinical records reveals an unexpected finding. The majority of the cases are being diagnosed during the neonatal period. This implies that only a relatively small number of the children with undescended testes would actually remain undiagnosed by the age of 6 weeks—the time when community screening could potentially find them. Dr Brooke therefore looks in more detail at the community screening records of these remaining children to see if they attended for their 6-week and 12-month check and whether they were actually detected by the screening test.

An unsatisfactory situation is revealed—although the screening test, when applied, missed few of the remaining cases, the positive screening tests did not lead to definitive action. Indeed, nearly all the older referrals were through clinical (demand-led) presentations.

Summary
Important characteristics of screening tests

- Eligible population excludes those previously diagnosed

- Screening tests that rely on physical examination produce unreliable results

This illustrates some *important characteristics of screening tests*. First, the 'eligible' population for a screening test excludes those who have been previously diagnosed. This can lead to rapidly diminishing returns when a sequence of screening tests is offered for the same condition. Second, screening tests that rely on physical examinations and clinical judgement often produce equivocal results. Without clear criteria to discriminate screen positives and protocols for subsequent action, there is a tendency to recall 'doubtful' children rather than refer them. This can raise anxiety levels and lead to subsequent non-attendance of high-risk children for reviews.

Dr Brooke decides that the emphasis within the comprehensive service should be to improve neonatal efforts at detection. The small number of residual cases, the inefficiency of community screening, and the possibility of generating unnecessary anxiety do not justify the substantial logistical effort required to ensure that every child is rescreened at later ages. This service is therefore withdrawn.

Setting priorities in this case has been facilitated by a relatively simple examination of the available medical statistics concerning the operation of a typical screening programme. Relating orchidopexies (the numerator) to their correct denominators (those babies eligible for screening at different ages) has not revealed an 'unmet' need. On the contrary, it has demonstrated a 'not-needed provision'.

Acute morbidity: accidents

Dr Sue Ingleton, the director of public health, has been presented with a challenge. She is given the lead responsibility to address the strategic policy aim of reducing accidental mortality among local children. Resources of £50 000 per year for three years are available, but how and where should they be used to most effect?

The director of public health in a health authority is a member of the senior executive team responsible for spending around £100 million per annum. Strategic decisions relating to new developments may result from local consumer demands, from deficiencies noted by health care providers (primary, secondary, or tertiary), from in-house service reviews, or from central government policy guidance.

Dr Ingleton finds in the published statistics (*Public Health Common Data Set*) that over the past five years the local child accident mortality rate is considerably higher than the national average (11 v. 8.5 per 100 000 per annum). However, she also notes that this figure for the local region is based on only 30 deaths per year. A more detailed database of all child accident deaths for the past 10 years in her own health authority is available to her and, on investigation, reveals the information shown in Table 2.2, for 1986–90.

The data in Table 2.2 are derived from the 'death tapes'—a local extract from the national system of death registrations which are collated by the OPCS (Office of Population Censuses and Surveys) from the returns of Civil Registrars of Births, Deaths, and Marriages. The 'Cause' column represents the underlying cause of death as indicated by doctors on the death certificate. The numerical code comes from the International Classification of Diseases (ICD 9) and, in this case, refers to the 'external cause' of death (e.g. E code 814 = pedestrian road traffic accidents), and 'enumeration district' (ED) refers to the district to which the home postcode relates (see below).

Dr Ingleton, although pleased that there are relatively few fatalities, is also conscious that informed and accurate planning will need greater numbers of injury events. She moves over to a separate database with records of all hospital admissions. She now has a much larger body of material to work with (see Table 2.3).

Additional Information
Public Health Common Data Set

- This annual publication by the Department of Health is a rich source of comparative health data across districts in England

- As well as a range of mortality indicators, there are additional data concerning abortions/conceptions/fertility, congenital malformations, cancer incidence, prevalence of smoking, excess alcohol, and high-fat diets

Table 2.2.
Local deaths from accidental injury (1986–90)

Postcode*	Sex	Age (yrs)	Cause (ICD 9 code)	Year of death	Enumeration district
NE17 9DG	1	8	814	1986	O6CH AL08
NE21 3BP	2	1	916	1986	06CH AN06
NE9 8EN	1	14	910	1986	06CH AE19
NE21 6BG	1	0	911	1986	06CH AK02
NE21 2RS	1	3	916	1987	06CH AN05
NE10 7UJ	1	15	987	1988	06CH AQ08
NE10 9DP	1	5	913	1989	06CH AA12
NE9 0PU	1	3	988	1989	06CH AJ27
NE10 4JR	1	4	814	1989	06CH AP01
NE17 6DA	1	1	890	1989	06CH AL01
NE17 8DA	1	2	890	1989	06CH AL09
NE21 1NP	2	14	850	1989	06CH AN23
NE9 3QG	1	9	814	1989	06CH AJ36
NE21 8QS	2	2	814	1989	06CH AN13
NE11 2DX	2	0	994	1989	06CH AW08
NE10 6EA	1	8	984	1990	06CH AA10
NE39 4DN	1	13	816	1990	06CH AM02
NE8 0AB	2	9	814	1990	06CH AG16

* These are fictional postcodes.

Summary
The National Census

- Every 10 years since the mid 19th century there has been a complete census of the population

- The census is supervised by enumerators who each supervise a group of about 150 households— an enumeration district (ED)

- As well as population denominators, the Census offers a variety of sociodemographic data about residents and the households they live in for each ED

WARD INJURY RATES - ALL ADMISSIONS <16 YEARS

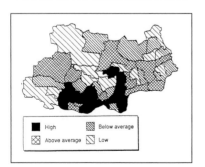

Legend: High ■ | Below average | Above average | Low

Fig. 2.3. Ward injury rates: all admissions of children under 16 years of age. (From Walsh, S. and Jarvis, S., 1992. Measuring of the frequency of "severe" accidental injury in childhood. *Journal of Epidemiology and Community Health*, 46, 26–32.

She also has rather more detail of the nature (i.e. type) of injuries involved, but unfortunately the external cause (E codes) are incomplete.

She is able to obtain some information of where in the city the children live as the home address postcode is recorded in the hospital record. She plots them on a large-scale map which reveals some obvious clusters. However, this may just reflect the fact that these are areas where a high proportion of the population are children.

To deal with this, Dr Ingleton uses the *National Census Counts* to establish the number of children living in each of her local authority wards (about 10–12 census EDs). This allows her to calculate rates of injury admissions per ward and to group the wards into high, above average, below average, and low frequency bands (Fig. 2.3). This confirms that there are indeed some obvious concentrations of higher injury frequency amongst wards along the riverside, an area of serious social and physical deprivation.

Accidental injury is strongly associated with *social deprivation*. For instance, burn fatalities are some 20 times more frequent among children from poorer *social classes*.

Table 2.3.
Number of admissions of children under 16 years of age (1990)

Type of injury	Boys	Girls	Total
Fracture of skull, neck, & trunk	21	7	28
Fracture of upper limb	80	36	116
Fracture of lower limb	17	18	35
Dislocation	6	2	8
Sprains/strains of joints and adjacent muscles	2	1	3
Intracranial injury, excluding those with skull fracture	106	75	181
Internal injury of chest, abdomen, and pelvis	1	1	2
Open wound of head, neck, and trunk	14	12	26
Open wound of upper arm	7	8	15
Open wound of lower limb	2	0	2
Injury to blood vessels/late effects of injuries, poisonings, toxic effects, and other external causes	0	2	2
Superficial injury	6	5	11
Contusion with intact skin surface/crushing injury	6	6	12
Effects of foreign body entering through orifice	19	12	31
Burns	11	9	20
Injury to nerves and spinal cord	16	16	32
Poisoning by drugs, medicaments, and biological substances	13	26	39
Toxic effects of substances chiefly non-medicinal as to source	7	9	16
Other and unspecified effects of external causes	17	21	38
Total	351	266	617

Summary
Social class

- I   Professional (doctor, barrister)
- II   Managerial, administrative
- IIIn Clerical and skilled non-manual
- IIIm Clerical and skilled manual
- IV   Semi-skilled labourer
- V   Unskilled labourer

(The classification for children is based on the occupation of the head of household)

Additional Information
Social deprivation/Social class

- Social classes I–V are arbitrary divisions of families, based on the occupation of the head of the household. Many of the data collected by the OPCS include this attribute (e.g. death and birth tapes, census statistics). These are brought together every 10 years as a 'decennial supplement' to the census in analyses of mortality rates by social class

- 'Social deprivation' is a concept usually used to describe areas rather than people (e.g. enumeration districts—see National Census above). A combination of census-based indicators of material deprivation among the component households are used to provide a single numerical score for each small area

The hospital admission data used to count injury events are collated within each health authority for their own residents, irrespective of place of treatment. It is necessary to bear in mind that presentation and admission to hospital is partly determined by distance, clinical judgement, and the availability of empty hospital beds. A way of avoiding these biases that distort the true injury event counts is to concentrate attention on the more serious injuries (e.g. long bone fractures, admissions lasting more than one day, etc.)

From these data and from other evidence in a sample of casenotes from more severely injured children, Dr Ingleton decides not only that the focus of attention should be on the socially deprived wards, but also that the priority issue is injury of pedestrians and cyclists in road traffic accidents (RTAs). She is not yet clear what can be done to prevent such events (or to improve the outcome once they have happened). A search of the relevant literature, however, reveals two important recent publications on the subject. First, there has been a large-scale survey of head-injury deaths among local children, which not only confirms the importance of pedestrian and cyclist RTAs, but also identifies a number of avoidable deficiencies in the acute management of these children on the roadside and subsequently in hospital. Second, the national Health Education Authority has published reviews of the world evidence concerning effective methods of preventing the occurrence of accidental injuries among children (and child RTAs in particular). From these, it is clear that area-wide traffic 'calming' schemes concentrated in those parts of urban areas with high RTA rates can have a marked effect in reducing the frequency of child pedestrian and cycling injuries.

Additional Information
NHS R&D Programme

- The NHS Executive has a research and development directorate, whose general remit is to improve the scientific and research base of the health service

- Strategic programmes of research to fulfill this aim are managed through outposted regional directors (e.g. the cardiovascular programme is co-ordinated by the Northern and Yorkshire Region)

A strong focus in the *National Health Service Research and Development Programme* has been on the production of definitive reviews of the evidence concerning effective interventions to improve health at both individual and population levels. Knowing the 'cause' of accidental injury, or even documenting poor outcomes from treatment, is not sufficient to initiate action. There should be firm prior evidence that the proposed interventions have worked under controlled conditions.

Summary
Principles of child injury prevention

- Education
- Environmental improvement (e.g. safe play areas)
- Engineering (e.g. safety integrated into house design)
- Enforcement of change by law (e.g. car seat belts)

Fig. 2.4. Child crossing a road.

Dr Ingleton discusses the material with colleagues in local hospitals and in nearby authorities and agrees to help underwrite the development and dissemination of *protocols* for acute management of head injury. At the same time, an *audit nurse* is appointed to undertake *clinical audit* and monitor compliance with the guidelines as well as the speed with which children with intracranial bleeding are diagnosed and treated. A similar investment is also made by the Health Authority in a full-time child accident prevention co-ordinator. This person could produce more specific data concerning the frequency and causes of local child injuries and also promote *collaboration between sectors* in implementing effective preventive strategies.

Many forms of health promotion relevant to children will rely heavily on collaborative work with other agencies—especially those responsible for the quality of the environment, both social and physical, in which children live and develop, for example, the *Healthy Cities Programme*. Similarly, successful attempts to modify health-related behaviour may depend on skillful interaction with peer groups, opinion leaders, the media, and marketing professionals, rather than didactic health education.

In this case study, the health events in question (accidental injury) are a national strategic priority for *preventive action*. The issue is not the setting of this priority but the investigation of how this might be best realized using local data and effective advocacy.

Perinatal mortality

Dr Harvey Johns is concerned that the rate of perinatal mortality in the district where he is a consultant in public health has been persistently ranked among the highest in the region. This is in contrast to the infant mortality rate, which is average despite the relative social deprivation of local families.

High infant mortality is often associated with poverty. Many of these excess deaths are caused by respiratory and gastrointestinal infections. Infant mortality is therefore seen as a key indicator of the effectiveness of health services aimed at mothers with small babies (both hospital and community). Data on *perinatal mortality and infant mortality* rates by District in England and Wales are published annually in the *Public Health Common Data Set* (see p. 33).

Additional Information
Clinical audit

Clinical audit is the process of routine examination of professional practice. In this case, the method being used is to develop the guidelines or protocols using a consensus of best clinical practice and then to disseminate these widely. The outcome is monitored in terms of compliance with the procedures in the guideline and also change in process targets. In a full audit cycle, the results will then be fed into further revision of guidelines

Additional Information
Collaboration between sectors

- Health authority and local authority boundaries usually coincide
- Promotion of population health requires action by other agencies than the NHS (e.g. traffic engineers, environmental health officers, planning departments, housing officers)

Summary
Principles of the Healthy Cities Programme

- Inter-agency working: health/local authority/voluntary sector
- High priority given to health issues by local authority
- Involvement of local people in planning
- Localized (small area) data collection on health
- Advocacy around health issues

Summary
Perinatal mortality and infant mortality

- Infant mortality: deaths < 1 year old per 1000 live births

- Perinatal mortality: still births and deaths < 7 days per 1000 total births (still births + live births)

Additional Information
Sources of birthweight data

Civil registration of birth has included a record of birthweight since 1981. These data are available in a variety of formats from the OPCS

Local notification of birth is made by the attending midwife to the district director of public health (often via the community unit) and includes birthweight and, occasionally, gestational age at birth. These notifications form the basis for the district child health register from which tabulations may be obtained

Hospital-based patient administration systems usually include birthweight and gestational age at birth in their computerized records for those births occurring in hospital (about 97% of all births)

As an initial approach to the question of where the priority should be placed in attempting to improve the perinatal mortality rate, Dr Johns decides to look for evidence concerning the underlying health of babies born in the district. First, he examines routine data concerning the growth of babies before delivery. He obtains figures on the proportion of births in his district which weigh less than 2.5 kilograms and compares these over a period of several years to the equivalent regional data. These analyses suggest that there has been a tendency for local babies to weigh less than expected.

Obtaining *data on the birthweight of babies* is fairly easy as this is recorded by several sources. However, this is not a true measure of growth as babies may be born at different degrees of maturity (i.e. duration of gestation). This 'gestational age' at delivery of babies is less readily available, as there is not always a clear record of when pregnancy started (e.g. last menstrual period). Alternative methods to estimate the duration of pregnancy are increasingly used, such as ultrasound-based measures of fetal size or examination at birth for developmentally sensitive physical signs, (see Ch 8, p. 153).

Dr Johns decides to examine more closely the individual causes of death among babies dying in the perinatal period to try to identify any unexpectedly frequent occurrences locally. To do this, he contacts the manager of the regional survey of perinatal mortality which collates a standard set of data from all local obstetric units.

Every area in England now has such registers—the material being returned centrally as part of the CESDI initiative (Collaborative Enquiry into Stillbirths and Deaths in Infancy—started in 1992). Among other variables, each perinatal death record includes the birthweight, gestational age, and a judgement of the cause of death—usually based on a case conference with pathological evidence.

Dr Johns obtains a listing of local perinatal deaths data covering a 9-year period. An example of the contents of the records (made anonymous) is shown in Table 2.4. He then compares the distribution of the causes given in his listing for the whole nine years to those he would expect if the regional pattern had applied.

In Table 2.5, the analysis has been scaled to an annual equivalent to show which specific causes of death appear to be contributing to the local excess of 4–5

Table 2.4.
Local perinatal deaths (1990)

DOB	Weight (g)	Gest. (wks)	Order/ Fetuses	Mat. Age	Soc. Class	Parity	Age at death	Immediate cause	Obstetric class	Post code
/90	1870	32	1 of 1	29	3	0+2	AP	AP anoxia	APH (PP)	NU10
/90	2925	38	1 of 1	33	2	0+0	AP	AP anoxia	Unk. (Other)	NU40
/90	2100	31	1 of 1	22	6	2+0	AP	AP anoxia	Unk. (Other)	DH3
/90	0750	31	1 of 2	23	3	0+0	AP	AP anoxia	Unk. (M. Abort.)	NU16
/90	2155	?35	1 of 1	42	2	2+0	AP	AP anoxia	Unk. (Other)	NU17
/90	1530	32	1 of 1	25	3	0+0	AP	AP anoxia	APH (Abrupt)	NU8
/90	3430	?42	1 of 1	17	0	0+0	IP	IP anoxia	Mec. (cord pro.)	NU10
/90	0660	31	1 of 1	24	5	0+0	AP	AP anoxia	Unk. (Other)	NU10
/90	2693	?40	1 of 1	18	6	0+0	AP	AP anoxia	Unk. (IUGR)	NU39
/90	2180	?33	1 of 1	30	0	3+0	AP	AP anoxia	Mec. (cord)	NU11
/90	1190	?28	1 of 1	22	4	1+0	AP	AP anoxia	APH (Abrupt)	NU10
/90	3240	41	1 of 1	30	6	0+1	1	IP anoxia	Unk. (Other)	NU10
/90	0838	28	1 of 1	38	3	0+0	4	HMD + IVH	Mat. (Hypert.)	NU9
/90	0440	?22	1 of 1	21	6	1+1	< 1	Pulm. immat.	Mat. (Infect.)	NU9
/90	?1000	?28	1 of 1	31	6	1+1	< 1	HMD	Unk. (Prem. lab.)	NU8
/90	0750	23	1 of 1	35	3	0+0	< 1	Pulm. immat.	Unk. (Prem. lab.)	NU9
/90	3415	?40	1 of 1	25	0	0+0	1	IP Anoxia	Unk. (Other)	NU8
/90	2400	?34	1 of 1	37	2	9+1	< 1	Rhesus	Rh (anti D)	NU8
/90	1450	29	2 of 2	23	0	0+1	< 1	Misc. (perinat.)	Misc.	NU8

DOB, date of birth; AP, antepartum; IP, intrapartum; HMD, hyaline membrane disease; IVH, intraventricular haemorrhage; Pulm. immat., pulmonary immaturity; Unk., unknown; APH(PP), antepartum haemorrhage (placenta praevia); Mec. (cord pro.), meconium (cord prolapse).

Table 2.5.
Analysis of perinatal mortality by cause of death among births (1982–90)

Immediate cause of death	Perinatal mortality (rates/1000)			Ann. no. perinatal deaths		
			Difference (%)	(Obs. locally)	(Expected if regional rates applied)	(Obs. – Exp.)
	Local (a)	Regional (b)	(c) = (a) – (b)/(b)	(d)	(e)	(f) = (d) – (e)
Malform.	1.67	1.84	–9.40	4.33	4.78	–0.45
Other AP–SB	5.38	4.17	29.10	14.00	10.84	3.16
IP anoxia/trauma	2.01	1.37	46.67	5.22	3.56	1.66
Prematurity	1.71	1.72	–0.84	4.44	4.48	–0.04
Infection	0.30	0.26	16.91	0.78	0.67	0.11
Misc. + Rh	0.60	0.51	18.34	1.56	1.31	0.24
Total	11.67	9.87	18.26	30.33	25.64	4.68

AP, antepartum; SB, still birth; IP, intrapartum

perinatal deaths (4.68) which are leading to the elevated perinatal mortality rate. The analysis appears to suggest that as well as a substantial excess of 'other antepartum still births' there is also a specific excess of perinatal deaths associated with 'anoxia/trauma' intrapartum (during delivery). Dr Johns decides to investigate this further by exploring whether these extra deaths are concentrated among babies of any particular birthweight.

In Table 2.6 we see why birthweight is so important to perinatal mortality. Small babies have very high mortality rates (col. a and b) so comparative analysis must take into account the distribution of total births by birthweight before rates can be compared between places. Even after adjusting for the local birthweight distribution it is still clear that the overall perinatal mortality rate is significantly raised ('*standardized*' *perinatal mortality rate*).

The analysis reveals that this raised 'standardized' perinatal mortality rate is due to a specific excess of deaths at medium to high birthweights in the local district (col. c). Dr Johns considers that these two pieces of evidence taken together form a

Additional Information '**Standardized' perinatal mortality ratio**

Table 2.6 shows that the (birthweight) standardized perinatal mortality ratio is 120.37. This is calculated from the column totals (32.5/27 and transformed into a percentage) and represents the ratio of the 'observed' number of deaths to that which would have been 'expected' if the reference birthweight specific rates (col. b) had been applied to the average annual number of births in the District

Table 2.6.
Analysis of perinatal mortality by birthweight among births (1983–88)

Bwt (g)	Ann. total births	Perinatal mortality (rates/1000)			Ann. no. perinatal deaths	
		Local (a)	Regional (b)	Difference (%) (c) = (a) − (b)/(b)	(Obs.) (d)	Expected if regional rates applied) (e)
< 500	0.67	1000.00	972.60	2.82	0.67	0.65
500–999	9.33	571.43	596.56	−4.21	5.33	5.57
1000–1499	18.67	294.64	249.09	18.29	5.50	4.65
1500–1999	38.83	115.88	101.76	13.88	4.50	3.95
2000–2499	120.17	33.29	28.74	15.82	4.00	3.45
2500–2999	464.00	8.62	7.50	15.02	4.00	3.48
3000–3499	997.33	4.68	3.07	52.20	4.67	3.07
3500–3999	715.00	4.01	2.29	75.41	3.83	2.19
4000–9999	240.17					
n–s	0.83				32.50	27.00

Standardized Perinatal Mortality ratio = 120.37

picture. At least part of the excess perinatal deaths locally appear to be among relatively heavy babies and may be associated with mechanical difficulties in their delivery—a classic obstetric problem. He decides to present this evidence to his obstetric colleagues for discussion of potential reasons and solutions.

This example is perhaps the most complete documentation of priority-setting using epidemiological data. This is because perinatal death is a relatively frequent and well-documented health event in childhood. Furthermore, there are excellent denominators.

Conclusion

Priority-setting using epidemiological techniques applied to population health data is a detective story with all too few clues! Although there appears to be copious routine data concerning the health of children—birthweight, congenital malformations, cancers, surgical operations, immunizations, screening coverage, even death (although fortunately very few)—it will be clear that these do little to illuminate the bulk of acute and chronic ill health amongst children. Furthermore, there is nothing to describe the extent of exposure of children to known health risks in their environment.

Imaginative use, however, can be made of these and other data, but several key resources are required:

1. A well-documented *population denominator* (census data, etc.).
2. Fully ascertained *numerators of different health-related events* in this population (health event databases).
3. Access to *detailed casenotes* for individual health care contacts.
4. Capacity to conduct *population (sample) surveys* to extend the range of health indicators and to answer specific questions.
5. *Reference population data* for comparative rates, etc. (e.g. Public Health Common Data Set).
6. *Reviews of best clinical practice* for definitive evidence concerning benefits from health services (Centre for Reviews and Dissemination, University of York).

Further reading

Hall, D.M.B. (ed.) (1996). *Health for all children*, (3rd edn). Oxford University Press.

Public Health Common Data Set (various years since 1988). Institute of Public Health, University of Surrey for the Department of Health. (Papers or spreadsheet formats).

Woodroffe, C., Glickman, M., Barker, M., and Power, C. (1993). *Children, teenagers and health—the key data*. Open University Press, Milton Keynes.

Rose, G. and Barker, D.J.P. (1986). *Epidemiology for the uninitiated*. British Medical Association, London.

3

The normal child: Child-centred health care

The normal child

How do children see the world?

The child's view of the world is very different to yours and mine. In part, this is physically because of their smaller height. But children also have a profoundly different mode of thinking to that of the adult. It is almost as if children assume that they live in a parallel universe to that of adults, which interconnects, but which to them is as foreign to adults as the adult world is to the child. From the young child's point of view, interpreting the world is exceedingly difficult. The adult world can be bizarre, inexplicable, and inconsistent. The child's universe gradually converges with that of the adult as he or she grows up.

The importance of being aware of a child's different world lies in the fact that children have different priorities to adults, and cannot be expected to relate to adult views and needs. In dealing with children we have to rediscover the viewpoint of children (so carefully grown out of in our later school days), and forget our adult view, if we are to be successful at understanding, diagnosing and treating children, preventing them from coming to harm, and encouraging them in habits and behaviours conducive to their future health.

There are a number of models of the ontogeny of children's thinking which will not be further discussed: of most help to the reader are likely to be those of Piaget, and of Erikson.

Why do we need to consider normality? And what do we mean by 'normal'?

First, it is a reference point for abnormality. It is all too easy to assume a child has an abnormal characteristic, particularly if this is how the parent perceives it, if you do not know what is truly normal and abnormal. Both parents and practitioners need to be able to answer: is this child normal?

Almost every biological characteristic—including everything measurable about children—varies. The extent of that variation can be ascertained fairly precisely, and a 'range' can be defined, corresponding usually to the 95% of observations centred on the average (the mean or the median). This gives a statistical definition of normal, but it neither implies that all children within the range are necessarily clinically 'normal', nor that children outside it are necessarily 'abnormal'.

Furthermore, normality may be viewed both as a static phenomenon and as a dynamic process. At any given age a child may have particular characteristics, such as height, weight, or developmental progress, which can be related to those of his/her peers at the same age. But growth and development are both processes, and their progress, or velocity, may also be either normal or abnormal (Fig. 3.1). The assessment of growth and development, and the diagnosis of their disorders, is dealt with in Chapters 7 and 16.

An awareness of normality is also necessary to guide practitioners as to the level of expectation of a child's understanding and behaviour, since this has implications

Fig. 3.1. An example of a growth and development centile chart.

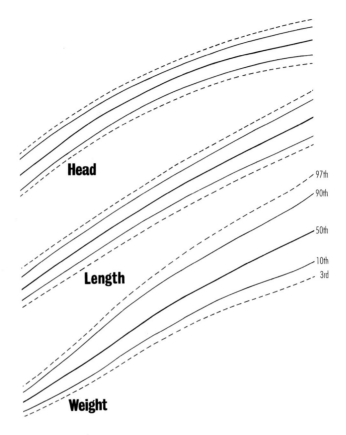

for communication, adherence to treatment and the ability to give consent for procedures, or treatment (see p. 61). Finally, children only maintain their normal growth and development if their emotional needs are recognized and met: their need for security and love within a family, and their social need for stimulation, and the chance to respond to it both inside and outside the family. Security implies a reasonable ability to predict the behaviour of the rest of the world, and requires defined limits on behaviour provided by parents and others.

Why are normal children of any concern to paediatricians?

There is a continual challenge in trying to keep children healthy. In part this is the work of epidemiology (Chapter 2) and prevention (Chapter 4). But it is also true that growth, development, and behaviour influence manifestations of illness, and conversely that illness affects growth, development, and behaviour; while a child's family and social environment have an impact on both physical and mental health. Everyone who deals with children has to be aware of this interconnectedness.

Who needs assessment as to whether they are 'normal', how should this be done and by whom?

All children who come to the attention of a medical practitioner need to have their developmental status taken into account. Part of the initial assessment of any child is a view of the appropriateness or otherwise of his/her developmental stage relative to his/her chronological age. To make such a judgement, the clinician must remember how the child might be expected to perform at a given age.

'Milestones' are commonly referred to for this purpose. However, they can be misleading because they are generally quoted as the average age at which a particular skill is expected to develop, so by definition, 50% of all children will achieve the skill later than this. Of much more value in routine practice is knowledge of the upper limit of normal age for the skill to develop: this might be the time by which 95% of children have acquired the skill. (See Fig. 3.2)

Formal developmental assessment is only required if there is a specific reason to believe that a child may be at high risk of developmental delay or cerebral palsy; this may be suspected because of obvious failure to develop normally, or because of problems in early life, such as preterm birth or neonatal convulsions (see Chapter 8).

How does normal development take place?

Development is discontinuous. This means that there will not necessarily be a smooth progress from one skill to the next. For example, a toddler who has just learned to walk will frequently do nothing else but explore his/her world for the next two weeks, and ignore skills requiring fine motor control. New skills build on previous skills, and a child who for any reason is prevented by disease or circumstance from developing one skill will necessarily be retarded in developing those which come after. Multiple or prolonged admissions to hospital also retard the acquisition of new skills.

For some skills, there are developmental 'windows' when learning is programmed, so that if the 'window' is missed for any reason, learning the skill is

Summary
Development

- is discontinuous
- new skills build on previous skills
- admissions to hospital retard development
- for some skills, there are developmental 'windows'
- first-borns are more advanced verbally than subsequent siblings
- girls are quicker than boys
- children from social class I walk later than in social class V

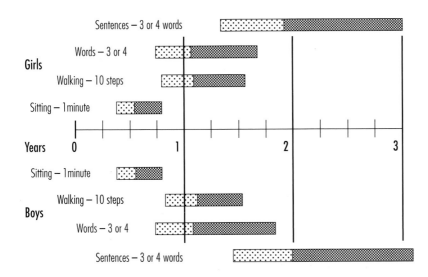

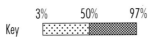

Fig. 3.2. Ages at which different skills were attained in Newcastle children. Note the similarity of ages for sitting and walking, but the general tendency for boys to attain words and sentences a little later than girls.

very difficult. An example of this is the difficulty a child may have with handling food in the mouth if there has been a prolonged requirement for nasogastric tube feeding from birth.

Rates of development are also influenced by other factors. For example, first-born children are more advanced than subsequent children in verbal skills (although not in achieving walking), and girls are quicker than boys.

Babies born prematurely should be judged in their early months with allowance made for their prematurity. However, between one and two years of chronological age the need to make this allowance disappears. The developmental quotient is often used as an assessment of performance. It is the ratio of the child's achievement on a standardized test or scale to that predicted for age. The Griffiths score is standardized to a mean of 100 with a standard deviation of 10. Another widely used system is the Bayley scale. For preterm babies the developmental quotient at the age of two (and the rate of severe disability) is best predicted by gestational age at birth; by school age, cognitive function is independent of gestational age, and more closely related to the social environment and maternal stimulation which the child receives.

Can children lose previously acquired skills?

The loss of skills constitutes developmental regression. This is unusual, and can never be regarded as normal. It requires immediate investigation, which is discussed in Chapter 7.

Nutrition: weaning and after

Weaning is the time of changeover from exclusive milk feeding to a diet of mixed solids and milk. In effect, the baby is making a gradual transition from a high-fat food (milk) to a diet richer in carbohydrates (solids). Milk continues to be a very important food for babies throughout infancy, but from the moment of introducing solids it increasingly forms just one part of a balanced diet. When to commence weaning is a matter of fashion among parents and health professionals, with relatively little scientific basis; there are several current recommendations, Figure 3.3 shows the discrepancy between professional recommendations and what mothers have actually done since 1950.

The gut of a baby is ready to accept other foods than milk by the age of 3 months, and exclusive breast-feeding beyond about 6 months places the baby at risk of iron deficiency because most babies' reserves are exhausted and demand for iron exceeds supply unless there is additional dietary provision.

Weaning diets are often short of iron, so education to encourage meat and fish in babies' diets is an important function of health visitor. This is especially important where breast-feeding is likely to be the baby's main source of milk for the next few months, or in families who choose to ignore the advice not to use unmodified cow's milk as the main milk source. Babies who continue to receive infant formula (or a 'follow-on' formula) as their main milk source are at little risk of iron deficiency at this time. Vitamin C or orange juice given with food increases the absorption of iron.

Additional Information
Committee on Medical Aspects of Food Policy (COMA) Recommendations on weaning 1995: major themes

- The majority of infants should not be given solid foods before the age of 4 months and no later than 6 months

- Breast milk is stressed as the best early nutrition

- Pasteurized whole cow's mild (not skimmed milk) should only be used as a main drink after one year, due to its low iron and vitamin content. But dairy products such as yoghurt, cheese, milk-based sauces, and custard may be given from 4 months, and unmodified whole cow's milk can be used on cereal from 6 months

- Iron deficiency is identified as an important public health problem and its prevention should be linked to dietary advice. Therefore, meat and fish should form an important component of the weaning diet

- Promotion of fluoridated water, and avoidance of sugar in some meals (backed up by more explicit food labelling), are key recommendations for improving child dental health

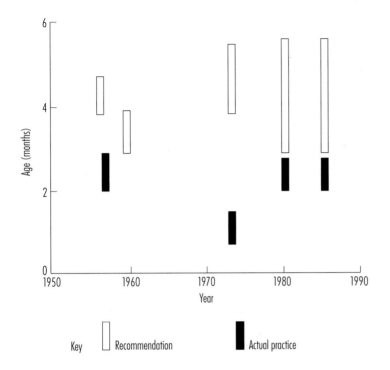

Fig. 3.3. Professional recommendations for weaning and actual practice (1950–90). (From Paul, A.A., Davies, P.S.W., and Whitehead, R.G., Infant growth and energy requirements: updating reference values. In Walker, A.F. and Rolls, B.A. (ed.) 1994. *Infant nutrition*. Chapman & Hall, London.)

At around the time when solid food is becoming an important part of a baby's diet, the first teeth (commonly the lower incisors) appear. Much lay significance is attached to teething, which is blamed for every kind of minor ailment which may coincide with it. As the first teeth start to come through, a baby will chew his/her fingers, salivate copiously, and sometimes appear to be in discomfort which is relieved by paracetamol. Since this is also the time at which a baby is taken about more, has waning humoral immunity, and therefore tends to contract upper respiratory infections, it is not surprising that these are blamed on teething.

From the end of the first year, it is reasonable to allow babies to have unmodified (but not unpasteurized!) cow's milk. This should always be whole milk: semi-skimmed or skimmed is suitable only for health-conscious adults, but young children need the fat.

Control of diet is initially entirely the parents' responsibility. However, even the introduction of a limited range of solids allows the infant to express likes and dislikes for each new taste as it is presented, and gradually these preferences modulate what the parent offers in the way of diet. Young toddlers are notorious for their apparent ability to thrive on very little food, and an apparent low food intake is quite a common reason for bringing a child to the GP or asking advice from a Health Visitor.

On a population basis, the most common *dietary deficiencies* in the UK are iron and vitamin D.

Summary
Weaning

- Weaning diets are often short of iron
- The best iron source is meat and fish
- Babies fed on formula are at least risk of iron deficiency
- Milk remains the main source of calcium

Summary
Dietary deficiencies

- Are most prevalent where there is socioeconomic deprivation
- Deficiencies of iron and vitamin D are commonest
- Vitamin D deficiency is a hazard for dark-skinned children

Early learning: playgroup and school

The neonate

Human babies are almost blind in the first few weeks, aren't they?

No. The newborn baby has a much wider repertoire of sensation, expression, and comprehension than she is generally credited with. But is the neonate almost blind? She can recognize her own mother's voice from others, and her own mother's milk by its smell, within two or three days of birth. The newborn preferentially responds to higher-pitched sounds, corresponding to the frequency content of a female voice. She prefers oval shapes to other shapes, and responds best at a distance of 20 to 30 centimetres, corresponding to the distance from her mother's face to her own when she is feeding. The effects of fetal learning can be demonstrated postnatally. The newborn can vary her cry to express different needs, and her mother quickly learns to interpret this simple 'language'

The process of *learning* probably starts before birth. It is now known that the fetus can learn, and remember, and that behaviour postnatally can be affected by intra-uterine experiences. However, there is no justification for the transatlantic fashion for deliberately stimulating the fetus prenatally with music or words in the hope of creating hyper-intelligent babies.

The first year

The world of the infant gradually expands so that within a few months the activities of other people in the same room will often capture her attention. Around 3 to 4 months she starts to coo and this encourages her carers to talk to her even more, and to start to create pretend conversations. As she starts to babble, this pattern develops, and is a necessary precursor to the start of real language.

Learning abilities expand dramatically with the development of an upright posture, and walking. At around the age of one year the twin abilities of enormously increased potential for communication and physical movement together transform the world of the child because this world becomes susceptible to manipulation and exploration. A child's constant fascination with experimentation in this expanding world is what adults call 'play'. (Adults do this as well, of course, but commonly give it less pejorative names!)

Preschool

Many parents like to take their child to parent-and-baby groups, or parent-and-toddler groups. The major function of these is to maintain social contact for the parents, rather than to have any positive developmental effect on the child. Children between 1 and 2 years seldom relate meaningfully to each other anyway. Working mothers may place their baby in a day nursery as soon as they start back at work.

Summary
Early learning

- starts in fetal life
- occurs through the interaction of baby and parent
- is impaired by maternal depression
- is impaired in deprived families
- is enhanced by nursery education

Additional Information
Child development normal ranges

Skill	Age
smiling	5–8/52
reaching for objects	5–6/12
transferring hand to hand	7–9/12
sitting unsupported	7–10/12
walking unsupported	15–18/12
single words with meaning	15–18/12
Speaks in phrases	22–30/12

Older children can benefit from attending a preschool playgroup. The environment here is more social, with group activities such as music, and gradually increasing structure to the session. A parent may accompany their child initially, then leave them for the duration of the session, giving themselves some welcome time of their own. Some playgroups are formally attached to the primary school which most of the children will subsequently attend. Children who attend a well-structured nursery, make more rapid progress in school than those who do not.

The child at school

School dominates the life of children, just as work (when it is available) dominates that of adults. For the child it is the main source of friends outside the family, and a major source of praise and criticism, success or failure. Importantly, it can be a source of friends for the parent through contact with other parents. Parents are increasingly encouraged to take an active interest in their child's schooling, although this has been a slower process than for child health care (see later).

School is not just about the teaching in the classroom. It is a community, and children learn many things about socializing, making and keeping friends, activities and interests in common, and relating to adults who are in authority but are not their parents.

Paediatricians, general practitioners, and other primary care professionals have to relate to children and their schooling in a variety of ways.

Fig. 3.4. The child at playgroup.

Case studies
Problem behaviour

Bobby is nearly 6 years old. He is notorious at school for poor concentration and disruptive, attention-seeking behaviour. He is often kept off school for upper respiratory illnesses. When assessed by the school doctor, at the request of the school nurse, he is found to have bilateral glue ear with scarred tympanic membranes. Audiological assessment confirms the clinical suspicion of hearing loss (25–30 decibels bilaterally), and he is referred to an ear, nose, and throat surgeon for consideration for grommet insertion. Surgery not only improves his hearing, but he starts to make educational progress and his behaviour and concentration improve.

Vision and hearing screening programmes are common at school entry, to identify those children with deficiencies in either sense, who will otherwise not be able to gain by their school experience and may be wrongly classed as slow learners.

Somatization

Michael is a big, shy, rather slow 13-year-old lad, and is referred to a consultant paediatrician with recurrent episodes of excruciating abdominal pain which have resulted in his missing 1 or 2 days of school every week for the past few months. A slow learner, this is his second term in a mainstream class to which he was promoted from a remedial group. Medical history and examination yield no clues, and he does not admit to any difficulties at school, either of bullying or through pressure of work.

Children presenting with somatized complaints often have an underlying problem relating to school.

Both paediatrician and parents are sure that the problem lies in the school, but are frustrated that Michael will not divulge his problems; physical investigations are avoided and some symptomatic treatment prescribed. Two weeks later he breaks down in tears at school, and it emerges that he simply cannot cope in the mainstream classes. He is returned to the remedial group, and his symptoms disappear at once.

In Michael's case the problem was to do with educational pressure that he could not handle, but bullying and other forms of abuse can present in the same way.

Infection control

Ann (age 8), is admitted to hospital with a diminished level of consciousness, a rapidly spreading purpuric rash, and neck stiffness. She is treated for meningococcal meningitis, and makes a good recovery although she loses two toes from vascular occlusion.

As soon as the diagnosis is confirmed the consultant in communicable disease control, liaising with school nurses and the local general practices, organizes for courses of rifampicin to be given to all members of Ann's class, and her immediate family. The schoolteachers are briefed by the school doctor about early symptoms of meningococcal disease. Ann is referred for audiological assessment because sensorineural deafness is an important complication of meningitis (see Chapter 18).

Because schools are communities in which people are in close contact, infectious diseases (influenza, chicken pox), parasites (threadworms), and infestations (head lice) can all spread rapidly. Of great importance, although rare, is meningococcal meningitis: classroom contacts of cases may be offered prophylaxis with rifampicin.

Special needs

Kylie received a severe head injury at the age of 4 years after running out in front of a car. She gradually recovered, and managed to regain her ability to walk, but she seemed much less bright and her parents became concerned about her ability to manage in mainstream school.

A child identified in the preschool years as having special needs (such as cerebral palsy, learning difficulty, hearing loss, visual impairment) may or may not manage in mainstream school.

Kylie received a full neuropsychological assessment in a child development centre, and a *statement of educational needs* (see Ch 7, p. 126) was generated as a result of this assessment. She was initially placed, with full agreement from her parents, in a school which catered for children with learning difficulties; but she did very well there and after a year she transferred successfully to mainstream primary school.

A statement of educational needs has to be prepared for that child with input from all the relevant professionals, and the parents also have a substantial say in the recommendations of the statement.

The school can also provide *other health functions*.

Many chronic illnesses have an impact on school life, and can be particularly hard to deal with in mainstream school where most children are, by definition, free of illness. Important examples are diabetes, asthma, and epilepsy. Children on regular medication who require a dose during school hours may need special arrangements to be negotiated. Serious but rare illnesses such as cystic fibrosis, leukaemia, and cancer can result in considerable spells of school absence, and close liaison is required between the hospital teaching service, the parents, and the school to minimize the educational effects of such serious illnesses.

Children and hospitals

How can a hospital be made 'child-friendly'?

Much has been learned in recent years about how to make the hospital less of a hostile environment for children, but much of the credit for this must go to the

**Additional Information
Other health functions of school**

- Schools are a useful place for organizing mass immunization with BCG for tuberculosis at around the age of 12 years

- There was a mass measles/rubella immunization programme for school age children in 1994 to prevent a predicted epidemic of measles in 1995 (see Chapter 4)

- Teaching on healthy eating, sexual health, smoking, drug and alcohol abuse takes place in most secondary school. Means of making this more effective remain controversial (see Chapter 4)

Summary
Child-friendly hospitals

- incorporate appropriate design and furnishing

- have play areas with toys and activities

- enable a parent or relative to accompany young children in hospital

- have relaxed clothing for staff

- do not separate children and parents, even for short times or minor procedures

- respect the various coping strategies for pain or discomfort which children have

- have a high awareness of pain prevention and analgesia

- have adopted patient-centred nursing care

- child-centred therapeutics

- have a routine for preparing a child for elective admission

- have a hospital teaching service

lay groups, such as the National Association for the Welfare of Children in Hospitals (now called Action for Sick Children), which helped to revolutionize professional thinking.

Some of the principles which underpin the health care of children both in hospital and in primary care environments are as follows.

- *Appropriate design and furnishing to cater for the relevant age groups*
  Getting away from dour functionality is not only pleasant for staff, it also signals to children and their parents that the institution is trying to be *child-friendly*. This means lively art work, perhaps including recent work by children attending that department; small chairs for young children; carpet on the floor (see Fig. 3.5).

- *Provision of toys, activities, and play areas*
  Outpatient and clinic visits always mean waiting, and waiting means boredom. This is greatly helped by toys appropriate to a child's developmental stage. In hospital, the older children need an area to themselves, and can often use a pool table or watch videos rather than daytime children's television, which is mostly aimed at younger children. Play workers can not only help to entertain the children, but can also help them to work through hospital or illness-related anxieties by means of play.

- *Young children admitted to hospital should, whenever possible, be accompanied by a parent or relative*
  This simple principle, first adopted in the Newcastle Babies' Hospital after World War II, took around three decades to achieve universal acceptance. It is now recognized that hospital admission, where illness is accompanied by enforced separation of children from their parents, can be immensely damag-

Fig. 3.5. A pleasant children's play area.

ing psychologically, and the corollary to this is the need to avoid admitting children wherever possible, and to keep hospital lengths of stay as short as possible. 'Visiting' by parents has itself become an obsolete concept: they come and go as they and their child feel they need.

- *Relaxed clothing for staff*
  It is now commonplace for child health doctors not to wear white coats, and for nurses to be less formal, often with colourful tabards over their hospital uniform. Interestingly, surveys of children's interpretation of clothing have shown, at least for doctors, that while the less formal look is appreciated as friendlier, the white coat is still associated with greater competence!

- *Awareness that it is inappropriate, unnecessary and sometimes counterproductive to separate children from their parents, even for short lengths of time or for minor procedures*
  It can be a major professional hurdle for staff to overcome their discomfort at having to work with parents present. Appropriate training, experience, and professional role models have all contributed to the recognition that parents can play a useful role in helping their children cope with procedures, including accompanying them to the anaesthetic room prior to induction of anaesthesia for surgery.

Fig. 3.6. Hospital scene with mother talking to a nurse.

- *Recognition that children have different coping strategies for painful or uncomfortable procedures, and that these need to be understood and respected*
  Whether it is an immunization or changing a dressing, some children cope better with distraction, and others with attention. Helping children to cope with a short-lived adverse experience is not assisted if staff carry preconceptions (from their own childhood or their own children) of how a particular child should manage. Children, like adults, need an individualized approach to coping with pain and discomfort.

- *Attention to strategies of pain prevention and analgesia*
  The development of skin anaesthesia with eutectic mixture of local anaesthetic (EMLA) cream (Fig. 3.7(a)+(b)) has revolutionized procedures such as insertion of intravenous lines, and drawing blood samples. At the same time, reliable methods for assessing pain in children are beginning to refine the process of giving optimum pain control after surgery, and intramuscular injection (a source of great fear for many children) for analgesia or any other purpose has become virtually obsolete.

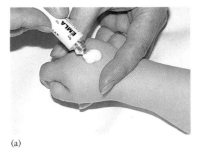

(a)

- *Patient-centred (as opposed to task-centred) nursing care*
  It was in paediatric nursing that the concept of patient-centred nursing first appeared, spreading rapidly to other specialties. It is a professional model which attaches nurses to groups of patients, the emphasis being on all the nursing needs of a group of patients, rather than a concentration on tasks independent of patients such as changing beds, doing drug rounds, and checking infusions.

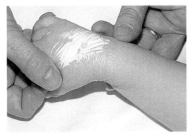

(b)

Fig. 3.7. (a) and (b) EMLA cream on a hand.

- *Child-centred therapeutics*
  Young children cannot manage inhalers, pills, or tablets; they may cope with liquid medicines but refuse them if the taste is unpleasant. Therapeutics is

always the art of the possible, and drug delivery systems and paediatric prescribing must take account of this. Syringes are now often used for administering oral liquid medicines as they are more convenient and accurate than the traditional 5 ml spoon. Dosage is generally calculated on the basis of body weight (which is easy to measure), although under certain circumstances surface area (calculated using a nomogram based on weight and height) is used in preference. Intramuscular injection is generally avoided, as few drugs need to be given by this route in preference to orally or intravenously (immunizations being the unfortunate exception). Oral drugs should taste pleasant, but it is important to avoid sucrose because severe caries can result if a child is on long-term treatment with a sucrose-containing liquid.

Impaired renal or hepatic function may alter the rate of disposal of drugs, in children just as in adults. The immature neonate is an extreme example of this, but awareness of the routes of disposal of drugs is important throughout paediatric therapeutics. Allowance must be made for immature function, blood levels measured, or alternative drugs used (e.g. cephalosporins instead of aminoglycosides in a child with renal impairment). Conversely, drugs used in epilepsy, a relatively common area of paediatric practice, may induce enzyme activity or interact with each other or with other drugs. Measuring the levels of these drugs (sometimes possible in saliva), can assist in prescribing and assess adherence to treatment.

Adherence to a prescribed treatment is generally the responsibility of parents, but must become the responsibility of the child as he or she gets older. Particular difficulties in ensuring adherence arise in adolescence in connection with the management of chronic conditions requiring continuous treatment, such as diabetes, asthma, and epilepsy.

- *Preparing a child for admission to hospital*
  Most children with medical conditions are admitted as emergencies, but for surgery most come electively. For elective admissions there is an opportunity to bring the child for a visit prior to admission, when staff can meet the family and the family can become familiar with the staff and the environment. Most children have fears about hospital, and any opportunity to alleviate these helps the child, and family, to cope with what is always a stressful experience. In addition, there are several good children's books about hospital which can be read with the child and which explain in outline what the child should expect.

- *Hospital teaching services*
  Median lengths of stay for children with medical conditions are very short—of the order of two days or less. However, there are still some children who may stay for many days or even weeks: those with orthopaedic conditions, children with rare illnesses such as cancer, leukaemia, cystic fibrosis, inflammatory bowel disease, or osteomyelitis, and some with psychiatric disorders such as anorexia nervosa. School-age children may have their education seriously disrupted by prolonged periods in hospital, and continuing their school work is potentially therapeutic for the child, mimicking to some extent the reality of the outside world, and preventing boredom. The hospital teach-

ing service exists to provide liaison with the child's school, and to maintain some measure of educational input. Occasionally, the hospital teachers find themselves invigilating for a public examination.

The paediatric consultation: history and examination

For those without much experience of working with children and families, getting full value from a *consultation* can be daunting. Some of the main uncertainties are set out below.

How should I set up the consultation?

Be welcoming. You will need to introduce yourself to both parents and the child. Position the parents and/or child at an angle to the doctor, not with a desk in the way. Children often want to go off and explore the toys in the room, of which there should be plenty. Notice, and comment, if the child has a toy of their own with them. It is helpful to explain how you will structure the consultation.

How can I form a relationship?

Use encouraging remarks to help the parent express the problem, which avoids the trap of abrupt interrogation. Try to express verbal and non-verbal empathy, nodding your head in response, and making encouraging noises. Parents and children will only open up if you are warm and friendly in manner. Condescension is to be avoided at all costs. Maintain eye contact throughout. Always include the child, if only tangentially (a young child may have gone off to play in a corner of the room). Don't write anything down initially. Find out the parents' names and use them: remember that the mother's, father's and child's second name may all be different!

How do I actually obtain the information I need?

Use open-ended questions, avoid leading questions and use appropriate non-technical expressions and language. You will often need to ask for clarification. Allow time for problems to be stated before asking for further details; give time for the parent or child to think of their answers. Remember that if all you do is ask questions, all you will get is answers: you must allow the parent and child to tell their own story. If you need to make notes, create pauses in the consultation.

You will need information about pregnancy, delivery, early feeding practices, and immunization, though the first three become of less importance in the school age child. To understand family relationships and family history of illness it is a good idea to draw a pedigree (see Fig. 3.8 for an example). The so-called 'systematic enquiry' of adult histories is seldom relevant to young children, but information about school performance, peer group attitudes, and sporting activities is essential for school-age children. Information about parental employment, finances, and housing is also important, and the *feeding history* in an infant, (*see also Ch. 8, p. 157*).

Summary
Consulting with children

- Set up the consultation with a welcome and introduction
- Empathize to form a relationship
- Use open-ended questions where possible
- Pause to make notes
- Explain why sensitive questions may need to be asked
- Parents, and older children, may need a private conversation
- Examine appropriately for age
- Use toys and games, an indirect approach, and a warm stethoscope
- Use an anatomical approach and always explain what you are doing
- Make a clearly defined end, organize follow-up, and include the child in a courteous parting

Additional Information
Infant feeding history questions

- What milk? (Has the milk been changed frequently?)
- How much in each feed?
- How often are the feeds?
- → calculate how many ml/day and ml/kg/day. You may find gross overfeeding or underfeeding
- How is it made up? (get exact description)
- How long does each feed take?

Fig. 3.8. An example of a pedigree.

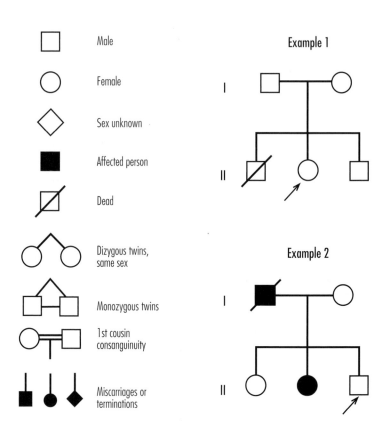

How can I handle personal or delicate issues?

Explain why such information is necessary: this prepares the parent or child for the questions. Unless it is necessary, you should not ask it. Don't be afraid to allow the consultation to pause while a reply is being considered; you should not feel uncomfortable with silences.

What about issues which may not be best discussed in front of the child?

There may be issues which you feel unhappy to discuss in front of the child. There will very often, with a teenager or adolescent, be things not best talked about in front of parents. You may need to know: does the child perceive the same problems as the parent? What are the child's attitudes to the problems? There may be specific questions (e.g. sexual behaviour, substance abuse, attitudes to parents) that can only be dealt with without the parents present.

How do I set about examining a child?

Babies can be examined on a warm soft flat surface; toddlers can be examined on their mother's lap; school-age children can be examined on a couch. Toddlers and preschool children are the most challenging group to examine. Use toys, playthings and games; use an indirect approach to the child; examine proxies, such as the child's teddy bear or doll before you examine the child (Fig. 3.9). Warm your stethoscope, and your hands. Get down to the level of the child, never tower above her.

Use an anatomical not a systematic approach: move from the top to toe in the baby, the periphery to the centre in the toddler, and any approach in school-age children, *but always explain what you are doing.* Also, leave potentially less pleasant items till the end: hips and head circumference measurement in a baby; ears, nose, and throat in a toddler; blood pressure in a child.

The examination of newborn babies is covered in Chapter 8. In infants and children there are also several *examinations* which are often less relevant to adults.

When should I write it all down?

Writing may take place at intervals during the consultation, after taking the history, or wholly at the end. As writing precludes eye contact and gathering non-verbal leads, it should be minimized while the parent or child is speaking, so create pauses for noting material and use writing time for further clarification or to summarize.

What then?

Summarize the consultation up to that moment: do you and parent perceive the same problems, ranked in the same way? Is there a single, firm diagnosis, a range

Summary
Examination of infants and children

- Feel the fontanelle in infants
- Check for femoral pulses (their weakness, absence or a radio-femoral delay suggesting possible coarctation of the aorta)
- Examine for lymphadenopathy in the anterior and posterior triangles of the neck, and the post auricular nodes
- Look down the throat
- Examine the tympanic membrane with an auroscope
- Measure the weight, height and (particularly in infancy) the head circumference, and plot these on a centile chart
- Measure blood pressure (often forgotten in children)

Fig. 3.9. Paediatrician examining teddy.

of possibilities, or the original problem as previously stated? Explain in appropriate language, without condescension, the nature of the diagnoses or problems, and do the same for the child. Recommend, or negotiate, a suitable plan for their management. If there are to be any investigations, these need to be explained to the child as well.

How do I bring it to an end?

Make a clearly defined end to the consultation, focusing on any follow-up contact (when, how), and make a courteous parting that includes the child.

Emma is $2\frac{1}{2}$ years old. She has come to outpatients with minor bowel symptoms. In the waiting area she clings to her mother, and her mother is getting frustrated.

The most important thing is to spend time introducing yourself to the mother and child, keeping up a patter of small talk as they come into the room. At this age, too much immediate attention to the child can be counterproductive, as it is often perceived at a threat. Inviting the mother to let the child do what she wants, rather than confine her to a chair or her own knee, allows the child to explore the room. She will find the toys in her own time, during which you can observe her developmental progress by watching her play, listening to her talk, and interact from time to time with her mother.

When the time comes to examine her, she may be best on her mother's lap, or may have the confidence to lie on an examination couch with her mother close by. Either way, it is useful to warm the end of the stethoscope, to make the examination a game, to distract from the examination with toys, and to do peripheral things, like feeling a radial pulse, before touching chest or abdomen.

Parents of the ill child

Susan is 3 years old. She has been admitted to hospital following her first febrile convulsion, and has been found to have a urinary tract infection. A couple of hours after admission the nurse in charge finds herself confronted by Susan's father, who angrily demands to see a doctor.

Susan's father does not realize that the person he has already seen is the paediatric senior house officer, because he was not wearing a white coat, his identity label had got mislaid, and he did not introduce himself. But the real reason for his anger is his terror: many parents who witness a simple febrile convulsion for the first time are convinced at that moment that their child will die, and it is probably the worst feeling they will ever have. The state of high anxiety carries over into the hospital, and quite often spills out in diffuse anger.

Having a sick child in hospital is inherently stressful, no matter how 'child-friendly' and 'parent-friendly' the environment. Under such conditions, explanations can easily be forgotten, or scarcely heard in the first place. It is vital to remember basic courtesies, such as introducing yourself clearly, and to spend a little time sympathising with the experience the parent has undergone. Most of what you say to a parent will need to be repeated later, and carefully written fact sheets can assist this (but can never substitute for a personal discussion).

When serious or life-threatening illness occurs, a great deal of time is needed for explanation, counselling, and helping the parents to come to terms with the situation. They often need help to find the most appropriate way to help and support their child. Nomination of particular members of the nursing and medical team to provide continuing support for families is helpful, and close liaison and outreach with community services, primary care, and school is vital. Formal or informal parent support groups can also be valuable for some families.

Ethical issues

It is important to distinguish between legal requirements and ethical practices. While it is unlikely that an illegal act is ethically justifiable, an unethical situation may nevertheless be perfectly legal. As practitioners, we have a civic duty to remain within the law, and to work to change bad or anomalous laws; and we have a professional duty to practise ethically.

The legal framework for child health (see also Ch. 1, p. 21) revolves around the Children Act (1989), all legislation involving primary and secondary education, the Infant Life Preservation Act (1929), the Human Embryology and Fertilization Act (1994), and a variety of other measures in relation to road traffic law, employment law, and much else.

Some *cardinal principles* of ethics are set out in the additional information box.

Mary, a 4-year-old girl, is admitted for insertion of grommets for refractory otitis media with effusion and bilateral hearing loss. The operation is explained to her parents, who then sign the hospital's 'consent' form.

Difficulties arise in child health because the younger the child, the less *autonomy* the child has. While an adult, or older child, can give informed *consent* for a treatment or procedure, for the young child or baby the parents can only *assent* on their child's behalf. There is no clearly defined cut-off between the dependent younger child and the independent older child. Consequently, the extent to which a child of less than 16 years can be forced to undergo a procedure or treatment which they do not want is neither legally or ethically clear, (see the Gillick principle, Ch. 4, p. 70).

Additional Information
Cardinal principles underpinning ethical considerations

Autonomy: respect for the individual to determine his/her own destiny

Beneficence: the intention to do good

Non-maleficence: the intention not to do any harm

Justice: the requirement to be fair. Some authors separately distinguish the need for *equity*.

Kylie has been on the 'at risk' register for two years. Now 3 years old, she is failing to thrive and showing signs of emotional abuse and physical neglect; she is in and out of hospital for relatively minor illnesses which invariably present in the late hours of weekend nights. At a case conference, the paediatrician forcibly argues for Kylie's removal from the family. The health visitor argues for further work with the family for another few months to try to improve the mother's parenting skills.

The conflict here is between the *autonomy* of the parents to bring up Kylie as they think fit (championed by the health visitor), *beneficence* towards Kylie (championed by the paediatrician), and the very real issue of *non-maleficence*: would Kylie be more harmed, at the age of three, by the removal from the family she knows, first to a foster home, and then perhaps for adoption, and the prolonged uncertainty which could arise from the inevitable legal wrangling? The legal framework is that of the Children Act (1989) whose cardinal principle is that the child's interests are paramount. The issue here is how to translate this into practice, which really means, what are Kylie's 'best interests'?

Serious ethical debates need to be informed by accurate facts whenever these can be found. For instance, legitimate concerns about the ethics of resuscitating very premature babies must be based on the facts of neonatal outcome among the survivors.

Sam is born with Down syndrome to a low-risk mother, and the family is devastated. Within hours, it is clear that Sam cannot tolerate feeds, and the diagnosis of duodenal atresia is made. Resection of a duodenal atresia is a well-established operation with very low morbidity and mortality. Sam's parents readily assent to this and he has the operation and does well thereafter.

Could Sam's parents reasonably have refused him surgery? Not so long ago, not all babies with Down syndrome would have been offered this operation. Yet the procedure is safe, it is certainly life-saving, and without it the Sam would have died of dehydration, modulated only by sedation (which would have had to be given parenterally). It would be very difficult to counter the argument of *beneficence* towards Sam with the suggestion that his parents have the *autonomy* to refuse him a simple and life-saving operation.

Two days postoperatively, Sam's colour becomes a little dusky. Echocardiography reveals a large atrioventricular septal defect, which is a well-known complication of

Down syndrome. His parents wonder whether they can justify putting him through an operation for this as well.

Fortunately, no one has to make a decision about this very quickly. An atrioventricular septal defect, if not repaired, usually results in irreversible pulmonary hypertension, which might be expected to limit Sam's life to his teens or twenties. A repair, if undertaken, would normally be done at the age of a few months. Clearly, if the mortality for this procedure locally was around 40%, a good case on the grounds of *non-maleficence* could be made for not undertaking it. But if the newly appointed cardiac surgeon with a special expertise in this condition could offer a mortality of only 4%, how might we view the parent's reluctance then? They could still argue that major cardiac surgery, with even a small chance of a poor outcome, would not alter Sam's quality of life for many years, and much else might have happened to him during that time. This would be the same argument of *non-maleficence*. The paediatrician might counter this by saying that any procedure offered to a normal child should be equally available to one with special needs: this is an argument on the grounds of *equity*.

Difficult ethical questions cannot have easy answers. For clinicians, the important thing is to be able to consider the issues within a conceptual framework, and the principles and examples given above provide a way in to this.

Further reading

Erikson, E.H. (1950). *Childhood and society.* Norton, New York.

Illingworth R.S. (1991). *The normal child: some problems of the early years and their treatment*, (10th edn). Churchill Livingstone, Edinburgh.

Piaget, J. (1926). *The language and thought of the child.* Harcourt Brace, New York.

4

Prevention in child health

- Models of prevention: social and medical

- Definitions

- Health of the Nation targets

- Case studies
  prevention of infection by immunization
  prevention of undernutrition and malnutrition
  prevention of deafness
  prevention of smoking

- Conclusion

The prospects for prevention in child health are very great. As shown in Chapter 1, there is still a high toll of morbidity from preventable conditions, including infectious disease, congenital disorders, accidents, and smoking. There is also a wide disparity in morbidity and mortality across the social classes with a two to threefold difference between social class I and V. Not all conditions are equally preventable, and for some (e.g. poor nutrition, accidents) prevention will need to come mainly from outside the health service. This field is challenging and there is much need for innovation, for working together between health and other sectors, and for evaluation of effectiveness.

Models of prevention: social and medical

It is important to be aware in describing prevention that the term is often used in relation to the *medical model* of causation.

A good example of this is *Haemophilus influenzae* meningitis. The bacterium causes illness, the doctor diagnoses the nature of it, prescribes an antibiotic, and the child gets better (if treated in time). Immunization prevents a person getting the illness. The medical model which is very widely taught to medical students is applicable mainly to those diseases in which there is a single cause. In many conditions, there is not one specific cause hence prevention is 'multifactorial'. It is equally important to be aware of the *social model* of causation. Lung cancer due to smoking is a good example.

Medical model

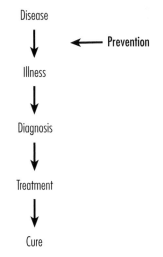

Social model

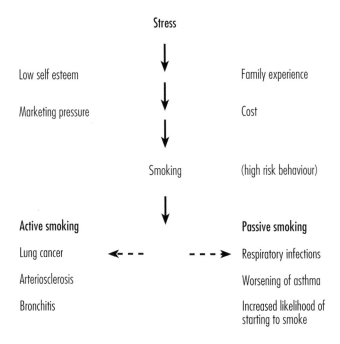

Additional information

Medical model
- Measles
- Pneumonia
- Meningitis
- Phenylketonuria
- Haemophilia

Social model
- Failure to thrive
- Accidental injury
- Child abuse
- Teenage pregnancy
- Behaviour disorders

Summary
Advocacy

- Advocacy is a long-established tool in public health

- It means speaking out on behalf of a child, a family or an issue (e.g. the need for day nursery provision or legislation on tobacco advertising)

- It may be undertaken on behalf of individuals, communities, or nationally

- Advocacy requires a good case, supported by evidence, persistence, and the avoidance of party political bias

- Advocacy requires an awareness of how the political system and the media operate

Summary
Public health responsibilities

- Monitoring and describing the health of the population.

- Identifying those groups most in need of health care of all kinds.

- Identifying the social, economic, and environmental factors that impinge on health

- Taking action to promote and improve health

- Assessing the impact of health care intervention on the health of the population

A social model approach to prevention requires attention to be paid to societal and behavioural factors as well as to disease. Here, it is clearly harder to apply a preventive approach than it is in diphtheria. Most illness is multifactorial in origin and there are few conditions where there is just one causal agent. Even in infectious disease, environmental and emotional factors play a part. Example of conditions where a mainly *medical model* and those where a mainly *social model* approach is necessary, are given is the additional information box.

In the medical model, doctors are likely to be able to organize prevention as a single intervention (e.g. immunization or screening). This can be readily done within the health service. In the social model, it is much less easy for doctors or the health service to organize prevention—in the case of smoking, cost is perhaps the most important factor and this requires government action. Hence, in the social model conditions, doctors can only develop preventive initiatives by working with other disciplines ('healthy alliances') or by influencing the government to alter its policy (e.g. to ban tobacco advertising, or increase the price of cigarettes). This type of lobbying or '*advocacy*' is appropriate for doctors to take part in, and there is a long history of advocacy in public health medicine.

Definitions

Primary prevention: reduction in number of new cases of a disease, disorder, or condition, (e.g. immunisation, accident prevention).

Secondary prevention: reduction of prevalence of disease by shortening duration or diminishing impact through early detection and prompt intervention (e.g. screening).

Tertiary prevention: reduction of impairment and disability, and minimizing suffering (e.g. team management of cerebral palsy).

The above terms tend to assume a medical model and are less easy to apply to a social model.

Health promotion is any planned measure which improves physical or mental health or prevents disease, disability and premature death. Child health promotion can encompass interventions in curative medicine, in education, in the environment or in public policy.

Health education is any activity which promotes health through learning.

Child health surveillance has been used to cover the oversight of the physical, social and emotional health and development of all children. *Health for all children* (Hall 1996) suggests that child health surveillance be regarded as synonymous with secondary prevention.

Public health can be defined as 'the science of preventing disease, prolonging life, and promoting health through the organized efforts of society'. The *responsibilities of public health* are shown in the summary box.

Health of the nation targets

In 1992, the UK government published a White Paper entitled the Health of the Nation. This proposed a series of targets for health improvement which would be supported by the Department of Health as well as all other relevant government departments. This publication was in line with World Health Organization policy as well as with progressive thinking in public health circles. The paper proposed targets in the following areas: coronary heart disease and stroke, cancers, accidents, sexually transmitted disease and sexual health, and mental illness.

Some of the targets relevant to children are unlikely to be reached by the time scale indicated and the government has been criticized for not taking strong enough action itself (e.g. by banning tobacco advertising). It was also pointed out that there was no target relating to a reduction in inequalities in health (e.g. the marked social class gradient in deaths from childhood accidents). Nevertheless, the concept of targets for health improvement was generally thought to be a welcome development.

There now follow four case studies which examine infectious disease, nutrition, deafness, and smoking to illustrate the approaches which could be taken in prevention.

Case studies
Prevention of infection by immunization

You are the chief medical officer for a developed country which has a relatively low incidence of measles: last year there were 1000 cases and two deaths in a population of one million children: both of the deaths occurred in immunosuppressed children. However, recently, there has been an increase in notifications from areas where immunization rates are not high. The national average is 90% but in some parts of the country it is only 80%, which is not high enough to prevent transmission. There has been a press campaign against measles immunization on the grounds that there are serious side effects which have not been publicized.

You decide to implement a national programme of eradication of measles.

Only one disease has been eradicated in human history and that is *smallpox*. Eradication is theoretically possible for measles but much more difficult than for smallpox since there are subclinical cases and also the disease is highly infectious in the prodromal phase (before the rash appears).

Prevention of measles uses the *medical model*. However, although immunization is a medical technology, obtaining a high uptake rate requires an efficient delivery system and a high measure of confidence among parents.

Additional Information
Health of the Nation targets for children

- To reduce the overall suicide rate by at least 15% by the year 2000
- To reduce the incidence of gonorrhoea by at least 20% by 1995, as an indicator of HIV/AIDS trends
- To reduce the death-rate for accidents among children aged under 15 by at least 33% by 2005
- To reduce the death-rate for accidents among young people aged 15–24 by at least 25% by 2005
- To reduce smoking prevalence of 11 to 15 year olds by at least 33% by 1994 (to less than 6%)
- To reduce the proportion of men and women aged 16–64 who are obese by at least 25% and 33%, respectively, by 2005

Summary
Immunization and vaccination

Immunization = Process of conferring immunity

Vaccination = Introducing the vaccine into the body

Vaccination does not always lead to immunization!

Additional Information
Smallpox eradication

Features that allow total eradication:
- No subclinical disease
- Contagiousness low and accompanies skin lesions
- Easy to identify contagious patients
- No prolonged carrier state
- No animal reservoir
- Effective and inexpensive vaccine with no need for cold chain

The campaign required the following:
- Feasibility demonstrated in individual countries before campaign began
- Collaboration of general public and mass media
- Strong links between field-workers and research scientists
- Indentification and containment of smallpox foci rather than 100% vaccine coverage
- Continued surveillance until certification of eradication

Additional Information
Consent in under-16s: the Gillick principle

Children under the age of 16 can give legally effective consent to surgical or medical treatment, independent of their parents' wishes, provided that they have sufficient understanding of their condition and what is proposed

It is agreed by the national immunization and vaccination committee that a two-track approach will be used: first, a national campaign based in schools to prevent a resurgence of measles; and second, a local approach through immunization co-ordinators to maintain a high uptake and ensure that all cases of measles are detected.

The national campaign is intended to achieve a very high level of immunity in the population at risk. It is aimed at the school-age population as children now over 8 were immunized at a time when uptake nationally was < 70%, so the level of protection is low. A mass campaign is planned to immunize every child aged between 5 and 16 years, in school, over a one-month period. School is chosen because the population is captive and it is possible to combine the campaign with an educational exercise about measles.

The injections are given in school by school nurses. It is agreed that consent forms will be sent out by the schools and brought back by the children. At a headteacher meeting, the question is raised what will happen about children who forget to bring their consent forms?

A reminder will be given through the school in the second week of the campaign. It is agreed that secondary schoolchildren may give their own *consent*, provided that the parents themselves have not withheld it (see also Chapter 3, p. 61).

Parents start phoning their GPs with queries about side-effects of the injection, and questioning the need for vaccination in a child who had already had the disease. How should these queries be dealt with?

Anxieties by parents over the side-effects of vaccines are known to be significant in reducing the uptake and have been stimulated by adverse press publicity and campaigning groups who oppose vaccination. Pertussis vaccine uptake has been particularly affected, following concern over possible 'brain damage'. This was analysed in the *pertussis encephalopathy study* in 1981.

It is important to provide full information to parents on *side-effects of vaccines*. A phone helpline is set up to operate in office hours during the duration of the campaign. Parents' queries are dealt with sympathetically and more detailed information is sent out on request.

The campaign thereafter proceeds smoothly and the eventual uptake is 95% of the target population.

The *high uptake* now has to be maintained. It is agreed that this will be done through a local strategy aimed at achieving a high level of coverage of all vaccines.

Most vaccines are given in the first 2 years of life by primary care teams, following a *vaccination schedule*, and GPs are paid according to whether they have achieved a target of 90%.

A successful programme depends on the enthusiasm and motivation of GPs, health visitors, practice nurses, clerks, and computer handlers who input the results.

A district immunization group is set up to include the district co-ordinator (a paediatrician), the consultant in communicable disease control, a health visitor, a GP,

Additional Information
Pertussis encephalopathy study

- The history was studied of every child admitted to hospital with neurological illness between 1976 and 1979

- Of the first 1000, 17 had recently had pertussis: 1 died, 1 became disabled, and 1 had speech delay

- 35 children had had DPT within 7 days before their illness: 8 of 11 children damaged had no other explanation

- The risk of persisting neurological damage was calculated at 1 in 300 000 immunizations

DTP = diphtheria, tetanus, pertussis

Summary
Side-effects of vaccines

Oral polio
Vaccine-induced polio: one recipient and one contact case per two million doses

Diphtheria
Redness and swelling at injection site; small painless nodule; malaise; transient fever; and headache

Tetanus
Redness and swelling at injection site; headache; lethargy; malaise (uncommon); acute anaphylactic reaction (rare)

Pertussis
Local reaction; fever, crying; encephalopathy; prolonged convulsions—less than 1 in 300 000 doses

Measles
Fever; rash—7–10 days after injection; convulsion (rare)

Mumps
Parotid glandular enlargement in 3rd week after injection; mumps meningoencephalitis 1 per 300 000 doses using Urabe mumps virus vaccine

Rubella
Arthropathy; thrombocytopenia (rare)

Hib vaccine
Local swelling and redness at site

Hib = Haemophilus influenzae type B

Summary
Achieving a high uptake

- Immunization co-ordinator: with responsibility for a district

- Regular communication, updating of professionals

- High level of motivation among primary care team

- Parent and child-friendly baby clinics held at accessible hours

- An efficient computer system for recording results and identifying non-immunized children, with feedback of uptake rates to practitioners

- Close attention to notifying cases of immunizable diseases

- Facility for home immunization for regular defaulters

- Media publicity promoting immunization

Summary
Schedule of vaccination in the UK

2 months
1st diphtheria, tetanus, pertussis (DTP), 1st Haemophilus influenza type B (HiB), intramuscular (IM)
1st oral polio

3 months
2nd DTP/Hib
2nd polio

4 months
3rd DTP/Hib
3rd polio

12–15 months
measles, mumps, rubella (MMR) (IM)

3–4 years
diphtheria, tetanus booster, oral polio
MMR 2

11 years
BCG (intradermal)

15 years
diphtheria (low dose), tetanus booster, oral polio

and members from the health education and child health computer departments. It is agreed that the following means will be used to increase uptake:

- audit of vaccination returns to computer;

- training of health visitors in vaccination; and

- provision of an improved information leaflet and video, and local press coverage of immunization.

A year after these measures were instituted, the rate for primary vaccination has increased to 95%.

Prevention of undernutrition and malnutrition

As a GP interested in child health and working in an inner city area you have noticed that there seem to be many underweight young children in your practice as well as many school children who are eating a diet largely composed of chips, hamburgers, crisps, and fizzy drinks. There is a high prevalence of dental caries and iron deficiency anaemia is also common.

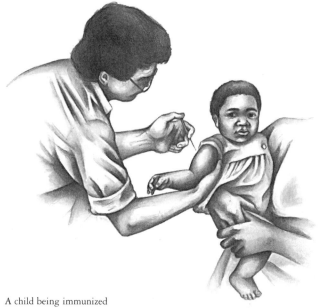

Fig. 4.1. A child being immunized

Poor nutrition is common in children in the UK (see Chapter 1). The main *nutritional problems* are a high intake of sugar, animal fat and salt, and a low intake of fibre. Children are also taking less exercise and this contributes to obesity. Other consequences of a poor diet are undernutrition and short stature, dental caries, and coronary heart disease and hypertension in adult life.

Prevention of poor nutrition—perhaps better thought of as promotion of good nutrition—requires a *social model*, as does prevention of accidental injury. Factors that influence eating patterns are education, cultural views, peer pressure, food marketing, food availability, and cost. The last, cost, is all important for low-income families.

Summary
Nutritional problems in UK children

- High intake of sugar
- High intake of animal fat
- High intake of salt
- Low intake of fibre
- Obesity
- Undernutrition in infancy is also present
- Iron deficiency is toddlers is common

You decide to find out what the nutritional problems are in the families in your practice. In discussion with the health visitor, you agree to concentrate on preschool children. Together, you will find out the breast-feeding rate, assess the extent of iron deficiency anaemia in toddlers, do a dietary survey of 3-year-olds, and find out how common obesity is at school entry (as defined as children who are over the 97th centile on a weight for age chart).

In your practice population of children under 5 years, you find the following:

Breast-feeding at birth	Breast-feeding at 6 weeks	Iron deficiency anaemia at 12–18 months*	Obesity at school entry (> 97%)
40% (63%)	20% (47%)	15% (15–20%)	10% (13%)

* This required a special survey using a haemoglobinometer, at the time of measles, mumps, and rubella (MMR) vaccinations. National figures are given in parentheses.

The dietary survey shows that 3-year-olds have a high intake of fizzy drinks, a high intake of sugar and saturated fat, and a high intake of additives in their food.

You decide to concentrate first on prevention of *iron deficiency anaemia* (see also Ch. 3, p. 48, Ch. 17, p. 295). Your practice has a mixed population with a significant number of families of Asian background (mainly Pakistan and Bangladesh). The main dietary problems appear to be early introduction of doorstep milk (around 6 months of age), late weaning, and unbalanced weaning diet with insufficient iron-containing foods.

You plan to develop a screening programme for children at high risk of iron deficiency, and an educational programme which the health visitor will deliver.

Summary
Prevention of iron deficiency anaemia

- Screening of those at high risk
- Case detection by high index of clinical suspicion
- Primary prevention through health education:
  - use of infant fomula up to a year
  - use of cereals containing iron
  - vitamin C containing-drink at meals increases iron absorption
  - encourage use of iron-containing weaning foods
- iron fortification of weaning foods

Screening for iron deficiency anaemia has been considered for inclusion in a child health promotion programme but there are considerable difficulties. It is doubtful whether iron deficiency anaemia fulfils the *Wilson and Jugner criteria* (Table 4.1) for a screening programme. Hence, there is more support for the concept of *high risk, and 'opportunistic' screening*.

Your screening programme will be targeted at those thought to be at high risk but will also be offered to any parents who wish it for their children. The education programme will include the use of a video and also participation in practical feeding sessions attended by a paediatric dietician. You plan to evaluate the benefit of this approach over the next 2 years, using parental knowledge, dietary intake studies of iron-containing foods, and haemoglobin levels.

Additional Information
Distraction test

- Contented infant, quiet room
- 'Tester' 1 m to side and behind infant's field of vision
- 'Distractor' fixes infant's attention
- 'Tester' signals to 'distractor' to avert gaze
- 'Tester' uses high pitched rattle in line with ear
- Infant turns to sound
- Repeat for opposite ear

Table 4.1
Wilson and Jugner criteria for screening programmes

Condition should be an important health problem
- There should be an acceptable form of treatment or intervention
- Natural history should be known
- There should be recognizable latent stage
- There should be a suitable/acceptable test
- Facilities should be available for diagnosis and treatment
- There should be agreed policy on whom to treat
- The treatment should improve the condition
- Cost-effective
- Case-finding should be continuing process

Additional Information
High-risk and 'opportunistic' screening

High risk screening: carrying out a screening test on a population thought to be at high risk. Thus, a screening test for IDA could be carried out on all children from Asian families, or on all children starting raw cow's milk before a year. This strategy is successful only if the high-risk population includes a very high proportion of those with the condition—otherwise more cases will be in the low-risk population

'Opportunistic' screening: carrying out a screening test when a patient presents. Thus, a screening test for iron deficiency anaemia (IDA) could be carried out on all children attending the surgery with a problem between the ages of 9 months and 2 years

Evaluation of health education is necessary but not easy. The expected outcomes are change in knowledge, change in attitudes, and change in behaviour. The first two can be achieved relatively easily but change in behaviour depends as much on environmental factors (e.g. income level) as on knowledge. It is important to assess knowledge and attitude change as not all forms of education are equally effective.

Prevention of deafness

As the paediatrician responsible for audiology in a large rural district of England, you are concerned that children with sensorineural hearing loss (the main congenital type of deafness) are picked up too late, often not until they start school. You decide to review and audit the programme for early detection of sensorineural hearing loss.

Sensorineural hearing loss is present in 3 per 1000 of the population and 1 per 1000 is congenital (see also Chapter 7, p. 132). Primary prevention is not possible as yet and secondary prevention is normally applied (*screening*). The commonest approach is the *distraction test* usually carried out by a health visitor when the infant is 6–9 months old. Often, this is used in combination with parental observation, which means asking parents to report early any anxieties about their baby's hearing. If parents do have concerns about this, they are usually right.

The distraction test requires rigorous quality control if it is to be carried out effectively on a total population. There are several *questions* which must be asked about a screening programme to test its efficiency and effectiveness.

You arrange to meet with the key people in the district, namely the paediatrician responsible for the surveillance programme, a health visitor, the ear, nose, and throat (ENT) surgeon who has an interest in deaf children, and the child health computer manager. You decide first to review the information collected on the child health computer from the surveillance programme.

The child health surveillance or child health promotion programme is a national programme designed to detect impairments early, to provide support and assistance to parents, to monitor growth and development, and to promote health in young children. A national standard is set by a working party between the professional groups involved (the Hall report, 1996, *Health for all children*). The Hall report sets out a *Child Surveillance timetable*. There is a national computer system whereby data is collected on immunizations, on screening tests, and also (usually)

Summary
Methods of screening for hearing loss

- Active parental and professional education programme
- Selective ('high risk') neonatal screening plus education programme
- Distraction test at 6–9 months
- Universal neonatal screening plus education programme

or other combinations of the above!

Additional Information
Questions to ask about a screening programme

- Is the coverage known and does it approximate 100%?
- Are the test conditions satisfactory?
- Do the testers follow a standard protocol?
- Is the hearing of the testers checked regularly?
- Is there a standard referral pathway following the test?
- Is there a suitable 'secondary level' hearing test available?
- Is data kept on the number of children referred?
- Can the final outcome be linked to the screening test to determine specificity?
- Is there a way of detecting false negatives (to determine sensitivity)?

Additional Information
Child surveillance timetable

Neonatal examination (see also Ch. 8, p. 154)
Full physical examination, including weight and head circumference
Check for congenital dislocation of the hip (CDH) and testicular descent
Inspect eyes, view red reflex
Test hearing if in high risk category

6–8 weeks
Physical examination
Weight and head circumference
Hearing if in high-risk category
Health discussion

6–9 months
Look for evidence of CDH
Distraction test of hearing
Discuss development, feeding, injury prevention

18–24 months
Confirm walking with normal gait
Confirm speech produced and comprehension
Haemoglobin level if local policy
Height and weight
Discuss behaviour

36–48 months
Height
Orthoptic screen if local policy
Health discussion

School entry
Height
Vision
Hearing

Summary
Sensitivity and specificity

Sensitivity
Ability of a test to give a positive finding when the individual screened has the condition

Specificity
The ability of the test to give a negative finding when the individual does not have the condition

on height and weight. Normally, the attendance of the child, the result of the test, and whether a referral is made are recorded. However, with the distraction test, if the child fails the test, he/she is usually recalled for a repeat and then may be referred to the GP to check whether otitis media is present before being sent on to the ENT/audiology clinic. The more re-tests are carried out the more chance there is of a deaf child falling through the net, as the parents may not attend for genuine reasons but then may not be recalled.

You agree to look at the statistics available on the distraction test. These show that the coverage is 83% in a population of 4000 children under 1 year of age, hence 680 (17% of 4000) children were not screened. Only 5 children were referred from the 3320 children screened whereas the expected prevalence is 10 (3 per 1000), hence you suspect that there are more cases among the unscreened population.

Another possibility is that the screening test is missing children with the disease (false negatives). Four of the children referred actually had a hearing loss, hence there are few false positives and the test has *sensitivity* of 80%.

Unless a screening test has a high coverage (over 95%), cases may be missed and it is well known that the population screened may have fewer cases than those unscreened. Hence, every effort should be made to screen the whole population. It is also known that if test conditions are poor (e.g. a room which is not sound-proofed, or distracting materials on the walls) or if the tester is not properly trained, then cases will be missed.

The group decides that greater efforts will be made to ensure high coverage by giving better information to parents and notifying health visitors if a parent fails to attend for an appointment. GPs will be asked to permit health visitors to refer directly to audiology after 2 failed tests. Funds will be made available to upgrade test rooms and provide refresher courses for all health visitors (and also to test the health visitors' hearing!).

As a result, the coverage goes up to 95% and two years later, 12 children are referred of whom 9 have hearing loss.

Prevention of smoking

You are the school doctor for a large secondary school in England and have been aware for some time of the high prevalence of smoking in teenagers. Each time you visit the school at dinner time there are groups of young people smoking in small groups at street corners. Smoking is banned in the school grounds but it is well known that many teachers smoke and their common room usually smells strongly of tobacco. Pupils are known to obtain cigarettes singly at nearby tobacconists.

The newly appointed headteacher has expressed his concerns about the high level of smoking and has suggested that the school undertake a campaign to make the school a genuinely non-smoking environment.

Reduction in teenage smoking is one of the targets in the UK government's 'Health of the Nation' White Paper, 1992. The current prevalence nationally is 22% of 15-year-olds. There is particular concern over smoking in teenage girls which appears to be rising. Figure 4.2 shows current prevalence figures for smoking in UK. Teenagers are heavily targeted by tobacco advertising, which is still legal in the UK, despite much pressure from doctors and other health professionals. It is illegal for tobacconists to sell cigarettes to children under 16 years of age, but this law is widely contravened. Educational campaigns aimed at children to encourage them to stop smoking are not very successful although there is a high level of awareness of risk in young people. However, many teenagers are not averse to risk-taking, indeed revel in it. Teaching about the effects of smoking is mandatory in all schools under the national curriculum, but the quality and enthusiasm of teaching on this subject varies widely.

You decide together to adopt a 'healthy school' policy with a target for '**no smoking school**' in three years time. All members of the school staff will be involved.

You decide to start by carrying out a survey of smoking habits in pupils and staff, and then hold a workshop for staff to present the results and discuss means of cutting down smoking. You seek a group of sixth-formers to co-operate with the campaign.

The type of prevention in operation here is primary prevention. It will require collaboration between health and other disciplines. *Questionnaires* are regularly used in school to determine the prevalence of risk-taking behaviour and are generally thought to be reliable, provided that they are anonymous.

Health promotion campaigns with young people are very unlikely to be effective without their co-operation, as they are jealous of their independence, have strong views on life, and are resentful of the interference of adults. Evidence shows that they are health-conscious and prepared to undertake health-promoting measures but often these are too expensive or too difficult. They are prepared to co-operate in planning as long as their views are respected: this is an important principle of UN Convention on the Rights of the Child (see also Chapters 1, p. 22 and 24 p. 398).

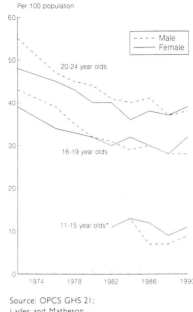

Source: OPCS GHS 21;
Lader and Matheson
1991

*England only

Fig. 4.2. Trends in smoking by age, Great Britain 1972–90. (from Woodroffe *et al.* 1993).

Additional Information
'No smoking school' strategy

- All staff to adopt the strategy and develop a personal target for stopping

- Pupils to be consulted about the ways to discourage smoking

- System of monitoring and feedback to pupils set up

- Help to be offered to pupils to give up

- Active education programme in classrooms

- Parents and governors to be involved

Additional Information
Factors influencing smoking in teenagers

- Peer pressure (seen as 'cool' to smoke)
- Teachers who smoke
- Children in school selling single cigarettes
- Easy availability of cigarettes from nearby shops and ice-cream vans
- Advertising in women's magazines
- Boredom

Additional Information
Measures used to combat teenage smoking

National
Ban tobacco advertising and smoking in public places

Local authority
Police under-age sales
Promote non-smoking pubs and restaurants
Promote non-smoking schools

School
High profile to non-smoking
Offer help to smokers to give up
Good quality anti-smoking education throughout the curriculum
Work with pupils
'Healthy School' award

A questionnaire survey is carried out of years 9 and 10 (13 to 15-year-olds) and shows that 15% of boys and 21% of girls in this age group smoke regularly (1–6 weekly). You also discover that 25% of staff smoke but most of these would like to give up. Discussion with a group of sixth-formers provides the information shown in the additional information box.

Prevention of teenage smoking is a major problem. Education alone is unlikely to be effective in preventing it. Some health promotion experts hold that all young people will wish to experiment whatever happens and preventive measures are useless—it is better to help young people give up. However, this ignores the highly addictive nature of cigarettes, and the wide variation in prevalence in different countries. It is agreed that a combination of *legislative and environmental measures* are likely to be successful.

A planning group is set up by the health education co-ordinator with the school nurse and a group of sixth-form pupils. Meanwhile, it is decided to offer smoking cessation sessions to teachers and to include smoking education in the curriculum for all pupils. The sixth-formers recognize that many pupils start smoking as a challenge to teachers' and parents' authority and the low achievers are most prone to take up the habit. They agree to offer peer support to young people who would like to give up, and to request an increase in extracurricular activities, sports, and clubs so that boredom is not such a major factor.

But what about the sale of cigarettes to under-age children in the local sweet-shops and newsagents?

According to the Department of Health, 250 000 children are sold cigarettes illegally each week—some 17 million cigarettes. The *Protection from Tobacco Act* (1992) gives local authorities greater power to stop this happening.

The best deterrent to illegal sales is by making test purchases with the help of children, usually aged between 10 and 13. The volunteer is accompanied by a Trading Standards Officer, who fines the shopkeeper if the law is contravened.

The sixth-formers agree to help to identify shops known to sell cigarettes to under-age children. Some of the younger pupils also volunteer to make test purchases in other areas of the city, where they are not known. A 'Smokebusters' group of 9- to 13-years-olds is set up with the assistance of the school nurse. This provides peer support for non-smoking and also raises awareness among the younger pupils. A further evaluation will be carried out to test the effectiveness of the interventions, using a control (non-intervention) school for comparison.

Conclusion

All doctors working in the field of child health will be able to undertake some aspect of prevention, which may range from the conveying of health information to advocacy. It is important to be consistent in the approach that is taken, to work with colleagues both within and outside the health sector, and to use techniques which have been subject to evaluation. Screening programmes should not be set up unless there is a high likelihood that they will do more good than harm!

Further reading

Department of Health (1992). 'Health of the Nation'. HMSO, London.

Department of Health (1996). *Immunisation against infectious disease*. HMSO, London.

Hall, D.M.B. (ed.). (1996). *Health for all children*, (3rd edn). Oxford University Press.

James, J., Evans J., Male P., Pallister C., Hendrikz J.K., and Oakhill A. (1988). Iron deficiency in inner city pre-school children: development of a general practice screening programme. *Journal of the Royal College of General Practitioners*, **38**, 250–2.

Summary
Children and Young Persons (Protection from Tobacco) Act 1992

- It is the duty of the local authority to consider, once a year, taking enforcement action

- It is an offence to sell cigarettes to any child under 16, regardless of how old they look.

- Special notices warning of the law must be displayed at all points of sale

- The provision of cigarette vending machines if used by a child may lead to court action

- Fines for breaking the law are increased

5

Child abuse and neglect

There are few situations in the life of a medical student or doctor, which are more difficult or painful, than those arising when deliberate injury or preventable misery of a child is presented within a health context. The physician should be aware that his or her contact with a child or family may prove to be the watershed event that helps to change the future for both.

An open-minded and non-judgemental approach is essential. Anger, horror, or disbelief are natural and unavoidable emotions that may be overwhelming, but may impede the development of the rapport and compassion, which accompany all other dealings with families of sick children. It is important that the physician does not feel isolated or burdened by an undue sense of responsibility. Extensive societal structures have been created, precisely because such situations are rarely simple and none may be wholly resolved by medical intervention alone.

Children at risk

Child abuse and neglect occurs in all sectors of society. Studies show that it is more prevalent where there is disadvantage or stress within communities. These situations include those where mothers are young and unsupported, or have had unrewarding or damaging childhoods themselves. Parents who have endured abused childhoods may lack role-models from which good child care comes; they may have poor insight into the needs of their child, or into their own needs. They may choose unsuitable or violent partners and may not have the resilience to cope with the violence, poverty, and neglect which their own partners bring into their households. Poverty, debt, poor housing, and loneliness may be part of a spectrum of stress that lead to falling standards of care.

An understanding of these complexities is essential if a child's well-being is to be established within the family and solutions found. There are very few parents who seek to harm their children deliberately, and it is always important to view the presentation of abuse in a child as a sign or marker of serious family distress. Evaluation of the child alone will not bring the answers to that child's needs. Criminal and civil legal processes should only be invoked with care and forethought.

Case studies
A bruised swollen eye

John is a 2-year-old boy who is brought to your attention, as the paediatric senior house officer because of a bruised ear. It is 10 p.m. at night and John's 19-year-old mother, Lisa, has attended with two children. She also describes persistent crying in the younger sibling, Maria, who is 10 months old. Maria's father, Gary, who is not the father of John, does not come into the department; he waits outside in the car park. You examine Maria and find she is physically healthy, but looks thin.

Her clothing is somewhat scanty and she has severe nappy rash. John appears tired and pale. You note further linear yellowish bruising over his buttocks and upper thighs.

The scenario you face suggests that John and Maria's mother has concerns about her children and is, in essence, presenting a 'cry for help'. Approach to this parent should include a full history of both John and Maria's problems, which should include a gentle exploration of the family history and a direct question about the origin of John's injuries. Given the lateness of the hour an admission to hospital is appropriate for Maria and John with their mother. It is important at this stage to consult the *Child Protection Register*.

Summary
The Child Protection Register

This register is maintained by social services and is a record of children who are at risk of child abuse. A child may be entered on the register as a result of a decision by a case conference (see below). The register may be consulted at any time by a health professional who is worried about the possibility of abuse in a child

Summary
Protection orders

Emergency Protection Order
Under the Children Act (1989), section 44, a court may make an order if satisfied that there is reasonable cause to believe that a child is likely to suffer harm. It gives parental responsibility to the local authority in addition to the routine carer. Such an order may only last for 8 days. The court may extend it for a further 7 days

Police Protection Order
Section 46, allows the police to prevent the child's removal from hospital for a period of up to 72 hours

John's mother tells you that John bumped into a door a week ago—'He is always bumping into things and bruises easily'. She is anxious to leave and tells you that her husband will not countenance her stay. You attempt to engage with Maria's father and bring him into the department. Gary is hostile, verbally abusive, and tells you that unless the children need emergency treatment he will not let them stay.

The findings in John also indicate a recent injury over the ear incompatible with the history given, the linear bruising is also compatible with contusion from a belt repeatedly applied, and the mark on his arm looks like a human bite mark (see Figs 5.1 and 5.2). Maria is poorly cared for and appears to be failing to thrive. Both children appear to be children at risk and there is strong suspicion of non-accidental injury in John. Immediate referral should be made to the social services and police. *Protection* of the children could be sought through a variety of orders.

Following the children's admission to hospital a child protection investigation will take place. You should contact the family's GP and health visitor to give and collect information. John's injuries should be further investigated with a skeletal survey, and coagulation profile and his injuries should be properly documented through photography (see Figs. 5.1 and 5.2 below). Evidence about Maria's growth and care will be provided by the health visitor and GP. Her personal health record should be requested as it will provide valuable information on growth and visits to the doctor. The hospital doctor will provide information about Maria's eating and weight gain in hospital and, if appropriate, will institute an investigation of her growth.

The social worker will interview the family and examine the dynamics, explore strengths and weaknesses, and also look for more appropriate safe placement (e.g. with another family member). The police will investigate the source of John's injuries, and may interview each parent.

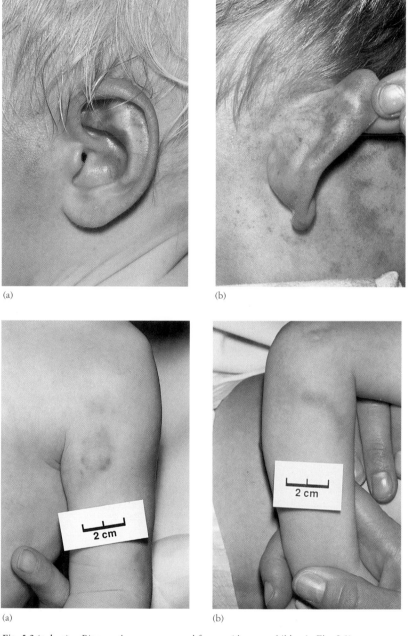

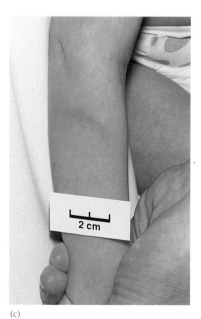

Fig. 5.1 (a, b). Ear bruising in a 2-year-old.

(a)　　　　　　　　(b)

(a)　　　　　　　　(b)　　　　　　　　(c)

Fig. 5.2 (a, b, c). Bites on the upper arms and forearm (the same child as in **Fig. 5.1**).

Not all children need to be admitted to hospital for investigation of suspected abuse: indeed, admission should be avoided if at all possible as it adds to the child's emotional distress. The investigations can be carried out as an outpatient, and if a place of safety is required outside the family then a foster placement can usually be found at short notice.

Summary
Case conference decisions

1. To place the children on the *Child Protection Register*
2. To remove the children from their parents under a *Care Order*
3. To place the children in foster care with parent's agreement with *no Order*
4. To return the children to their parents under a *Supervision Order*
5. That the children return to their parents with *no order*

Additional Information

Physical abuse is defined as actual bodily trauma to a child which has been deliberately inflicted. Such an injury may include bruising, fractures, bites, burns, and scalds, or the administration of toxic substances

Non-accidental injury is sometimes used initially when a child presents with an injury for which there is not an adequate explanation

The outcome of all these processes is shared in a *case conference*. The children's parents should be kept closely in touch with all that is happening and should be invited to participate in such a conference. Certain decisions will be made at the *case conference* (see also p. 95).

At the case conference it was agreed that the injury to John was caused by trauma, and Lisa states that he had been in the care of her partner the night before. She says that she has split up with him now because of his violence and is living alone. She has always found Maria difficult to feed but is managing better with the help of the health visitor. It is agreed to place her children on the Child Protection Register on a Supervision Order, and review the situation in three months.

Clinical presentations of injury

Non-accidental injury.
Physical abuse to children is the most overt form of abuse to a child, detected by the assessment of visible injury and the finding of hidden trauma, (e.g. fractures). There are several features which should lead the doctor to suspect *non-accidental injury*.

Accidental injuries
to children are common, once a child is mobile. The rough and tumble of play is likely to account for a small number of minor bumps and bruises. Such injuries usually are seen over bony prominences in the young child (e.g. the parietal bone, chin or nose, the knees and shins of more mobile youngsters). Bite marks may be inflicted by another child, scratches and abrasions during play with pets or sharply contoured toys.

Inflicted injuries
should be suspected if multiple soft tissue sites are involved, particularly the buttocks or back of the thighs. Injuries to the abdomen, genitals, chest wall, inside the mouth or to the ear lobes should cause concern.

Instrumental trauma
(or adult human bites) will be recognized by their characteristic shape and pattern. Tears to the upper lip and frenulum occur either by direct blows or by forced feeding, or rough jamming in, or a pacifier. Pinch marks and squeeze marks often appear as multiple oval bruises resembling fingertips. Bruises to the neck may be due to choke injuries, linear petechial haemorrhages due to twisting of a child's collar, or pulling a child up off the ground by the clothing. There may also be linear bruising (see Fig. 5.3).

Munchausen syndrome by proxy (MSP)
or induced-illness syndrome is a rare and unusual form of abuse which should be suspected when a child presents with features of illness which have no apparent cause.

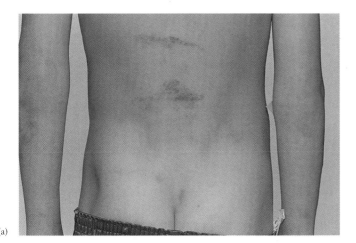

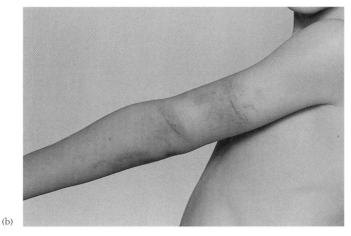

(a)

(b)

Fig. 5.3 (a, b). A 9-year-old—beaten with a billiard cue. The pattern of injury is linear, supporting the history.

An ill baby with conjunctival haemorrhage

Jenny is 8 weeks old. You are the family's GP and are called urgently to the home at 2 a.m. in the morning. Jenny's father tells you that he woke to offer Jenny her feed, and found her pale, unconscious, and limp.

Jenny's parents have been married for 4 years. They have a healthy elder daughter aged 3 years. Both parents are in professional jobs, and Jenny is cared for during the day by her grandmother. You have had no previous worries about her care.

Summary
Features suggesting
non-accidental injury

- The appearance of the injury does not reflect the history given because of:
  - its appearance and age
  - the developmental stage of the child
  - the injury represents a known pattern of deliberate injury (e.g. a hand or belt mark)
- There are multiple injuries of several ages
- The history is inconsistently offered
- The injury is not explained, 'I just found him like this'
- There is delay in seeking medical care
- There are other worrying factors or clinical signs.

Summary
Munchausen syndrome by proxy (MSP)

This is a complex form of child maltreatment, in which a child's carer (usually the mother) deliberately and repetitively fabricates or induces symptoms in the child to bring him/her to medical attention. Examples include deliberate administration of salt, or the introduction of blood or sugar into urine or stool samples. Children may suffer prolonged hospitalizations before diagnosis. They may also undergo painful tests and procedures and be required to accept unnecessary and potentially harmful drug treatments

When you arrive at the home, you find Jenny very ill. Focal fits have begun 10 minutes before you arrive. On examination, Jenny is not febrile, she is twitching on the left side, and looks very pale. There is nothing else to find, apart from a small conjunctival haemorrhage in the right eye.

Her parents tell you that she has been fractious all evening, but perfectly well and there were no other symptoms.

Summary
Shaking injury

- Conjunctival and retinal haemorrhages
- Rib fractures
- Metaphyseal fractures
- Subdural haematomas
- Bruising to chest wall
- History of persistent crying
- Unconscious infant with convulsions or anaemia

Clearly this baby is ill and requires urgent transfer to a paediatric unit. The only clue to her altered status is the haemorrhage in the eye. You should consider the possibility of a *'shaking'* injury. It would be entirely inappropriate to explore this issue at this stage and Jenny's immediate safe transport to hospital should be your paramount concern.

In hospital, Jenny is found to have a right subdural haematoma and on skeletal survey a metaphyseal fracture of the right distal tibia is found. The paediatrician has involved social services and the police are investigating the origin of these injuries. Jenny is in intensive care and she is making progress, but is likely to have residual neurological handicap. There are signs of a developing left hemiparesis.

The social services department arrange a *case conference* which you attend. Information about the family is shared, and you are told that after initial denial by both parents, Jenny's mother acknowledges shaking Jenny roughly on the night of her admission because she had refused a feed and was crying incessantly. She has told the investigating workers of her own exhaustion and feelings of depression and is asking for help.

The case conference has now learned who is responsible for Jenny's injuries. Both parents have been open about their needs. As their GP you will have a major role in mother's future care, and management of her depression. The conference is likely to place the child on the Child Protection Register and to discuss ways in which rehabilitation of Jenny with her parents may occur. This is likely to include a full assessment of the family's needs and of Jenny's future care needs.

The willingness of parents to acknowledge their role in a child's injuries and their willingness to change and accept help usually indicates a good prognosis for therapeutic intervention.

The shaken baby syndrome

Infants and children under the age of 2 years may present with multiple injuries associated with violent shaking events. Many of these injuries may not be evident on initial examination. These injuries are commonly the result of the shearing forces of rapid unco-ordinated accelerating and decelerating forces. The child is picked up around the chest, shoulders, or neck. The rib cage is firmly held or squeezed. Violent shaking in a young baby with poor muscle tone results in repeated hyperextension of the neck and spine and unco-ordinated flailing of the limbs. These forces results in rupture of vessels in the conjunctiva and retinae and bridging veins in the brain. Tears of the falx cerebri may occur. Lateral rib fractures because of compression, or posterior rib fractures because of shearing forces, are also noted. The growing ends of the long bones are particularly susceptible to these forces and chip or separation fractures occur at the metaphyses. Such fractures are virtually pathognomonic of deliberate injury and are rarely, if ever, seen in accidental injuries.

A scalded child

James is 22 months old. He is brought to the local clinic where you are the community doctor. He is accompanied by both parents, grandfather, an uncle, and three other children, the eldest of which is 9 years old. James has a superficial scald over his left shoulder and upper chest. There is no blistering and most of the area of the scald is dry and in areas scabbed over. He is miserable and in obvious discomfort.

This little boy has been brought to medical attention after a significant delay (the scalds are not fresh). It also seems clear that he has not had pain relief and has been distressed for several hours. A full history of the circumstances of the family, living conditions, and any previous encounters in the past should be taken from

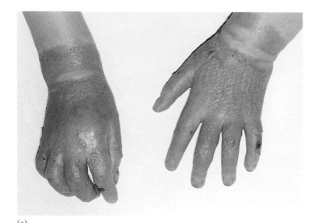
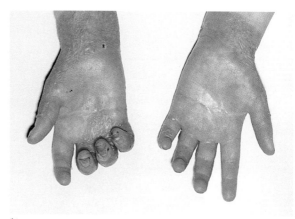

(a)   (b)

Fig. 5.4 (a, b). Classical pattern of immersion scalds (a 22-month-old). Classical pattern of deliberate immersion scalds. Note the clean demarcation lines and bands of dorsal sparing where the wrists were held hyperextended—the expected reaction to such a painful process.

Summary
Burns and Scalds (see also Ch. 9 p. 182)

- Accidental burning in childhood is not uncommon and up to 80% occur within the home.

- They are more frequent in winter months and at times of the day when stress is at its highest in the family

- Accidental injuries relating to spills of hot liquids from kettles, pans, or cups may be extensive and severe, but most often are small, involving the upper limb, chest wall, and shoulder

- Flow is rapid and cooling simultaneous with the spill; the findings are therefore of scalds of variable depth with splash marks and 'trickling' effect

- Immersion scalds are characteristic (see Fig. 5.4) with a clear demarcation line and the involvement usually of the distal part of the limb

Summary
Neglect

Failure to provide essential care of a child (e.g. food, warmth, or exposure to danger, indifference to or denial of a child's physical, medical, and educational needs)

the family and then from the health visitor and the GP. The Child Protection Register should be checked.

The GP tells you that less than 6 weeks ago, Jamie sustained a fall against an unguarded fire in the home and was treated in hospital. The health visitor tells you that the home is a two-bedroomed flat belonging to the grandfather and that Jamie's parents, who are in considerable debt, have moved in. Four children and four adults have lived in the flat for several months. The older three children are rarely in school and standards of hygiene are poor. Neighbours have reported that the children, including Jamie, are out in the street, late in the evenings and often poorly clad.

You have a significant amount of information to make a decision that this child and his siblings are at risk in the home. There is no evidence that Jamie's injury was deliberately inflicted, but there is strong evidence for lack of care and poor insight into Jamie's needs. Therefore there is an issue of **neglect**. Although a formal child protection case conference may not be necessary, an inter-agency planning meeting to examine the needs of the family should be held urgently. Some of the risk to Jamie could be reduced by rapid rehousing and education of his parents in home safety. An assessment of their ability to accept such intervention will be needed.

Growth patterns

One of the most objective markers of the health of a child is the physical growth of the child. Careful and serial analysis of weight and height progress in children provides important clues to poor or deteriorating health.

'Failure to thrive' was a descriptive term developed to describe apathetic, wasted children living in long-term state institutions. The same term is often now used to describe children who fail to grow for no apparent reason of illness ('non-organic failure to thrive') and where psychosocial factors are suspected to be critical (see also Chapter 14, p. 256).

However, it is most important to recognize that growth in children is influenced by calorie intake, calorie absorption, rest, exercise, hormonal influences, and metabolic requirements in disease and health. Trauma and neglect in children will impact on their growth through a combination of these influences. The withholding of food from an infant will result in a clearly observed failure to gain weight, rapidly correctable by adequate replacement. On the other hand, an abused child may fail to grow because of abnormal food intake mediated by apathy and anorexia, difficult to resolve without therapeutic intervention other than food. The unhappy child may be stunted in height without necessarily demonstrating a serious fall in weight achievement. Neglect may coexist with organic disease states, and severely neglected children are predisposed to more low-grade infection and ill-health.

The chart (Fig. 5.5) is taken from 'A child in trust', a public enquiry report into the death of Jasmine Beckford, who died on 5 July 1984. She was then $4\frac{1}{2}$ years old and was in the care of the local authority from the age of 20 months. She was admitted at that age with a femoral fracture. After a period of 7 months in foster care she was returned to the care of her parents. At her death over two years later, she weighed less than she did in foster care and had suffered numerous soft tissue and bony injuries in the weeks and months leading up to her death.

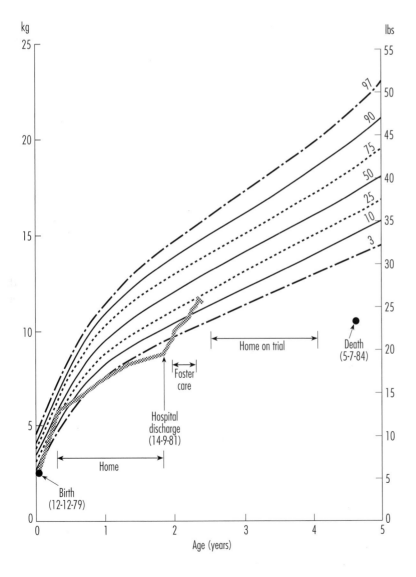

Fig. 5.5. The weight chart of Jasmine Beckford, who died of neglect in 1984. It shows good weight gain in hospital and foster care, and weight fall off at home. (From 'A Child in Trust', public enquiry report into the circumstances surrounding the death of Jasmine Beckford. London Borough of Brent, 1985.)

Case study
An anxious withdrawn child

Sarah, a 6-year-old, has been noted at school to be withdrawn and isolated. Previously bright and bubbly, she has tended to be clingy to her teacher recently. She has wet herself on two occasions and has once fallen asleep during a lesson. Her single-parent mother has recently become pregnant to a new boyfriend, Stuart. Sarah shows anxiety at the end of the day and sometimes shows fear and refuses to go home when her mother's boyfriend comes to collect her. Her teacher asks you, the school doctor, to see Sarah and an appointment is made. However, Sarah's mother does not bring Sarah to school on the day of the appointment.

Sarah is presenting worrying signs suggesting that something untoward is happening in her life. To discover more about her you could speak to her family GP, and the health visitor from the practice. Make a telephone call or visit the practice. Share your concerns.

The family's GP tells you that Sarah's mother is not attending antenatal appointments. She is 21 years old and was 15 when Sarah was born. She has lost the support of her own mother because of a family row. The health visitor reports that Sarah's mother and Stuart drink heavily and that the home is visited by several of Stuart's acquaintances, who sleep nights and sometimes take care of Sarah.

You now have a lot more information to suggest that Sarah is at risk. You should make a further attempt to see Sarah with her mother at school. You and the school nurse could visit the home to talk to mother about Sarah's behaviour. You should also consult the Child Protection Register and discuss the situation with the consultant community paediatrician. The meeting with Sarah's mother should be properly planned and you should have some idea of what you wish to offer Sarah and her family. Social worker support to the family and/or assessment by a community team (e.g. the child and family guidance psychologist could be helpful). A meeting involving yourself, the GP, teacher, social worker, school nurse, Sarah's mother, and Stuart could be arranged so that discussion is open and Sarah's problems can be properly aired. The family may have explanations for Sarah's upset and ideas about the way forward.

When you visit Sarah's mother and Stuart they speak to you on the doorstep and tell you they have no concerns. They are angry that the school staff have involved you and tell you of their intention to keep Sarah away from school and find another placement for her.

You now face a difficult problem. The family deny a situation you know to be present and it is also likely that Sarah is going to suffer from:

(1) a loss of her peer group;
(2) loss of educational time; and
(3) loss of an adult (teacher) whom she trusts.

This situation should be reported to the local social services as a child protection matter, as you have evidence of potential harm. You must also consider the possibility of sexual abuse.

A planning meeting with social services can be held. This meeting could have a representative from the police present. The police and social services may have information about Sarah's carers which is indicative of risk to Sarah.

However, there may be no further information. With an open, frank and non-judgemental approach, especially from a non-threatening professional (e.g. the family GP), Sarah's mother may be willing to acknowledge there is a problem and that Sarah's interests are paramount. If she is unable to do so, consideration may be given to legal proceedings. At this stage, however, there is little specific evidence for abuse, and it may be that Sarah's interests are best met by a non-confrontational friendly approach and by sharing information with the new school health and educational staff so that there is heightened vigilance. She may be given special attention. Staff must be alert to any distress and should listen to her with care.

Several months after Sarah's problems were brought to your attention, the police investigate an allegation made by another young child, the daughter of Sarah's mother's boyfriend in a previous liaison. She is 11 years old and has felt able to talk about sexual molestation by Stuart which went on for several years before he left her home.

The strategy is now much more clear. Given the previous concerns about Sarah's behaviour, a case conference should be called. This will provide a forum where Sarah's mother's views may be sought again. Stuart will be advised to leave the home, at least for a period, while investigations are underway, and this period should be used to involve Sarah and her mother in therapeutic work, directed at uncovering any fears Sarah may have.

There are many situations where a suspicion of abuse may be not acted upon, because evidence is too slim for any statutory intervention.

Additional Information
Recognition of sexual abuse

- A child may make a direct statement to a peer or adult about abuse. Such statements should always be taken seriously

- Children rarely, if ever, lie about such matters; when indeed, such statements are thought to be unfounded the gravest concern should exist about the reasons and background to the child's statement.

- Children may present with changes in behaviour, disturbances in conduct, school progress, or peer interaction. These may be non-specific (e.g. depression, anger, anxiety), or more specific (e.g. self-harming or sexualized behaviour)

- Physical symptoms include urinary and stool incontinence as a direct outcome of trauma, and vulvo-vaginal infections including sexually transmissible disease

- On rare occasions, the situation may present with acute symptoms of genital bleeding, bruising, and swelling

Additional Information

Emotional abuse implies harm inflicted by acts of verbal or behavioural aggression which damages the child's self-esteem, cause fear and anxiety, and which prevent healthy emotional development.

Emotional neglect implies the absence of thoughtful, loving interaction with a child, periods of privation, isolation, and a sense of non-existence in the mind of the parent. These factors often coexist with the more direct abusive situations of chronic criticism, scapegoating, or severely punitive styles of management.

Sexual abuse

The existence of sexual exploitation of young children has been recognized for many centuries. In the 1980s and 1990s the issue has been brought increasing into public debate as health workers have learned to recognize the variety of presentations.

For child sexual abuse to occur, two preconditions need to exist. First, there should be an individual who is predisposed to sexually abuse children. Such a person needs to have access to a child and needs to overcome inhibitions. This is more likely to occur if alcohol or drugs are involved, however, as the second precondition is usually fulfilled such influences are unnecessary. The second precondition is the vulnerability of the child victim. Factors that influence a child's vulnerability are developmental immaturity, susceptibility to coercion, absence of a protective parent, and poor experiences of nurturing. Most child protection workers recognize that although isolated, unpredictable sexually traumatizing events occur in the lives of children, in the majority of children, sexual abuse occurs as part of a spectrum of adverse experience. It is also true that perpetrators of abuse are most commonly someone known to the child, often in a position of trust.

Behavioural markers

Children subjected to non-accidental trauma, sexual abuse, and those who live in emotionally depriving environments are highly likely to suffer significant developmental damage, even when there are no permanent physical sequelae. Observations of such children may reveal certain behaviours that appear characteristic. A withdrawn, apathetic state, or alternatively, a hyperalert, still child who smiles little, a baby whose feeding is anxious and gulping in nature, loss of appetite, and pallor are all features which should arouse the gravest concern. Emotionally neglected children may display self-stimulatory behaviours like rocking or head-banging. The sexually exploited child may display inappropriate boundaries of behaviour, for example, seeking to touch or stroke adult carers or engaging in compulsive masturbation or simulated sexual activity with toys or peers.

Evaluation of the abused child

An understanding of aetiology and family systems predisposing to abuse, should underpin the clinical approach to suspected child abuse.

Communication with the family should be intense and with the intention of obtaining a wide perspective. Pre-judgement, anger, or hostility will impede such an approach. The best way to ensure a proper evaluation is to adopt a systematic approach to the *investigation of child abuse*.

Management of child abuse

In 1991, the Department of Health produced a document entitled 'Working together' to guide professionals towards proper arrangements for child protection. All health districts, and individual hospitals and health units should have procedural guidelines accessible to all professionals. Each local authority has an area or district child protection committee. This is a body made up of representatives from health, social services, police, education, and voluntary bodies whose role is to design a strategy for child protection practice for the area which it serves.

The case conference

The case conference provides a forum for the development of an understanding of an individual child's needs and the way forward to secure that child's safety and future good care. The views and knowledge of various professionals and carers are sought in a discussion which should be open and frank and which should include the child's parents and other adults who may have contributed or wish to contribute to his or her upbringing.

The responsibility for calling such a conference rests with the social services department and is influenced by guidelines laid down by the Department of Health. When a doctor, health worker, or any other agent suspects abuse or is concerned about a child's care, notification of that concern is the start of the process leading to a case conference.

The objectives of the case conference are crystallized into outcomes described as 'decisions'. One of these is the 'decision to register'. The *Child Protection Register* is a central list held by the social services department in each district. The list records the names of children deemed to be 'at risk'. There are firm *criteria* for deciding that a child is a person at risk of harm.

The Child Protection Register (CPR) will give information to concerned professionals when requested (e.g. a doctor dealing with an injured child in an accident and emergency department). A child whose name is placed on the CPR with or without legal sanctions remains on it up to such time as sufficient change in his or her circumstances occurs. Once a child is on the CPR regular reviews occur to reassess the continuing need for registration and a care team meets regularly to provide for needs. Table 5.1 shows the number of children on the CPR nationally (England and Wales) in 1990 and 1991.

Table 5.1.
Number of children on the Child Protection Register in different categories (1990 and 1991)

Type of abuse	1990	1991	Ratio/1000 (1990)	Ratio/1000 (1991)
Neglect	3700	3300	0.34	0.31
Physical abuse	7100	6700	0.65	0.62
Sexual abuse	4200	3900	0.38	0.36
Emotional abuse	1200	1300	0.11	0.12

From Department of Health Statistics (1992). 'Registration data for years ending March 1990 and March 1991'.

Summary
Investigation of child abuse

- Full family history including social supports, physical and developmental examination, diagrams including measurement and dating of all injuries
- Coagulation profile
- Growth charting
- Skeletal survey: healed and fresh fractures
- Developmental assessment
- Swabs for semen detection (sex abuse)
- Swabs for trace material and items of clothing (sex abuse)
- Screening for sexually transmitted disease (sex abuse)
- Reconstruction of events (e.g. home visit may be helpful in non-accidental burns)

Summary
Criteria used in the Child Protection Register

- Actual physical abuse or future potential physical abuse
- Actual emotional harm or potential for future emotional harm
- Actual or future risk of sexual abuse
- Actual neglect or risk of neglect

Prevention strategies

Child maltreatment is a societal issue, rooted in complex interpersonal relationships and the often unpredictable coming together of a series of environmental factors. Prevention is, thus, intensely difficult and follows the social rather than the medical model (see Chapter 4, p. 67). At a simple level one may consider three strands to a prevention process.

1. *Primary*. Strategies to prevent abuse involve a clear understanding of potential danger. For an individual child, this could mean the early provision of support to a young mother, sensitivity to the stresses of poverty and unemployment in a family, material and befriending provision for children and parents in distress. At a community level, examples of primary prevention include programmes like 'Don't shake the baby' campaign, prevention of under-age pregnancies, and early parent-craft education in secondary schools. There is now much emphasis on programmes to promote positive (violence-free) parenting and health visitors are key health professionals in this exercise since they have access to every parent. There is an organization in the UK (EPOCH, End Physical Punishment of Children), which is seeking legislation to ban corporal punishment in the home, a move successfully carried through in five European countries. In the USA, a programme of teaching children about sexual abuse has been increasingly tested in institutions.

2. *Secondary*. This refers to early detection of abuse, particularly of emotional neglect, scapegoating, and neglect. All primary health care services have a role in such detection, and most child protection agencies are heavily involved in such work. The National Society for the Prevention of Cruelty to Children (NSPCC), established in the late 19th century, has done much to bring to public attention the fact of abuse in our midst and their creation of a facility for a member of the public to report anonymously is an example of secondary prevention.

3. *Tertiary*. This refers to the recovery of children who have been maltreated. Compensatory nurturing, high-quality tailored mental health services, early acknowledgement of damage and of need, may allow some children to grow into adults who can fulfil the needs of their own children. The availability of proper resources both within the criminal justice system and within psychotherapeutic health services and social service systems, could make it possible to treat some people who maltreat children and reduce their potential to harm in the future.

Further reading

Addcock M., White R., and Hollows A. (1991). *Significant harm*. Significant Publications, London.

Department of Health (1989) *Working together under the Children Act*. HMSO, London.

Hobbs, C.J., Hanks, H.G.I., and Wynne J.M. (1995). *Child Abuse and Neglect: a clinician's handbook*. Churchill Livingstone, Edinburgh.

Newell, P. (1989). *Children are people too. The case against physical punishment*. Bedford Square Press, London.

6

Emotional and behavioural problems

- Parenting and stress

- Simple behavioural management techniques for parents

- Case studies
  sleep problems in babies
  food refusal in toddlers
  'the terrible twos'
  school-age children: attention deficit
  hyperactivity disorder (ADHD)
  urinary and faecal incontinence
  conduct disorder
  pre-teen and teenage children: anorexia nervosa
  obesity
  anxiety and phobias

- Sources of help and indications for referral

If they grow up not to love, honour and obey us
either we have brought them up properly
or we have not:
if we have
there must be something the matter with them;
if we have not
there is something the matter with us.
R.D. Laing (1970) Knots. Tavistock, London

It is a depressing fact of life that, as parents, all of us pass on our emotional baggage to our children, no matter how hard we try not to. Good parents have learned to understand their own self and are also able to empathize with their children, that is, be able to 'put themselves in their shoes'; this, in turn, requires a good theoretical understanding of the stages of emotional development of children (which parents will not automatically acquire). Hence, it is rarely the case that any emotional disorder in a child can be dealt with in isolation from the dynamics of his or her family or group of carers.

Parenting and stress

John Bowlby is primarily credited with the recognition that early strong emotional bonding (initially with one adult) underpins secure normal development. Insecure attachment may result from such factors as maternal depression, lack of parenting and illness, or early institutionalization, although one or two traumatic separations from a caring adult early in life does not usually result in long term emotional damage. Some children are resilient to disadvantage, but frequent changes which prevent the formation of emotional bonds during the first two years (e.g. parent care followed by a series of foster carers) often cause serious emotional damage with an increased risk of later delinquent behaviour.

Chapter 1 has described the recent demographic and social trends responsible for increasing family disruption during childhood. All of these social problems may adversely affect a child's emotional development, although many are resilient enough to develop normally despite disadvantage.

What seems to be important is that the child develops a sense of their own self-worth and a feeling of positive regard for a significant other person (usually a parent). Even in adverse circumstances these develop through positive involvement with the parents, other persons outside the family, or in interests and hobbies.

The most consistent parts of a baby's or young child's environment are the home and parents. Children need to feel loved, secure and valued, and to know the boundaries they should not go beyond. This can be engendered in positive ways, through praise and encouragement in the main, with the occasional retribution for stepping 'beyond the mark'. Most people learn parenting by example from their own parents. Even though a young person may say 'I'm not going to smack my children as my father did' and carry this through, other features learnt from their own parents will persist (e.g. in their daily routines and patterns of interaction).

Deprivation and health

The combination of poor housing, low incomes, uncertain health, insecure employment (coupled often, no doubt, with limited knowledge of parenting skills) offers a prescription for low achieving, poorly behaved, disenchanted or alienated young people. It represents a 'prescription for anti-welfare' for children.
P. Wedge (1984)

Where there is a lack of routine and consistent management children can become difficult and out of control. This may occur in families where there is no social disadvantage and no pathological interaction. There may be lack of parental involvement because of excessive attention to work or television, or abuse of drugs or alcohol. Such families may care for their children physically, but do not give them 'quality time'. A child may not identify with such emotionally detached parents, will not want to please them, and may develop a conduct disorder. Conversely, some families apply extreme discipline but lack supportive warmth. These parents may be over-restrictive, hostile, ridiculing and sarcastic, causing the child to become anxious and withdrawn. The most serious sequelae occur where this translates into physical, sexual, or emotional abuse.

Simple behavioural management techniques for parents

Good parenting is very analogous to good teaching. The parent/teacher has a wealth of knowledge and experience to impart which might benefit the child, but there are some well established ideals of good parenting.

We *do not own our children*: they are their own (autonomous) person; they are only entrusted to our care and if we abuse this trust society has a right to protect them against us. They should be as autonomous as their stage of development allows. They should not be punished for things that they could not have been expected to know. Conversely, they must be induced into learning that they are not alone in this world; that they have to 'share and share alike' and adopt the general social values that distinguish a civilized society from an uncivilized one. Educational theory teaches us that people (even adults) do not learn well by simply being told what is the right thing to do. People 'learn by their mistakes'. The best way to help a child to learn is to allow him/her to practise and make mistakes in a supervised environment and to learn that the 'mistakes' result in a worse outcome for them.

There is no evidence that physical punishment of children is on balance in any way beneficial. On purely ethical grounds, given acceptance of the child as an autonomous person and their weak and dependent position in relation to adults, there are few or no circumstances in which physical punishment can be regarded as other than abuse. It also provides children with a model for bad behaviour that they are likely to copy. There are other ways of imposing order: even these must be used fairly and consistently.

The most important thing is that the parent does not use withdrawal of love as an instrument of punishment. This is not justified because it is not true: with rare exceptions we do not cease to love our children because of what they do and we should not play on their inbuilt insecurity by implying that. The first step is therefore to show affection and to reinforce that 'mummy/daddy still loves you' while asserting that the present behaviour of the child is unacceptable for some given reason.

In using these methods it is important to achieve a balance such that positive rewarding interactions are more frequent than negative ones. Too much of the latter can only lead to rejection or rebellion. Also, if the only attention a child gets

Summary
Behavioural management techniques

- Reward desirable behaviour. An increase in positive behaviours leaves little room for negative ones

- Ignore irritant but not serious bad behaviour

- Time out. The child is helped to understand that his/her behaviour is not socially acceptable. Not a withdrawal of love but simply an opportunity to 'cool off'

- Positive behavioural re-enforcement

- Give an incentive. If the incentive is excessive the child will perceive the desired behaviour to be of similarly excessive importance

- Stop a treat. Needs to be carefully used to avoid breakdown of trust

- Diversion. Particularly helpful and easier to achieve in young children

is through punishment then he/she will seek to 'be bad' more often. Hence, it may be important to concentrate on changing one or two aspects initially and rewarding behaviours that are part way towards the desired behaviour (putting away one toy in the box rather than all of them). This is referred to as 'shaping behaviour'.

Parenthood is about creating civilized, socialized, caring human beings who achieve whatever is their personal potential. It is essential that parents (and other adults) provide a good model. Parents are, however, only one (albeit the main) influence: at certain stages of child development the pressure on a child to conform to peer group culture may cause rejection of parental values.

Case studies

Under the age of 1 year it is a good general rule that if a baby cries there is a good underlying reason—it is his or her main way of drawing attention to physical and emotional needs. Emotional needs are important even in the newborn. The normal baby rapidly becomes highly bonded to the mother. A baby has no concept of 'coming back': separation is therefore the baby's greatest anxiety—as far as he/she is aware mother may have gone forever when she goes to the bathroom. The needs of normal babies for intensive physical and emotional attention (particularly demanding for lone parents) underlie many of the complaints that mothers present to the primary health care team. Understanding of these complaints demands a good knowledge of the range of normal child development.

Sleep problems in babies

A 22-year-old mother of Simon, a first baby of 10 weeks, complains to her GP that he is driving them mad because he won't sleep. He is difficult to settle and he usually cries and wakes two or three times a night.

As many as one in five mothers seek help for crying/sleeping problems but these rarely indicate physical disease. Although there is an obvious relationship between crying and night-waking video studies show that many babies are 'quiet wakeners' who do not disturb their parents: night-waking is therefore more the rule than is perceived by parents. A few babies have patterns of particularly intense and prolonged crying, again almost always not related to physical disease (although this must be excluded). As babies develop other ways of communicating, crying becomes a more selective response to, for example, pain or loss of dignity.

'Why won't he sleep through the night? My friend's baby is about the same age and he hardly ever wakes up now?'

The friend's baby may be a 'quiet wakener' because it is unlikely that any baby has an 'adult' sleeping pattern at this age. In the first 2–3 months an average baby sleeps for about 15 hours per day, almost equally divided between day and night, sleeping for at most 4 hours and waking for perhaps 3 hours at a time. Rapid adaptation to a more mature pattern takes place by 3–4 months when many babies sleep for a full 8 hours at night although there has not yet been much reduction in the total amount of sleep.

> 'Is it because he's hungry. Can I start him on solids?' Simon is bottle-fed but seems to be hungry all the time. His mother is not sure whether he 'just needs a cuddle', because sometimes this seems to be enough to settle him, or whether he needs to be weaned on to solid food although she has read that he is a little young for this.

The meal size of bottle-fed babies may be more constant because it is easier for the parent to control intake: conversely, being more demand-led, breast-feeding may reduce the risk of overfeeding and subsequent obesity. On average, breast-feeding mothers continue to give a night feed to the baby until about 4 months of age, whereas bottle-fed infants tend to be denied this from the age of about $2\frac{1}{2}$ months: consequently, bottle-feeding mothers are often forced to introduce solid food earlier so that the infant sleeps through the night (see also Chapter 3, p. 48).

Questions to ask. What is the parental level of understanding of normal child development and the source(s) of their information? What exactly happens—how do they settle Simon, what do they do if he wakes up? How alert are they to his needs? Is the physical environment conducive to sleep? What is the baby's overall sleep pattern including daytime sleep? What are family lifestyle patterns—are they expecting Simon to fit into an unreasonable pattern? Is there any indication of parental illness (especially maternal depression)?

Application of behaviour management techniques to this age group. The general technique of rewarding desirable behaviour and extinguishing unwanted behaviours is applicable even to young babies particularly if it is used to support natural developmental progress (e.g. the development of a mature sleep pattern). General aims would be to encourage natural development of a diurnal cycle by reducing daytime sleep, increasing daytime stimulation, decreasing evening stimulation and minimizing rewards for waking up (by meeting genuine physical demands (e.g. feeding, changing) with minimum fuss). Apart from grandparents and other lay sources, health visitors are the main professional resource. First-time mothers especially often find contact with peers through mother/baby groups extremely supportive. Referral for management of emotional or behavioural disturbance would be extremely unusual under the age of 12 months.

Food refusal in toddlers

In many aspects of his life the toddler, at a stage where he is acquiring many new capabilities and beginning to see himself as an independent person, can have this exciting new freedom curtailed by an adult. The toddler is dependent on adults for many things but they can't control the child's bodily functions over which he retains complete autonomy. Most children assert their independence through refusing to eat, sleep, or toilet train in the socially acceptable ways that their parents are trying to impose. This happens to a greater or lesser degree depending mainly on the personality traits of the child and the consistency with which parents 'set the limits'.

Darren is $2\frac{1}{2}$ years old. His mother (who is a few months pregnant with her second baby) says Darren is driving both parents round the bend. 'He won't do anything he's told, he won't eat his meals properly and throws his food around and at people.'

Questions to ask. Ask the parents to write down everything Darren eats for 2–3 days. This may show the child is actually eating reasonably or can reveal excess milk consumption or that large amount of sweets, crisps, and biscuits are being eaten.

Principles of management. Some children will eat only a very limited range of foods for a time without coming to any harm. If a child does not appear to be eating or is very fussy about food it is wise not to exacerbate the situation by spoon-feeding/getting cross, or in other ways giving attention to eating. The child should be fed at the same time and alongside other members of the family (preferably at the table) and offered small amounts of the foods the adults are eating (including finger foods and lumps or pieces of food). The child's eating should not be discussed in his presence but the mealtime should proceed normally. If the child cleans the plate he is briefly praised but, if not, the plate should be taken away after 10–15 minutes and the whole matter ignored. Coaxing and bribery are counterproductive and should not be used.

Summary
Food refusal in toddlers

- Children of 2–3 years often seem to have a poor appetite

- The rate of growth is much slower than in the first year

- Parents become anxious when the child eats less and attempt to bribe or force eating

- The child enjoys the attention and manipulates parents at mealtimes

- If growth is adequate as it usually is, parents should be reassured that intake is sufficient and that the child should be left to eat what he/she wants, as long as a balanced diet is offered (toddlers never starve themselves)

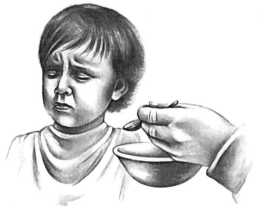

Fig. 6.1. Food refusal.

Food fads. Children sometimes develop a strong dislike for certain foods or textures (e.g. pieces of meat or carrot). If this is the case the child should be given more of the foods it is known he likes (provided it gives a fairly balanced diet) and may be offered the disliked food from time to time. However, the diet should always include some lumps and variation in textures.

The 'terrible twos'

Darren's mother complains that he is also always on the go, 'into everything' often breaking things as he goes. He never seems to settle for even a few minutes, just going from one thing to another all the time. He won't go to bed and hardly seems to sleep at all waking up at about 6 o'clock every morning. He used to bang his head on his cot for ages every night but he seems to have stopped that now. 'Is he hyperactive, doctor?'

Fig. 6.2. Child having a tantrum in a supermarket.

This is a trying stage for many parents. The normal child has attained high mobility and all sorts of new skills which he wants to explore and try out. At the same time, at this stage he is entirely self-centred and incapable of seeing another person's point of view. He therefore cannot understand why he should not simply do anything he wants to do and this can lead to apparently antisocial behaviour. Most parents have read about the so-called hyperactivity or hyperkinetic syndrome (see below) in one or other of the many child-care books for parents. At the toddler stage, in the absence of developmental delay or other abnormality it is usually impossible to distinguish true hyperactivity from normal extremes of variation and poor 'limit-setting' by the parents. Many toddlers are very active, needing almost constant supervision to prevent accidents, and their attention span is very limited. Finally, head-banging is a less usual variant of a number of comforting behaviours, like rocking, masturbation, and thumb-sucking, that the normal child indulges in when tired, bored, or unsettled. These behaviours can become abnormally fixed in retarded or autistic children but disappear in the majority by the preschool years.

'He also gets terrible temper tantrums. If he doesn't get his own way he just throws himself to the floor and bangs his head with temper. Once, about a year ago, he got so annoyed he held his breath until he went blue and passed out. We had to call the doctor then but she said it was all right and not to worry.'

This is still within the range of normality. After refusing to eat, sleep, or toilet train, refusing to breathe is the ultimate weapon in the child's armoury of assertive behaviours. Fortunately, few children carry independence this far and the few who do usually stop this extremely disquieting behaviour after a few attempts, especially if the parents can steel themselves to respond calmly without the panic which the child intends to provoke. The doctor must distinguish these breath-holding attacks from loss of consciousness due to other causes, especially when they mimic convulsions (see Chapter 20, p. 339).

Questions to ask. Are either of the parents (especially father) similarly hyperactive? Do his mother and father disagree about how to control Darren's unacceptable behaviour? How consistently do they apply their limits and rules—does Darren sometimes 'get away with it'? Are there any potential sources of insecurity for the child (e.g. father away a lot, marital disharmony)? Do they set aside 'quality time' for him (e.g. reading, interactive play), or is he left to watch videos most of the time? How does he behave with other children? What do they expect from the doctor?

Principles of management. The parents need a more thorough understanding of child development, the great range of normal variation, and 'how the child's mind works' (i.e. empathy). Once they see things from Darren's point of view they will realize that it is not in his interests to be inconsistent or overindulgent. At the same time, they must minimize any potential sources of insecurity for Darren, provide a constant background of affection and be equally consistent in recognizing and rewarding him with 'quality time' for good behaviours. This is a crucial stage for development of socially acceptable behaviour. Excitable, highly active children do not like to be restricted. The parents should concentrate on modifying the most unacceptable behaviours, like violence towards other children and excessive shows of temper, while trying to allow him as much freedom as they safely can. Consistency is essential: Darren will quickly learn to manipulate his parents if they react to his 'bad' behaviour in different ways. Finally, some parents have read about drug treatments or dietary restrictions (see p. 107). At this stage, use of one of these would distract parents away from the central problem of establishing consistent limit setting by affectionate and non-violent means.

School-age children: attention deficit hyperactivity disorder (ADHD)

Because a certain degree of overactivity is often normal in toddlers, diagnosis of abnormal hyperactivity is not usually considered unless it persists into the school years.

Ryan is 8 years old and has always had trouble at school. He doesn't concentrate on his work, is always out of his seat, interrupts the teacher and talks all the time. He can't keep himself to himself and annoys other children. At home he is constantly 'on the go' and doesn't seem to need much sleep.

A small percentage of children, mostly boys, may be unusually hyperactive, excitable, and destructive, with short attention span, constant restlessness, and poor concentration. Many adults (again, mostly men, and often successful people in business or professions) are similarly hyperactive and self-centred with 'butterfly minds' and these traits may be passed on to children whether through heredity or upbringing. Delineation of ADHD is therefore contentious and best defined by expert psychological testing. ADHD is much more commonly diagnosed in the USA than elsewhere in the world.

Ryan is always flitting from one thing to another, seeming to have a very short attention span. Even when apparently settled at for example, drawing or watching television he is always fidgeting. He seems impulsive and disinhibited, shown by angry outbursts or tearfulness for no apparent reason. He also shows lack of awareness of 'personal space', always intruding on other people, and will often take apparently high risks in climbing and exploring.

These features might indicate ADHD, but they could also be due to conduct disorder (e.g. due to emotional damage from insecure attachment, poor discipline, example, and limit-setting by parents, etc.).

Ryan's mother has read about food colourings being a cause of hyperactivity and she has stopped him having tomato ketchup and orange juice, but it doesn't seem to have made any difference.

There is some evidence that up to 80% of hyperactive children are affected by tartrazine (an orange food-colouring additive) and other substances such as cow's milk (64%), grapes (50%), wheat (49%), oranges (45%), and eggs (39%). Apart from the hyperactivity itself, other symptoms such as headaches or abdominal pain have responded to a change in diet in double blind studies conducted at the Maudsley Hospital in London. In severe cases an oligo-antigenic diet (lamb and chicken, rice and potatoes, banana and apple, a vegetable, vitamins and minerals) may be introduced under supervision of a specialist dietician followed by gradual re-introduction of other foods.

Questions to ask. Are there any antecedents of possible emotional trauma? Are the features of ADHD confirmed? Is there a familial pattern? Are there any suspected precipitants/sensitivities (e.g. to foodstuffs)?

Management includes the following:

- Behavioural management and/or cognitive therapy.
- Dietary changes (see previous remarks).
- Drug treatment.

Methylphenidate (Ritalin) is a stimulant that paradoxically, does not increase activity in these children, but may improve attention span, concentration, and co-ordination. It has little effect in children under 5 years of age. All cases of hyperactivity in children of 6 years or over should in any case be referred for specialist assessment. Methylphenidate should be prescribed only on specialist advice, usually as an adjunct to psychological treatment.

Urinary and faecal incontinence

Incontinence of the bladder or bowel at an age when most other children have acquired an adult pattern of control is the most common problem at this age. It is usually yet another example of normal variation although parents may find this difficult to accept. Even this normal delay in maturation may be a cause of emotional problems in the child. Conversely, gross or persistent antisocial behaviour (such as deliberate smearing of faeces on furniture or other people) may be a symptom of severe underlying psychiatric disorder.

Summary
Attention deficit hyperactivity syndrome (ADHD)

Hyperactivity is thought to affect 3–5% of the school-aged population in the USA (ratio 3 boys: 1 girl)

- Poor sustained attention, easily bored, constantly shifting activity
- Impaired impulse control or delay of gratification
- Excessive task-irrelevant activity (fidgeting, can't sit still)
- Difficulty in following through instructions
- Early onset (mean age 3–4 years)
- Problem less obvious in a one-to-one situation
- Normal IQ

Summary
Enuresis

- Most children are dry by age 3 but 10% of 5-year-olds have nocturnal enuresis

- 1 in 20 children are still not completely dry by age 10

- Male predominance

- Wetting by day does not imply greater likelihood of an organic cause

- Enuresis is almost always functional

- Organic causes are found in only 1/100 cases and can generally be excluded by urine testing and culture

- Management is by positive re-enforcement, 'star chart' for young children, desmopressin/buzzer in over-7s

- Tricyclic antidepressants should not be used because of side-effects and high relapse rate

Enuresis

Sam, a healthy 7-year-old, attends with his father. Although he can sometimes stay dry for several days, he wets his bed most nights and sometimes wets himself during the day. Sam himself finds this distressing and mother tends to be annoyed because the sheets have to be washed almost every day. Father is a bit more sympathetic because he himself was a 'bed-wetter' until about 9 years of age and is still embarrassed about it.

At school age, the indignity and constant sense of failure engendered by *enuresis* may cause serious emotional problems and hinder other aspects of mature development. A punitive, blaming approach by the parents is very unhelpful but, fortunately, increasingly rare with greater public awareness of the real nature of the problem.

Questions to ask. Have there been long periods of dryness (relapse might suggest possible urinary tract infection)? Is there a family history (unfortunately tends to predict delayed resolution)? How does the child feel about himself (problems of self-esteem)? What is the attitude of the parents (sensitive and understanding or punitive/disciplinarian)? What have they tried so far? How do they think the doctor can help?

Principles of management. It is most important to help the child maintain his self-esteem. The approach must be highly positive with no 'putting down' of the child. In preschool-age children the parents may try lifting the child to go on his potty before they themselves go to bed but little more should be done unless or until the problem persists into the school years. By the age of 5 or 6 it is worth trying 'star charts' where the child records his dry nights and is praised for them. Battery-driven pad or sensor alarms which buzz when electric contact is made by small amounts of urine are the most successful method. Tricyclic drugs are *not* now recommended because of toxicity, but desmopressin (anti-diuretic hormone) nasal spray may be used at night for up to two months if an enuresis alarm has not worked. Desmopressin is also useful for short periods where dryness is essential (e.g. holiday camps) and for helping initiate the use of buzzer alarms.

Faecal soiling (encopresis) (see also Ch. 14, p. 258)

Mark is a 7-year-old boy who soils his pants two or three times a day. According to his mother he was toilet trained at about 3 years but often had accidents. The problem became a lot worse when Mark started school and he's never really been clean since. He does not seem to know when to defecate and his pants are found to be dirty three or four times each day. When he was younger, he frequently screamed on defecation and streaks of blood had been seen in his motion.

Bowel control is still precarious in most youngsters and going to school often precipitates a lapse in control. By the age of 7, however, almost all children are fully continent. As with enuresis, delay is generally thought to be maturational and organic causes (e.g. Hirshprung's disease, see Chapter 14, p. 259) are exceptionally rare. Constipation, irregular bowel habit, and faulty training are all predisposing factors. Faeces being even less acceptable than urine to most people, the potential for emotional disturbance in the child is correspondingly much greater than with enuresis.

Mark doesn't seen to know when he is soiled and seems not to care. He won't even go and try on the toilet now.

If the soiling has gone on for some time it is likely that the lower part of the rectum has become stretched and insensitive. Parental anger or punitiveness may aggravate toilet refusal, leading in extreme cases to rebellious behaviour and smearing or other provocative behaviour.

Mark's mother doesn't think he is constipated or 'bunged up'. She says 'If anything he is really loose, it just seems to dribble out of him all the time'.

Leakage past a solid mass of faeces (constipation with overflow) is the most common cause of soiling: no real progress can be made until this mass is cleared (see Fig. 6.3).

Mark's mother tries to get him to eat fruit and vegetables, but really he prefers to eat chips and burgers.

There should be persistent effort to increase dietary fibre but by this stage laxatives (usually lactulose and Senokot) are commonly needed and occasionally even enemas or inpatient treatment.

Mark wants to go on summer camp with the cubs. His mother wonders if he will be better by then.

Additional information
Differential diagnosis of soiling

Chronic constipation
- History of constipation in infancy
- Often painful defaecation before potty training
- Large amounts of faeces in rectum
- No period of normal defecation

Primary behaviour disorder
- Constipation not a prominent feature
- Other behavioural disturbance
- Faeces often hidden
- Often soiling is secondary—follows period of normal defecation

Fig. 6.3. Mechanism of soiling in childhood. Leakage past a solid mass of faeces is the most common cause of soiling.

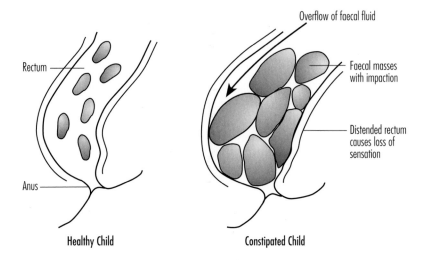

A major motivator may help, but rapid return to normality is unlikely. Prolonged and integrated physical and psychological management with a 'no blame' ethos and small rewards for daily achievement needs to be maintained often for as much as two years.

> Abdominal examination revealed palpable faeces, a loaded colon and many hard faeces could be felt in the rectum. Mark was treated initially with quite high doses of Senokot and lactulose which helped empty his loaded bowel and the parents were given instructions on trying to encourage him to go to the toilet regularly. A 'star chart' and incentives were used to encourage regular toileting and when successful he was praised strongly for his achievement. Over the next few weeks Mark started to go to the toilet and his soiling improved greatly.

Questions to ask. Has the child had a period of being clean and performing in the toilet? Is the child likely to be very constipated and require bowel clear-out? (Physical examination alone may not always determine this). What are the parents' and the child's attitudes to soiling and has it become an emotional issue? Is there any physical problem?

Principles of management. Initial bowel clear-out as necessary, followed by a combination of physical (laxative) and psychological (reward and praise) techniques. Normally, a routine needs to be established and maintained over a considerable period of time with gradual modification of laxative dosage, (see also Ch. 14, p. 258).

Conduct disorder

Mrs Reed, a young mother of 25, arrives with Rachel saying she is 'at the end of her tether' with this 8-year old girl who is increasingly defiant, rude, and scarcely does anything her mother asks.

Non-compliance and temper tantrums are common in toddlers, who are developing their sense of self and testing for limit-setting by their parents and preventive work can be done at this stage. If difficulties persist beyond the toddler stage, the child may become outwith parental control. Children with abnormal disorders of conduct are generally 'externalizers' who direct their behaviours towards the environment. Boys outnumber girls in the ratio of 10 to 1.

Mrs Reed can't understand how this has happened. 'I have always done everything for her, given her all the things she wanted and which I didn't have when I was her age and she just takes liberties with me. The other night she didn't come home after school and then when she did come in she didn't want the tea I'd made her even though it was her favourite.'

A common causal factor is that the parent has overindulged the child with material possessions while providing poor limit-setting and inconsistent management, with the parent giving in on some occasions but not others. The mother tries to win the child's affection by giving her what she thinks she wants but the child rejects the mother because she wants attention, even negative attention, rather than material goods.

When asked what she has done about Rachel's behaviour Mrs Reed says 'I have tried everything, putting her to her room, stopping her TV and smacking her, but it is all to no avail—she says she doesn't care anyway. At the moment she's gated for the next week'.

Almost always the parent has tried several punishments, but inconsistently, so that frequently the parent gives in to the child's demands for the sake of peace or the child receives a punishment out of all proportion to the crime. Play periods between parent and child can be used to increase the amount of positive attention for the child (rather than negative attention).

Summary
Factors that assist in the development of prosocial and moral behaviour

- Reasonable parental control and limit-setting
- Positive re-inforcement of 'good' behaviour
- Attention and affection without overindulgence
- Provision of a good model
- Non-punitive and non-aggressive approach
- Consistent management and ability to carry through threats/promises
- Ability to discuss and accept child's developing autonomy
- Giving the child responsibility

Mrs Reed states that she has little time for play. She has two other children younger than Rachel and they are also a handful. Her husband works long hours and is usually too exhausted at the weekends for the children. 'We're not badly off though—we've seen that we have a nice house and as I say the children have everything.'

Many parents with problem children have adverse social circumstances and are often single parents struggling to make ends meet. However, as in this case, there are also parents who, because of this struggle or perhaps because of their own experience of parenting, attend in the main to material well-being and are unaware of the concept of 'quality time' to be spent with their children. The father (if present) needs to be drawn in to the treatment plan. Frequently, parents feel that nothing can be done and it can be quite difficult to persuade them to look to the more positive aspects of the child's behaviour and build on this. A no-blame approach is needed, often beginning with an attempt to change some small aspect of behaviour while working on improvement of the parent–child relationship.

Questions to ask and information to gather. How severe is the problem and how long has it gone on? How much distress is it causing to parent and child? Is the child at risk for physical or emotional abuse? What are the parental management issues and is there a balance of positive over negative rewards, consistently used? Is there evidence of a good warm relationship between parent and child or does this need to be strengthened? Are the parents able to carry through any plans and to what extent might other problems interfere with potential treatment plans?

Principles of management. Generally, through positive re-inforcement of good behaviour and the development of a positive parent–child relationship. Non-re-inforcement (ignoring) and occasionally 'time-out' may be used for negative behaviours. There are agreed local and national guidelines for dealing with children considered to be at risk of emotional and/or physical abuse (see Chapter 5).

Pre-teenage and teenage children: anorexia nervosa

Jane's mother is very concerned about her daughter as she is getting very thin and hardly ever eats. Normally a quiet girl, Jane screams and shouts at her mother whenever she brings up the subject. At the moment 12-year-old Jane seems to exist on pizzas, salad, and coke.

Although Jane might seem young, 5% of all anorexics are under 13 years of age and the majority are girls.

Jane sees herself as quite fat, although she doesn't like to use the word. When her mother has mentioned her periods starting Jane just walks away and seems to abhor anything related to her body changing and growing up.

Jane shows many of the features of prepubertal anorexics (e.g. fear of puberty) as well as some features more common in older anorexics (e.g. distortion of body image, fear of fatness, excessive exercising).

Jane's mother is extremely worried and tries to get Jane the things she wants (e.g. new clothes) and Jane promises that she will eat more. But Mrs Brown says 'whenever I mention food she screams and starts running up and down stairs. I just don't know what to do'.

Jane has great control over her mother and the two of them are emotionally enmeshed to a high degree. This is typical of anorexic families, which are often dysfunctional, and family therapy (the most effective form of treatment for younger anorexics) is required.

Questions to ask and information to gather. Is this a temporary dieting phase or are the other features present? Questions pertaining to control and family issues.

Management. All cases of suspected anorexia or bulimia nervosa (binge eating followed by self-induced vomiting) should be referred to a clinical psychologist or child psychiatrist for specialist assessment and management.

Obesity

Judy is an overweight 10-year-old. She has always been rather chubby, but she has taken little notice of the attempts of her parents to change her diet. She eats enormous quantities at mealtimes and likes to snack on crisps and chocolate biscuits between meals.

About 5–10% of schoolchildren are 20% or more above ideal weight. Fat children usually become fat adults through a combination of inheritance and upbringing.

Fig. 6.4. Overweight child being teased in the school playground.

Summary
Obesity

- Becoming more common probably because children are taking less exercise

- Family factors (overweight, conflict between parents) and low self-esteem common

- Treatment is difficult: requires a family- and psychologically-oriented approach

- All the family should follow the same dietary pattern

- Increase the amount of exercise the child takes if possible

- Improving self-esteem is all important

- Aim to maintain the same weight so as to regain a lower centile with growth

Other children call her names and she has difficulty with PE (physical education). She has always been withdrawn and underconfident.

Many overweight children have a familial pattern of comfort eating. Major trauma (often separation from a parent or sexual abuse) may be a contributory factor and low self-esteem is a common feature.

Judy is really concerned about her weight now and wants to tackle it.

At this stage, with constant monitoring and high motivation, Judy may succeed.

Questions to ask and information to gather. Make height and weight measurements and compare with developmental norms. Investigate any possible endocrinological problems (highly unlikely). Ask about family diet and eating patterns and if there are any family problems.

Management. This needs to focus on both the child and the family. With high motivation, regular small targets and an initial focus on no weight gain (especially with younger obese children who will increase in height to compensate) some progress can be made, although relapse is common. Many fat children do not like to exercise (partly through embarrassment) but activity should be encouraged. Swimming and bicycling are usually acceptable.

Anxiety and phobias

Susan hasn't been to school for about six months. The problem began soon after she started secondary school when off with 'flu'. She was back in school again for a couple of days but complained of nausea and stomachache thereafter. There is nothing physically wrong with her although she feels sick most school days and has to stay at home.

This type of school phobia should properly be called separation anxiety as distinct from truancy. Truants usually show a bored and 'don't care' attitude, rather than overanxiety: often they are really lonely and miserable, tending to come from disorganized and uncaring families.

Susan has always been well behaved at school and is attaining well but her parents moved house just before she joined and she she was very clingy at first. Her mother doesn't work, is always around for her, and is involved with the Girl Guides, which Susan also enjoys.

This illustrates that Susan has always been an anxious, introverted type. This predisposes her to developing neurotic-type disorders (internalizing problem), such as panic attacks, specific phobias, habit disorder, or psychosomatic complaints. In cases of school phobia, the parent often has a problem with separation and is enmeshed with the child.

Susan's mother has often tried to take her to school but she has only gets as far as the school gate and runs home. The other children call her 'scaredy cat' and try to trip her up. Susan does her school work diligently at home and watches television with her mother in the afternoon.

The problem needs to be tackled on two dimensions: separation from the mother; and gradual re-integration into school. Other factors include bullying (a common problem for which most schools now have a policy), domineering teachers, or an unfriendly large building.

Questions to ask and information to gather. Is the problem out of all proportion to reality? Are the physiological responses to fear present (e.g. increased heart rate and breathing rate)? Is the child's daily life affected to a major extent? Are there family or other problems (e.g. bereavement)? How long has the child had this problem? Has the problem always been there and/or has there been a recent precipitant? What has been done already? Is there school or other involvement?

Management. Collaboration between the school, educational psychologist, clinical psychologist or psychiatrist is usually necessary for school phobia. For other types of anxiety, a combination of relaxation training and systematic desensitization would normally be carried out by a trained therapist. For psychosomatic and other physical complaints relaxation training, lifestyle management, and withdrawal of inappropriate rewards for avoidance behaviour would be recommended (e.g. in Susan's case watching television is re-inforcing avoidance of school). Specialist referral is recommended in most cases of school phobia and other anxiety neuroses, particularly if there are obsessive-compulsive elements.

Sources of help and indications for referral

Many professionals and lay people may be involved in the management of children with emotional and behavioural difficulties. In the majority of cases the primary health care team (PHCT) is the main source of help, mostly through GPs themselves or the health visitors attached to the practice. Health visitors have a particular role in helping babies and small children, mainly through

Summary
Depressive illness in children

- Feelings of hopelessness for the future and low self-esteem: children younger than about 8 cannot become depressed in this sense because they have not yet developed concepts of 'future' or of 'self'

- Older children and adolescents *can* develop adult-like depressive illness and this may be becoming more common

- The prognosis for emotional disorders in childhood is *not* as good as previously supposed—vulnerability to relapses of affective disorder often persists well into adulthood.

- Academic failure is common

- There may also be a real risk of suicide

- Parents often fail to recognize the signs of depression

- Depressed children usually have multiple problems

- There is a strong family tendency

- Most cases require specialist assessment

Fig. 6.5. An anxious child clinging to mother.

Summary
Drug abuse

- Abuse of drugs and other substances is a major health problem in adolescents
- Average age of initial abuse is declining and younger age of starting is correlated with higher likelihood of continued use
- In some settings abuse is almost 'normal'
- Although abuse may be experimental and short-lived it cannot be ignored because:
  - death can result even from initial use
  - psychologically vulnerable children may fail to develop healthy solutions to problems such as parental control, anxiety, and low self-esteem if substance abuse is used as a ready solution
  - drug use is systematically linked with other problem behaviours which are generally anti-social and damaging to health or welfare of the child (e.g. theft and prostitution)
- The example set by adults (tobacco and alcohol) may encourage experimentation with such substances

Additional Information
'Glue-sniffing'

- Inhalation of volatile hydrocarbons (e.g. sniffing glue or lighter fluid) is a common form of substance abuse affecting large numbers of young adolescents in all areas of the UK, both rural and urban
- The immediate effects, which last up to half an hour, are similar to alcohol intoxication
- Polythene bags are commonly used to concentrate the fumes
- Severe cerebral damage and death have resulted from asphyxia, and/or the direct effects of the fumes (e.g. cardiac arrythmia)
- Some solvents, particularly those used for the 'dry-cleaning' of clothes, may have severe acute toxic effects on the liver and kidneys
- Chronic abuse may lead to severe and permanent renal, hepatic, and neurological damage

advising and supporting the parents. Difficult toddlers and young children are most likely to be seen by paediatricians and/or clinical psychologists, often in association with other professionals (e.g. social workers, dieticians, speech therapists). School-age children are usually referred to child psychiatrists or clinical psychologists, but educational psychologists (who work in schools), social workers and behaviourally trained psychiatric nurses may also be involved. A child guidance service consists of child psychiatrists and psychologists working in a team with social workers. Child psychiatrists usually see children with severe problems such as hyperactivity or anorexia nervosa, especially where medication is likely to be needed. Voluntary agencies, such as Parent-Link and Home-Start, can be very helpful in supporting young mothers with small children.

Further reading

Bowlby, J. (1977). *Child care and the growth of love*. Penguin, Harmondsworth.

Buchanan A. (1992). *Children who soil*. Wiley, Chichester.

Douglas, J. and Richman, N. (1984). *My child won't sleep*. Penguin, Harmondsworth.

Green, C. (1992). *Toddler taming: a parents' guide to the first four years*. Vermilion, London.

Green, C. (1994). *Understanding attention deficit disorder*. Vermilion, London.

Myers, B. (1992). *Parenting teenagers*. London, Jessica Kingsley.

Herbert, M. (1991). *Clinical child psychology: Social learning, development* and *behaviour*. Wiley, Chichester.

Wedge, P. (1984). Disadvantaged children: the scale of the problem and the scope of the solutions. In *Towards a coherent policy for disadvantaged children* (ed. E. De'Ath and M. Hill). Environmental Department, University of Dundee, Scotland.

7

The child with a disability

Feeling beastly—nasty,
Many discuss me so—
Can he hear? Can he see?
Can fools fly? Evidently no!
Christopher Nolan (1980). *Damburst of day dreams*. Weidenfeld & Nicholson, London

Writing became Joseph Meehan's Word-World. Brain-damaged, he had for years clustered his
words, certain that some cyclops-visioned earthling would stumble on a scheme by which he could
express Hollyberried imaginings.
Christopher Nolan (1987). *Under the eye of the clock*. Weidenfeld & Nicholson, London

These two quotations from the writings of an exceptionally talented person with spastic quadriparesis poignantly contrast the exterior image with the interior illumination. His definition is his writing, not his cerebral palsy. Such talent is rare, it illustrates the importance of potential. The aim of those who work with children with disability is to help the child reach his or her potential what ever that may be. Picking up a Smartie is not too small a miracle. To achieve that potential depends on early identification of handicap, on careful assessment, on teamwork in treatment, and on support and commitment from parents and carers.

Definitions

The words *'impairment'*, *'disability'*, and *'handicap'* have different meanings. The dictionary definition is quoted followed by a definition in the context of this chapter. The summary box illustrates the differences between the World Health Organization (WHO) terms and the equivalent terms used in other medical conditions.

For example, a long eyeball (impairment) results in myopia (disability), but spectacles prevent handicap.

A *'special need'* used in the terms of the Education Act (1981) is an educational need of a child which is not normally met by the normal provision for a child of that age.

'Special provision' is what an education authority provides to meet special needs.

Early identification of special needs gives the most time to alleviate handicap, to reach potential; it also gives parents the best opportunity to come to terms with their child's disability. Early identification is based on sound knowledge of the normal development of the child set out in Chapter 3. Assessment of development is based on the appreciation of the 'normal' and its variations. There are common paths to development but some children are exceptions; most sit, crawl, then walk, some sit and then bottom shuffle.

Development does not progress in a linear fashion but stepwise with 'treads' and 'risers'. The staircase is uneven and some stages of development constitute a quantum leap, the achievement of which opens up new possibilities: erect

Summary

WHO Model	Medical Model
Impairment	Aetiology
Disability	Pathology
Handicap	Manifestations

Summary Definitions

Impairment: an injurious weakening: any loss or abnormality of physiological or anatomical structure

Disability: a want of ability: any restriction or loss of ability, due to an impairment in performing an activity in a manner or range, considered normal for a human being of that developmental stage

Handicap: a disability that makes success more difficult: a disadvantage for an individual arising from a disability that limits or prevents the achievement of desired goals (an uncompensated disability)

locomotion—a gross motor skill, frees the hands for the myriad of fine motor tasks that are subsequently learnt.

The different developmental 'streams', gross and fine motor, hearing and speech and social development are interlinked but run at different rates, experience is needed to determine when disassociation between them becomes pathological. Cultural and rearing practices influence developmental patterns. When 'kangaroo' pouches were popular, babies used to them developed head control earlier than usual for the UK population, and also developed a difference between flexor and extensor tone which, for them was normal, but which otherwise was sometimes a warning sign of cerebral palsy—disproportionate extensor tone. A baby who is stimulated will always do better than a baby of the same endowment who is deprived of stimulation.

The general health of an infant plays its part too, a baby seen during or after an acute illness may well have decreased tone and apparent delay, but will often catch up. Chronic illness will also slow down development, as will frequent hospitalization.

Fig. 7.1. Child with disability unable to get on a bus.

Finally, family position and pattern is important, but there are hazards of paying too much attention to it. A second child may be slower to talk than the first, but the diagnosis of many speech-delayed children has been delayed because of this. A late walker in a family of late walkers is less likely to be pathological than a late walker in an early walking family.

The case studies that follow illustrate the principles of management of the disabled child.

Fig. 7.2. A disabled teenager using a Maltron keyboard.

Case studies

Down syndrome

Down syndrome was first recognized as an entity in 1866 by Dr John Langdon Down in England. In 1959, Lejeune and his colleagues demonstrated that this syndrome was associated with an extra chromosome 21. It is one of the commonest causes of mental handicap.

> You are the paediatric senior house officer on call. The midwife calls you to say 'a Down syndrome has been born', what would you do?

First, remember that this is a baby who has an incidental Down syndrome. Not a Down syndrome with a baby attached. Gather more information!

> Baby John was born at term, weighing 2.9 kilograms. He was a normal delivery to Jim and Brenda Jones, a couple in their mid twenties. He is a first child. His parents are devout Christians and had opted not to have antenatal diagnosis because of the possible consequence of termination of pregnancy. Now what will you do?

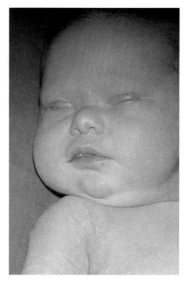

Fig. 7.3. Newborn with Down Syndrome.

Inform your senior or consultant. This is a diagnosis that has lifelong implications for the family, for their hopes and aspirations, and the parents deserve a senior opinion. You should arrange for both parents to see the consultant as soon as practicable.

> You go to explain the arrangements to John's mother and to carry out the routine neonatal medical. She asks 'What's wrong?' How would you answer?

Never lie. Down syndrome is very recognizable, parents may have suspected the diagnosis; if they are able to ask 'Is it Down syndrome?' an honest reply is required. Bad news is hard to give and receive. It can be broken well or badly. If done well then that experience gives a firm foundation for all that follows, if badly, the converse applies. Parents prefer to be told this sort of news as soon as it is possible, and together. The baby should be present as this will help the consultant explain. If the parents wish, a key relative could be present but this is not an occasion for crowds. This is a private matter, not one for the open ward and disturbances must be minimized. It is helpful to have someone known to the mother,

Summary
Telling parents about a disability

- Tell parents together
- Tell parents as soon as possible
- Tell them in a quiet place without interruptions
- Tell them without an audience of staff
- Tell them honestly and directly, with warmth and understanding
- Allow them privacy after the initial interview
- Meet them again to allow them to ask questions
- Give them time and space to consider their feelings

Summary
Down syndrome

Cause
Trisomy, translocation or mosaicism of chromosome 21

Overall incidence
1/600, Incidence increases at both extremes of maternal age (i.e. maternal age 45, 1/30)

Clinical features
Round face, flat profile, flat occiput, epicanthus, 'mongoloid' slant to eyes, Brushfield spots on iris, protuberant tongue, broad short-fingered hands, clinodactyly little finger, single palmar crease, wide-spaced big toe, short stature, hypotonia, and general learning difficulties

Associated problems
Congenital anomalies: heart disease; duodenal atresia; Hirschprung's disease; hypothyroidism; increased risk of leukaemia

Associated minor problems
Dry skin, feeding difficulties, recurrent upper respiratory tract infections, glue ear, feeding difficulties, hypermetropia, and myopia

Prognosis
Normal life span, increased incidence of Alzheimer's disease in later life, increased risk of myeloid leukaemia

Reacting to bad news
Stages

1. Shock
2. Anger
3. Disbelief, denial
4. Acceptance

perhaps a midwife or sister present, and providing numbers do not get too great, the social worker or liaison health visitor of the child development team may find it useful for future management to hear what has been said.

There are different ways of breaking the news. One effective approach is to examine the baby with the parents, gently indicating the features of the syndrome and explaining that although individually they can occur in normal people, the overall features indicate Down syndrome and that this will be confirmed by a blood test. This approach avoids raising a false hope that the test will be normal and that all this will have been a bad dream.

Once the words '*Down syndrome*' have been said, parents have told us that they *cannot* take in any more information.

All questions should be answered but this information will have to be reiterated on a follow-up visit when further questions can be answered. The inability to take in information after bad news is part of a well-described reaction to tragedy which parents take a variable time to work through.

After this initial interview, what would you do next?

A karyotype is necessary for accurate genetic diagnosis, and to assess the chance of recurrence of Down syndrome. 95% of all Down Syndrome is due to non-disjunction trisomy 21 (see Table 7.1 and Chapter 15, p. 264).

Send blood for a karyotype for assessment of the genetic risk. Arrange cardiac ultrasound because of the possibility of a congenital heart defect, which may be murmurless or acyanotic, and arrange for John's hearing to be checked.

Table 7.1.
Recurrence risk in Down syndrome

Type of syndrome	Parents' chromosomes	Chance of recurrence
Trisomy 21	Normal	1/100 if mother is under 40. If mother is over 40, twice usual chance for her age
Translocation with chromosome (13, 14, 15, or 22)	Normal	Usual chance for mother's age
	Mother a carrier	1/8
	Father a carrier	1/40
Mosaicism	Normal	Unknown, probably the usual chance for mother's age

From Selikowitz (1990).

Before the parents leave hospital, give them relevant literature and put them in touch with the Down's Syndrome Association which has local groups throughout the UK. After discharge, Mr and Mrs Jones may find it helpful to meet the local child development team who will be working with John.

John's parents go through the stages of grieving quickly and ask 'what can we do?'

Although there is not a lot of evidence that early intervention improves the final developmental outcome in a child with Down syndrome it helps them achieve that potential and certainly helps the parents. A learning support teacher or Portage worker (who provides special assistance with development in the home) could draw up a home-based, individual programme for John.

John is now 14 months old. His mother is worried as he seems to be slowing up.

Early on, the gap between a child with Down syndrome and a normal child is not very apparent. As the normal child develops the gap becomes more apparent, typically at this time. A co-ordinated team approach will help the family here.

What about John's education?

This worries parents. The most rational approach is to identify John's special educational needs, the extent of his learning difficulty, the presence of any other handicap (e.g. deafness), the presence of any physical problems, (e.g. motor problems due to hypotonia), or any other difficulty. This should be a team assessment together with colleagues in education. When John's needs have been identified the education authority matches the special needs with special provision (see below). This could be a 'normal school' with extra support or a 'special school'. John will need regular follow-up of his vision, hearing, height and weight, and thyroid function tests need checking. Down syndrome children are prone to recurrent upper respiratory tract infections which can lead to conductive deafness. Some develop hypothyroidism, some weight problems, and some long or short sight.

Cerebral palsy

The conditions known as 'cerebral palsy' were first described by William John Little in 1863 but it was William Osler who coined the term in 1886. This is the commonest cause of physical disability in the UK.

Summary
Cerebral palsy

Definition
Cerebral palsy (CP) is a disorder of movement and tone caused by a non-progressive brain lesion

Incidence/types
2/1000 live births
Spastic (hypertonic)   70%
Athetoid ⎤
Ataxic   ⎬ 20–25%
Mixed   ⎦

Pattern
Hemiplegic: one side body involved
Diplegic: lower half body involved
Quadriplegic: all body involved

Associated problems
CP alone: *c.* 30%
CP + epilepsy: *c.* 30%
CP + learning difficulties: *c.* 30%

Additional Information
Presentations of cerebral palsy

Neonatal
Hypotonia, head lag
Hypertonia
Disproportionate tone
Asymmetric reflexes

Infant
Persistent neonatal reflexes
Failure to develop mature reflexes
Hypertonia
Early 'handedness' (unilateral weakness)

Toddler
Delayed motor milestones
Late sitter
Late walker
Abnormal posture

Child
Excessive clumsiness

Beth is 9 months old. You are her GP. She is just beginning to sit up and is still wobbly. Her mother, who is a single parent, has observed that she is left-handed and draws your attention to this at a routine medical.

This child is not left-handed. Hand preference is not established until at least two years of age. Why is Beth not using her right hand? Examine the child thoroughly.

Examination shows increased tone on the right, the hand is clenched, the thumb adducted, there is weakness in that arm but the reflexes are brisk on that side with an upgoing plantar.

At nine months an upgoing plantar is not a sin by itself, but the picture is one of a right-sided weakness with hypotonia and upper motor neurone signs.

What do you do next?

This is a cerebral palsy manifest as a spastic hemiparesis. As the GP, you should explain that the child is not left-handed but may have a problem with strength on the right side and that a paediatric opinion is necessary. The paediatrician will confirm the diagnosis and explain it to Beth's mother. It is better to explain and use the words 'spastic', 'cerebral palsy', and 'hemiplegia' at an early stage, and written information helps. A common early question is 'Why did it happen?' Investigations such as a computerized tomograph (CT) or magnetic resonance imaging (MRI) scan and a genetic opinion may be useful. Physiotherapy will be arranged, directed at reducing spasticity, maintaining the range of movement, and improving strength. The local child development team should be involved. Orthotics may provided a light weight splint to prevent shortening of the achilles tendon (ankle foot orthosis, AFO) and a muscle relaxant, baclofen, can be used to reduce tone. The social worker on the team can discuss benefits and allowances.

Beth is now just over 3 years old, she is walking independently with an ankle foot orthosis on the right leg. She has regular physiotherapy and occupational therapy. Her mother wonders if she will cope in school?

When is Beth's school target date? It differs in England and Scotland. Age three is not too early to start to prepare a *Statement of Special Educational Needs* (England

and Wales) or Record of Needs (Scotland) in line with the Education Act (1981). There is a duty in law for doctors to notify the Education Department of children who have or are likely to have, special educational needs upon school entry. Beth's mother has a statutory right to be involved in the Record/Statement process and to receive a copy of the record or statement. An educational psychologist will take the lead role in this process. The record or statement describes Beth's strengths and weaknesses and sets out her needs. Beth's IQ is in the normal range and her needs are centred around her physical problems. She will be regularly reviewed.

Beth manages well, she is starting puberty and approaching secondary school transfer.

This can be a difficult time. Primary and secondary school ethics are different. The handover will be co-ordinated and her record or statement will follow her. A problem of puberty is that at the time of maximum growth there is often a spell of poor co-ordination and Beth's physical difficulties may temporarily increase. Extra physiotherapy, careful attention to orthotics, and possibly a muscle relaxant will help.

Beth indicates that she wishes to leave school at 16.

At school leaving, responsibility for Beth's continuing care will pass to the social services. Between the ages of 14 and 16 the child development team will have informed the social services of Beth and her difficulties and the latter will have offered to assess Beth's needs under the terms of the Disabled Persons Act (1986). Beth and her mother can decline assessment. The social services can decide that a particular child is not disabled within the meaning of the Act.

Cerebral palsy can present in many different ways. Usually, the earlier the diagnosis can be made the greater the eventual disability but also the greater the opportunity for remediation.

Delayed walking

A relatively common condition, but this case is a reminder that you must not assume that delay is normal without careful history-taking and examination.

Benjamin is 18 months old and has not yet started walking. His elder brother, William, walked at 12 months. Mrs Smith felt that Ben was 'just lazy' but is now concerned and has consulted her GP who has referred Ben on to you, his paediatrician.

Additional Information
Provision for special needs

- Regular professional physiotherapy
- Orthotic review
- Auxiliary support at PE classes
- Medical review
- Additional learning support
- Opportunities for one-to-one teaching
- Small group teaching
- Specialist educational provision

Additional Information
Statement of Special Educational Needs

- Co-ordinated by educational psychologist for the education department
- Requires input from health, education, and psychology as well as parents
- Community paediatrician writes 'medical advice'
- Placement is decided depending on needs: mainstream preferred
- Additional support is described in the statement
- Parents have a right of appeal

Summary
Differential diagnosis of late walking

- Extended normal range
- Familial late walker
- Bottom-shuffler
- Isolated motor delay
- Global delay
- Cerebral palsy
- Rarities, including muscular dystrophy

Summary
Bottom-shuffling

- An alternative developmental pathway
- Sit–Shuffle–Walk, normally but late
- Occurs in 2% of children
- Often positive family history
- Dislike of prone position
- Relative lower limb hypotonia
- Absent/decreased caudal parachute response
- Does not crawl

The history includes Ben's development, general symptoms and illnesses, neonatal and family history, with particular reference to walking.

Ben sits and crawls, he does not cruise but pulls himself up to the furniture. All other development is reported as normal. He has had no serious illnesses. He was a full-term, normal delivery. All the family members, except Ben's mother, walked at about 12 months, she bottom-shuffled.

There is evidence of a non-familial isolated motor delay. Physical examination includes general examination, a neurological examination, and a developmental assessment to confirm the mother's report.

Ben is well. There are no abnormalities on examination. Neurological examination shows normal power, tone, and reflexes. He appears well-muscled. Developmental examination shows normal fine motor development, normal speech, language and hearing development, and normal social and adaptive skills. Ben's gross motor development is as his mother described.

Global developmental delay and cerebral palsy can be eliminated on the developmental and physical findings. Eighteen months is rather late to be an example of a child at the end of an extended normal range. Ben's mother was a bottom-shuffler but there are no other clues to this. Isolated motor delay is possible. Rarities are possible. Medical problems can be important because they are either common and affect a lot of people or because they are rare and serious. Muscular dystrophy is important because it is serious. Unlike other disorders discussed in this chapter it is fatal. It must therefore be excluded in late-walking boys.

In all late walkers, particularly in boys, check the creatine kinase. If this is normal than the diagnosis is one of isolated motor delay. The prognosis is good and the management is one of expectant observation and physiotherapy.

Ben's creatine kinase is 10 000 units per litre! This is grossly elevated. Ben also demonstrates Gower's sign in that he 'climbs up his legs' when getting up from a prone or seated position on the floor.

A diagnosis of *Duchenne muscular dystrophy* (DMD) is probable. Ben requires referral to a paediatric neurologist with a team experienced with working with neuromuscular disorders. The parents will need to be told that a serious disorder is being considered and that a muscle biopsy will be needed. Duchenne muscular dystrophy was recognized by Guillaume Duchenne in 1861 when he noted the pseudo-hypertophy of the calves in affected children.

The muscle biopsy is positive, showing an increased variation in fibre size and an increase in rounded fibres staining with eosin.

This confirms the diagnosis. The poor prognosis of DMD is well known so that in addition to explaining the diagnosis and prognosis to Ben's parents they should be told that there is much that can be done to give Ben a stimulating and valuable if shortened life. Ben's diagnosis will have been made before Gower's sign, pseudo-hypertrophy and muscle weakness were apparent. This can pose problems. Parents prefer early diagnosis, but it may be easier to accept a diagnosis when the signs are apparent than before they appear. Ben's prognosis is progressive muscular weakness with loss of walking between the ages of 7 and 13 and death in the late teens or early twenties. Physiotherapy aims to maintain a plantar grade foot to help Ben walk as long as possible. Ankle foot orthosis (AFO) can also assist, an achilles tenotomy may be required. These measures require teamwork. Even short periods of bedrest can be associated with a decreasing power which may not be reversible. Ben will eventually need a wheelchair. Liaison between physiotherapists, occupational therapist, and orthotist is necessary to select the right chair, adapt Ben's environment, and enable Ben to get the best from it.

Genetic counselling is necessary for the family with an affected male, for although the condition can arise *de novo* it is usually transmitted down the female line. This has implications for breaking the news (see above, Down syndrome) as it can result in inappropriate feelings of guilt in the carrier. The DMD gene has now been localized on the short arm of the X chromosome and its product, dystrophin, identified. Specific therapy is not available now but may become available in the future, (see also Ch. 15, p. 269).

When Ben stops walking this is inescapable proof of deterioration and there may be a resurgence of the grieving reaction. Ben will come to realize his reduced life-expectancy and this may be brought home to him by watching other boys attending the specialist team clinic. He probably will not be afraid of death but may be afraid of the mechanism of dying. This may need careful exploration of the unspoken with Ben, at a time he chooses. He will need to know that death is not painful nor uncomfortable and that he will die with dignity, having been loved. The clergy team and the hospice movement can contribute. The family, Ben's brothers and sisters, should not be forgotten.

Summary
Duchenne muscular dystrophy (DMD)

Incidence
1/3000

Aetiology
X-linked recessive gene, carried by asymptomatic females, 50% of whose daughters will be carriers and 50% of whose sons will be affected. The affected boys lack the protein, dystrophin, in muscle. Antenatal diagnosis is now possible.

Diagnosis
Creatine kinase and muscle biopsy

Prognosis
Loss of walking at 7–13 years
Death in the late teens–early twenties

Associated problems
Lower than average IQ
Muscle contractures
Scoliosis
Progressive weakness
Respiratory problems

Summary
Speech delay

Incidence
2.4% totally unintelligible at age 7, according to teachers (National Child Development Survey)
4 male: 1 female

Associated problems
Non-reading
Poor number work
Motor learning difficulties
Impaired visual acuity

Differential diagnosis
Global delay
Hearing loss
Lack of stimulation
Specific language disorder
The autistic continuum
Elective muteness

Speech delay

This is an increasing problem in the UK and one of the commonest reasons for referral to a developmental paediatrician. According to their teachers up to 10% of 7-years-olds' are difficult to understand.

Richard is 2 years old, you are his GP. He is the second of two children to a professional couple. He has no words. He is prone to temper tantrums. His sister, Elizabeth aged 4, is a chatterbox. His mother asks, what can be done?

First, take a history. Has Richard ever had any words? Babble? What is he understanding? Seek examples or offer them. Is Richard hearing? Can his parents give examples; overhead aircraft, ice-cream van, sweetie wrapper at 10 paces? Is his development otherwise normal? Any history of illness or fit? Any history of feeding or chewing difficulties as a baby? Then examine. Check ears for signs of serous otitis media (glue ear), nose, throat, and mouth for normal anatomy and movement. A short frenum (tongue tie) does not cause speech delay. Check that physical examination is normal.

Richard babbled but got stuck at the 'da da da' stage after a bad cold. He does not seem to understand instructions, particularly the word 'No'! He is never the first to the front door when the bell goes. Richard's mother felt that Elizabeth was 'doing all the talking for him and he would come on'. Examination shows glue behind both ear drums.

You have raised serious doubts about Richard's hearing which may be the cause of his problem. You must now refer for hearing assessment, and take active steps to exclude deafness. If Richard's hearing is normal then he will be referred for a joint assessment by a speech and language therapist and an educational psychologist, the latter to establish that his cognitive functioning in non-language areas is normal.

The community paediatrician carries out a hearing test which shows a 60-decibel hearing loss across the frequencies in both ears.

Richard needs a specialist assessment of his hearing and speech and language delay. The object is to establish what part Richard's hearing impairment is playing in his speech and language problem, to treat the hearing impairment if possible and to provide speech and language therapy.

Further tests including evoked response audiometry establish that Richard's hearing loss is conductive, he has glue ear (see Ch. 12, p. 226). He has middle ear drainage followed by the insertion of ventilation tubes (grommets, see p. 226). Later, Richard's parents note that he is taking more notice, is less loud, does not study faces so intently, and that he increases the number of single words uttered.

Richard benefited from surgery. Glue ear can re-occur. He will need continuing supervision by his speech therapist, by an Audiologist, and his ENT surgeon.

His mother asks, why did he have temper tantrums?

There are two main reasons, first, the 'terrible twos', a developmental stage in the acquisition of personal autonomy (see Chapter 6, p. 104). This occurs at a developmental age of 2 years, which is not necessarily the same as the chronological age. Second, frustration due to the hearing/language problem can cause tantrums. Imagine trying to make sense of the facial movement in speech when you only hear parts of words! Speech and language delay is often associated with behavioural problems. Constant firm handling helps and clinical psychological advice may be needed.

Richard starts to progress and within a few months is starting to put words together. He is less aggressive and Elizabeth has ceased to be spokesperson. Richard's mother raises the question of a nursery place.

At the right time, increased peer exposure in a nursery can bring on speech and language; too soon and you can switch a child off. Be guided by the Speech Therapist. Checking the hearing, speech therapy, and increased peer exposure in a speaking environment are the main approaches to most speech and language delay.

Richard was lucky, his hearing loss was easily amenable to treatment. That is not always the case but even with sensorineural hearing loss the impairment need not constitute a handicap.

Richard's language delay resolved so that by the time of school entry his speech therapist felt, along with an educational psychologist, that he could manage in the first-year class in primary school.

Summary
Causes of deafness

Conductive	Sensorineural
Foreign body	Drug-induced
Wax	Meningitis
Serous otitis	Encephalitis
(glue ear)	Asphyxia
	Hyperbilirubinaemia
	Congenital infection
	Genetic

The deaf child

In the previous case study, conductive deafness was discussed. Deafness can have many causes, 5–10% of children will have a significant conductive loss and approximately 3 per 1000 a sensorineural loss. There are many genetically determined causes of deafness. Although screening of high-risk neonates by an acoustic cradle is possible, the most effective screening agents are the child's parents. If a parent thinks their child has a hearing problem, the child must have the hearing assessed and proven normal or otherwise.

Certain acquired medical conditions carry a high risk of deafness and all children who have had a meningitic or encephalitic illness should have their hearing tested on recovery. If a hearing child becomes deaf as a result of such an illness then a cochlea implant is possible but this is not appropriate for the majority of deaf children.

A team approach to the management of a hearing impaired child is important. Awareness of a child's problem and the correct classroom positioning will help. Deaf children do not just have a problem hearing and speaking, they can find it hard to learn written language which is a system of symbols, representing speech which is in its turn a system of sound symbols representing both abstract and concrete thoughts. Hence, the need for specialist provision whether integrated into a mainstream or special school.

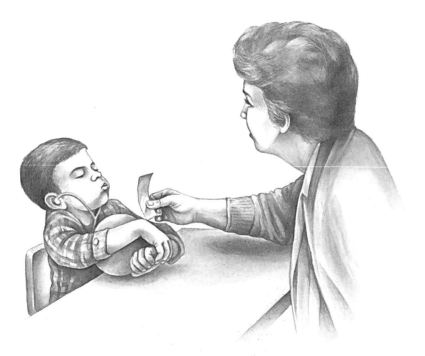

Fig. 7.4. Specialist provision is needed for the deaf child.

Visual impairment

In this section we are concerned with problems severe enough to affect daily life and education (see also Ch. 13).

James is 8 weeks old. At a 6-week check, his mother reported that he is not 'looking at her'.

This observation requires full follow-up and investigation, do not dismiss it lightly. Ask what the mother noticed. Take a history, including birth history. Examine the child, check the pupil reactions, the retinal light reflex, the eye movements, the media and the fundi if possible. Check fixation, using your face in the lower half of the visual field at a 20–30 cm range. Do not speak as you do this, it is visual fixation you want to test, not auditory. Attempt to elicit the 'threat blink response'; bring a closed hand to the outer canthus of each eye in turn, then suddenly open the fingers, avoiding the face. By doing this you avoid a draught which could elicit a corneal reflex blink. A blink to the threat implies a protective closure of the eyelid to the visual threat of the fingers. Examine other systems, pay particular attention to the neurological and developmental observations. Then refer the child to a community paediatrician or to a visual assessment team.

Always check for cataract and the retinal light reflex. A white or green retinal reflex demands investigation. It might mean retinopathy of prematurity or rarely, but importantly, retinoblastoma.

Within the team, the ophthalmic, genetic, and educational, as well as social aspects will all be assessed.

James was full-term, with no adverse history. He does not have cataracts or abnormal retinal reflex. His pupil reactions are present but there is no threat blink and no fixation. His other development seems normal. Further investigations, including fundoscopy under anaesthesia, electroretinograms (ERGs), and visual evoked responses (VERs), were carried out. Examination is normal, ERGs normal but VERs show reduced amplitude and delayed pattern.

This could represent delayed visual maturation (DVM). Just as there can be delay in gross or fine motor skills there can also be delay in visual maturation. If there is no associated mental retardation, a child with isolated DVM is likely to have useful vision by 6 months. If there is associated mental retardation then useful vision will also be delayed. A visual stimulation programme devised by the

Summary
Visual impairment

Definition in UK
Visual acuity 3/60 or less in better eye with correction.

Incidence of blindness
32/100 000 children

Causes in the developed world
- Genetic disease:
  congenital cataract, albinism, retinal dystrophies
- Perinatal disease:
  retinopathy of prematurity, birth hypoxia
- Non-genetic congenital malformation:
  microphthalmos, congenital glaucoma
- Congenital infection:
  rubella

Causes in developing countries
- Corneal scarring:
  vitamin A deficiency, measles, malnutrition, herpes simplex, ophthalmia neonatorum,
- Cataract:
  rubella
- Genetic causes

Associated problems
In the UK many cases of blindness are associated with other severe handicaps including cerebral palsy, epilepsy, and learning disability.

orthoptist and peripatetic teacher using aids such as fluorescent toys under ultra-violet light, or voice-activated disco lights can help.

> With visual stimulation James starts to respond, vocalizing to light and then to face, and by the time he was 1 year old his visual skills were normal for age.

James should still be followed up because some children with resolved DVM go on to have other later learning difficulties.

Beware of the word 'blind'; parents do not like it. Pause a moment, what would a blind person see? Blind is not 'blackness', or is so very rarely.

Parents tend to think the word 'blind' does mean blackness and also means forever. Instead, try to get over the concept of functional vision. If a child has a visually directed reach then he/she is not blind. 'Blind' is an administrative category or level of visually acuity below which certain benefits are payable.

Learning disability

The terms 'mental handicap', 'mental retardation', and 'general learning disability' are all synonyms. The existence of such synonyms and other older ones are an indication of a lack of acceptance of mental handicap within society and the progressive adaptation of technical terms as terms of abuse by the public: 'retard' in the USA and 'mental' in the UK.

Learning disability is different from the other problems discussed which cause special need; it is not a disease, not a disorder, not a syndrome nor a specific disability; it is a blanket or administrative term for a wide variety of different genetic, social, educational, and specific medical conditions with the common feature that 'affected' individuals repeatedly score below 70 on specific IQ tests. Learning disability is sometimes an indication of brain dysfunction, just as epilepsy is, or as cerebral palsy is, but often it is not.

IQ scores are not used as the sole delineator of learning disability. A comprehensive assessment by an educational psychologist acting as part of a team is required. To conduct epidemiological studies IQ testing is relevant.

Mental retardation with IQ consistently less than 70 can be further subdivided into severe, IQ less than 50; and profound, IQ less than 20. The prevalence of learning disability in recent studies from Scandinavia was between 7 to 10 per 1000.

Additional Information

Intelligence quotient (IQ) =

$$\frac{\text{Subject Score}}{\text{Population Mean}} \times 100$$

> Mary is just over 3 years old. You are her GP. She was a slow walker at 18 months, a slow talker saying her first words at 18 months, and linking words at 2 years 6 months. At her 3-year developmental surveillance, she is functioning at about the

2-year level. She is well. Her mother needed some remedial help in school and so was not particularly worried about Mary. Mary was a normal, full-term delivery. There is nothing obvious on examination. What do you do?

Refer Mary to your local community paediatrician with an interest in special needs.

On *paediatric review*, the history is confirmed. Physical examination is normal with no specific clues. The Denver Screening Tests confirm functioning at around the 2-year level, at a chronological age of 2 years 3 months. Mary's birthday is such that her school target date is in September, 13 months away. Thyroid function tests, urine amino acids, chromosomes, and DNA studies for Fragile X (see Ch. 15, p. 271) are all normal. Vision and hearing tests are also normal. What next?

Mary requires a co-ordinated assessment directed towards establishing a Record/Statement of Needs within the next year. That process requires input from her mother, the educational psychologist, speech therapist, physiotherapist, occupational therapist, and the community paediatrician. During the assessment a social worker can advise Mary's mother of the benefit entitlement.

Mary's IQ is found to be 60. Her speech delay is commensurate with her general delay, her fine motor skills are at the same sort of level, and there is no significant gross motor impairment.

Mary's special needs are identified as: a requirement for a highly structured developmental curriculum, for small group teaching, one-to-one teaching with extra classroom support, a speaking environment with regular professional speech therapy and occupational therapy review with advice as necessary. Meeting this package of needs with a package of provision is a matter for the local educational authority taking into account parental preference. It is not the job of the doctor to indicate a particular school.

In practice, there are two main ways of making suitable provision, a special school for pupils with learning difficulties or additional support to allow true integration in the local primary school. True integration is allowing the child to function as part of the school, educationally and socially, meeting her needs and thus allowing her to obtain her potential. It is not a inexpensive option.

There are arguments for and against 'special' and 'mainstream' schools. A compromise approach is a mainstream school with a special unit on site.

Summary
Investigation of learning disability

'Always'
Hearing test
Vision assessment
Thyroid function
Urine: reducing substances, ketones, amino acids, mucopolysaccarides, porphyrin screen
DNA Fragile X

'Often'
Chromosomes
Serology for antenatal infection
Skull X-ray for intracranial calcification

'Sometimes'
CT scan: head
MRI scan: head
Electroencephalogram (EEG)
White cell enzymes

Summary
Special school vs. mainstream

Special school	Mainstream school
May be far from home	Near to home
May be stigma to child	Good for mixing, but may be bullying
Specialized teachers	Specialized support may be lacking
Good for parental support	Enables child to be with siblings

After discussing the options Mary's mother opts for a special school 'like hers'. At school Mary continues to make slow but definite progress. At the age of 9 years, Mary's mother mentions that she is having problems with Mary's temper at home. No similar report was received from school.

Behaviour problems are not unusual in children with learning difficulties. There is a temptation to carers to 'make allowances'. Consistent, firm management is advisable and not allowing any undesirable behaviour to become entrenched. Mary and her mother will need advice on managing Mary's menstruation and her developing sexuality. Secondary school transfer, coming at about the same time, can also be a source of stress. Mary's needs and provisions will both have to be reviewed. At around the age of 14, with parental approval, information about Mary should be passed to the social services so that plans can begin to be made for a handover of continuing care.

The young person with learning difficulties can remain in school until the age of 19 as long as there is educational benefit. Education may stop at 19, but Mary will continue to require lifelong special help.

Dyslexia (specific reading difficulty)

A specific difficulty with reading and spelling not associated with any general impairment is present in approximately 5% of school-age children.

David is 8 years old. He is having problems at school with reading, spelling and written work. Numbers are not too bad. He likes drawing and PE (physical education). He has been receiving help from a remedial teacher. His father had some difficulties at first in school but there is no other relevant history. David's mother wonders if 'anything more can be done?'

This problem is different from the others discussed in this chapter. David is older than the other children, his problem is not a disease and its remedy is outside the medical arena. There are things that can be done; the most important being to get the diagnosis right. Review the history, look at examples of David's school work, examine him, pay attention to the neurological system, look for evidence of motor learning difficulty, ascertain hand, foot, and eye laterality, get David to draw and write, check for squint and visual acuity.

There are no historical pointers other than David's father's problem. David's writing is not yet joined, 'B' and 'D' get mixed up; words are frequently misspelt. Neurological

Summary
Features of dyslexia

- Early reading difficulties
- Slowness in reading
- Difficulty reading aloud
- Bizarre spelling, 'B' and 'D' confusion
- Confusion with sequencing
- Right–left confusion
- Positive family history

examination is normal, he is not overtly 'clumsy' and he is right-handed and right-footed but looks through the kaleidoscope with either eye. There is no squint, vision is 6/6 6/6, (for explanation, see Ch. 13, p. 236).

The only 'abnormality' you have uncovered is that David seems to lack a fixed reference eye and that there is prima-facie evidence of a specific learning difficulty with reading. Two steps are appropriate, referral to an Orthoptist and to an Educational Psychologist. Some children with dyslexia do not have a dominant eye and for some of these establishing eye dominance can help. For most, the Educational Psychologist will arrange specialist remediation, perhaps with a Record/Statement of Needs.

Reading difficulty is commonly seen in practice because it is fundamental to most of education. Dyslexia tends to be a socially 'acceptable' special need. It is necessary to distinguish specific learning difficulty from a more general learning difficulty.

Autism

Autism was first described in 1943 by Leo Kanner, a child psychiatrist. The current prevalence is between 4 per 10 000 and 20 per 10 000, depending on the criteria used.

Sarah is 3 years old. At her 3-year check her parents report that she has three words, 'mama', 'biccy', and 'no'. Her parents became concerned about her at 2 years. She ignores other children at her playgroup and spends a lot of time twiddling a piece of string or rocking on her bottom. She only wants cuddles on her terms and rarely makes eye contact. Her GP has referred her to you, the local Paediatrician. What would you do?

Your history-taking should include details of her behaviour, including any evidence of obsessions, communication skills should be explored, as in the problem in speech delay, but enquiries should be made of non-verbal skills. Physical examination should be thorough, checking ear, nose, and throat as well as the neurological systems. Any dysmorphic features or cutaneous stigmata should be noted. A cutaneous stigma is the external, dermal, manifestation of an internal disease (e.g. the *café au lait* spot of neurofibromatosis). Normal findings are likely. Developmental observations are important.

Autism is a condition where a team approach to diagnosis is necessary, involving as it does neurology, communication, the emotions, and education. *Hearing must be tested.*

Summary Autism

Prevalence
2–4/10 000 rising to 20/10 000, depending on definition

Sex ratio
3–4 males: 1 female

Features
Presence of abnormal impaired development before age 3
Impairment in social interaction
Impairment in communication
Repetitive and stereotyped patterns of behaviour

Associated problems
Mental retardation in 70–90%
Epilepsy in 25–30%

Sarah was referred to the child development team and it was their view that she had the features of a pervasive disorder of communication and behaviour required for a diagnosis of autism.

In most cases of autism there is no underlying cause, a few are symptomatic. The thorough examination should have eliminated neurofibromatosis, tuberous sclerosis, or genetic syndromes. Urine should be tested for amino acids—phenylketonuria can cause autism. Blood should be taken for chromosomes and DNA studies for Fragile X, (see Ch. 15, p. 271).

Sarah's tests were normal. The educational psychologist established that her non-linguistic IQ was low. What would you do now?

Sarah requires very specialized provision. In addition to the global communication disorder (i.e. not understanding speech, sign, or body language), autistic children often lack the concept of people as people and can not invest other people with thoughts and feelings. Try putting those problems together and then try to make sense of daily living! Obsessions, rituals, and sameness are the autistic child's way of coping in our world.

Clearly, a child such as this will require a very specialized educational approach, consistent behavioural management, and specialist input into all communication skills. This may involve residential provision.

Spina bifida

The term 'spina dorsi bifida' was coined by Professor Nicolai Tulp in Amsterdam in 1652. The prevalence in the UK is about 1 per 1000 and falling. It is salutary to remember that many UK schools for physical handicap were originally established because of this condition.

Peter and Jane are a couple in their mid thirties. This is Jane's first pregnancy. They have opted not to have antenatal diagnosis because this baby is much wanted. Michael is a full-term, normal delivery and is noted to have a spina bifida. You are the paediatric resident! What will you do?

This is an event of great importance to the family. Arrangements must be made for Michael to be seen by a consultant. He will require careful evaluation by a paediatrician or paediatric neurologist and a paediatric surgeon. The object is to allow the parents to participate in an informed debate about what should be done.

The policy of early surgery for all spina bifida babies led to the survival of large numbers of children with a subsequent poor quality of life. Certain *adverse factors* were recognized as being associated with a bad outcome are now regarded as relative contra-indications to surgery.

Michael's mother asks you to let her see his back. What would you do?

Let her see it in a controlled setting. The human imagination can conjure up far worse deformities than in reality. Seeing Michael's back will assist explanation later.

Michael's defect is a lumbar sacral myelomeningocele covering L4, L5, S1 confirmed on X-ray. There is no hydrocephalus, no spinal deformity, and no major defects.

Michael is a suitable candidate for surgery. This will involve closure of the back and almost certainly the insertion of a ventricular peritoneal shunt to prevent the development of hydrocephalus. This arises secondary to the Arnold–Chiari malformation, which is a congenital downward displacement of the cerebellum and medulla oblongata through the foramen magnum resulting in an obstructive hydrocephalus. Before the development of the Spitz–Holter valve allowing drainage of cerebrospinal fluid (CSF) in this complication only 10% of spina bifida children lived more than a few years and of these, 70% were grossly handicapped.

The early surgery goes well. Michael's parents want to know what the future holds.

With a lesion at the level of Michael's, walking with short calipers is possible. There may, however, be orthopaedic problems associated with muscle imbalance. Kyphosis and scoliosis can occur and will need to be watched for. There may be problems with urinary continence and infection, there can also be bowel problems. Learning difficulties are not uncommon. There may be sexual problems in adult life. Thus, there is a litany of possible problems but with the level of Michael's lesion the outcome is relatively good.

Michael's parents are worried about the risk if they have another child.

Additional Information
Adverse factors in spina bifida

- Thorax-lumbar lesions, motor level L3 and upwards
- Spinal deformity, severe kyphosis, or scoliosis
- Gross hydrocephalus, occipital frontal circumference (OFC) > 2 cm above 90% percentile
- Other motor defects

Once there has been an affected child there is an increased risk of neural tube defects in that family. The inheritance of neural tube defect is multifactorial. Genetic counselling is possible, as is antenatal diagnosis. There is good evidence that periconceptual vitamin supplementation with folic acid reduces this recurrence risk.

Support for parents

Parents are a very important part of the team. Most parents are a great asset to a disabled child; sadly occasionally, a parent can contribute to handicap. Support for carers is an important role for the team. Breaking the bad news and the stages of grieving have been discussed in the section on Down syndrome; the principles apply to all disabilities.

When a baby is expected, it is expected to be perfect. If it is not, many parental dreams and hopes are shattered. Some parents will adjust but not all. The quotations at the beginning of the chapters could not have been written if Christopher Nolan had not been read to and had not been fully involved in his family's love of literature.

Parents need information about causation, about the current situation, about future prospects and about genetic implications. They need practical help in alleviating their child's handicap; aids to easier living, access to appropriate appliances, and environmental adaptations.

They need the essential break, respite from care, but they may not know it at first. This may be short or long term or crisis relief and can take a variety of forms including hospital provision where there are very special needs. They need financial assistance and help to gain access to the statutory benefits that are available.

Parents need information about available services which may come from specialist social workers, liaison health visitors, books or support agencies.

They frequently benefit from the mutual support of other parents with children with similar problems. There are many different syndrome support groups in existence.

Conclusion and teamwork

The problems mentioned in this chapter all require a team approach if the children involved are to reach their full potential. The team consists of a core group of professionals who work together regularly, both helping individual children and planning and developing services for children with special needs as a group.

Teams need to meet regularly and most evolve their own pattern; weekly meetings about clinical problems and monthly meetings about broader issues are usual. Child development teams also require a clear lead and this should be provided by a consultant paediatrician.

Additional Information Allowances and financial support

Disability living allowances

Care component

Lower level:	small amount of care at specific times
Medium level:	attention or help needed by day or night
Higher level:	constant attention or help needed by day and by night

Mobility component

Lower level:	can walk, but needs more supervision than a child of the same age
Higher level:	virtually unable to walk

Family Fund, PO Box 50, York, YO1 2ZX

Funded by the government, administered independently by the Joseph Rowntree Trust. Provides help to the families of severely handicapped people in areas not the responsibility of other agencies

Further reading

Cunningham, C., Morgan. P., and McGucken, R.B. (1984). Down's Syndrome: is dissatisfaction with disclosure of diagnosis inevitable? *Devel Med Ch Neurol* **26**, 33–9.

Hall, D.M.B. and Hill, P. (1996). *The child with a disability*. Blackwell Science, London.

Selikowitz, M. (1997). *Down syndrome: The facts* (Second edition). Oxford University Press.

WHO (World Health Organization) (1980). *International classification of impairments, disabilities and handicaps*. WHO, Geneva.

Griffiths, M. and Clegg, M. (1988). *Cerebral palsy: problems and practice.* Souvenir Press, London.

Emery, A.E.H. (1994). *Muscular dystrophy: The facts.* Oxford University Press.

Shaw, C. (1993). *Talking and your child.* Hodder and Stoughton. London.

Baron-Cohen, S. and Bolton, P. (1993). *Autism: The facts.* Oxford University Press.

Section B

8

The newborn

- Measuring neonatal outcomes: epidemiological terms and their meaning

- The prenatal environment and its influences on the fetus and newborn

- Birth and adaptation to extra-uterine life

- Case studies
  resuscitation
  birth asphyxia
  cardiorespiratory adaptation
  birth trauma
  assessing gestation

- Normal term babies, their examination and common problems

- Feeding
  Immunization

- Important congenital abnormalities

- Babies born small for gestational age, or mildly preterm

- The particular problems of the very preterm baby

- Multiple pregnancy; the teenage mother

Measuring birth: definitions

How 'new' is newborn? And what is a neonate, apart from obviously being a newborn baby? Understanding early life is much easier when some definitions of these terms are adopted. In general, we define babies in terms of the length of their *gestation and their birthweight*, and the *events* which may occur in terms of the times at which they occur: the first week, the first month, the first year.

Early infancy is a hazardous time of life, even though it has become relatively much safer during the 20th century. So dependent is the welfare of babies on such extraneous factors as whether their fathers are in work, and the economic state of their country that death-rates in early and late infancy are used for international comparisons of socioeconomic success and quality of welfare provision, as well as more local factors such as the quality of a maternity service.

Why are death-rates used rather than those of survival?

Death is a 'hard' outcome: that is, it is relatively easy to ascertain and verify, and is also related to the rate of illness, more formally called 'morbidity'. To describe patterns of mortality in early life, statistics on rates of still birth, perinatal mortality, neonatal mortality, and infant mortality (together with further subdivisions) are routinely collected. The data may be collected over many years and so describe their change, and their improvement, with time. They may also be used to compare outcomes in different populations, either within or between countries.

Although all of these rates are useful to describe differences between populations, either geographically or over time, they cannot in their crude form be used to measure the relative safety of different hospitals or maternity units. This is because women with high-risk pregnancies are systematically referred from smaller hospitals to centres where fetal medicine services or neonatal intensive care are available. Thus, the women delivering in the smaller units tend to be lower risk, and on average have less chance of a perinatal death; their high-risk friends are more likely to have an adverse outcome by definition, and these adverse outcomes will tend to occur in the unit to which they are referred.

These measures are also less useful now for measuring the outcomes of maternity services in general for other reasons.

1. The rates of all perinatal and neonatal death are so low that just a few deaths more or less, from chance variation alone, tend to give large fluctuations in rates from year to year in a small population. This can be overcome in part by averaging the rates over a longer period of time, say three years.

2. Many more very preterm babies are now surviving. It therefore makes sense to aim to avoid fetal loss from mid gestation onwards, and hence to measure the outcome of pregnancy from 20 weeks. Conversely, some very preterm babies who do not survive do not die until they are several months old. The World Health Organization has therefore recommended that all fetal death from 20 weeks, and all postnatal death to 364 days inclusive, should be counted and expressed per 1000 births.

Summary
Definitions: weight and gestation

Weight

< 2500 g: low birthweight (LBW)

< 1500 g: very low birthweight (VLBW)

< 1000 g: extremely low birthweight (ELBW)

Gestation

≥ 42 weeks: post-term

37–41 weeks: term

< 37 weeks: preterm

Summary
Definitions: events

- Infant: 0–364 days
- Neonate: 0–27 days
- Early neonatal: 0–6 days
- Late neonatal: 7–27 days

Summary
Definitions: death-rate

- *Still birth* is said to have occurred when a baby, of at least 24 completed weeks of gestation, on being completely expelled from the mother, shows no sign of life

- *Still birth rate* is measured as the number of still births from 24 completed weeks of gestation, per 1000 total births. The typical rate in the UK population is around 4/1000

- *Perinatal mortality rate* has been used as the main outcome measure for maternity services for many years. It is defined as the number of still births from 24 completed weeks of gestation, plus the number of deaths from 0 to 6 postnatal days inclusive (early neonatal deaths), per 1000 total births.

- *Neonatal mortality rate* is the number of deaths from 0 to 27 postnatal days inclusive, per 1000 live births (*not* per 1000 total births). It is sometimes helpful to examine early and late neonatal deaths, and death-rates, separately

- *Infant mortality rate* is all death in the first postnatal year, per 1000 live births. It is usual to examine neonatal and postneonatal deaths separately, hence the *postneonatal infant mortality rate*: deaths from 28 to 364 postnatal days inclusive, per 1000 live births

3. Lastly, because there is an irreducible minimum of 'unavoidable' perinatal death due to lethal congenital malformation, a virtual certainty of death where birthweight is less than 500 g, and a high mortality for babies born under 1000 g, the International Federation of Obstetrics and Gynaecology (FIGO) recommend subdividing the deaths, and mortality rates, as follows: all < 500 g, all < 1000 g, and non-malformed babies ≥ 1000g.

The prenatal environment

Life does not start at birth, and the well-being and early life of an infant can only be understood in the context of events that predate birth, and may even predate pregnancy.

Pregnancy may be wanted or unwanted, planned or unplanned, whether or not it occurs within a conventional marriage. The extent to which a family tries to plan pregnancy is a function of the cultural context of that family, and the parents' beliefs about their control over their destiny. Whatever the planning, between 10% and 20% of pregnancies in the UK end in termination.

The concept of a 'standard nuclear family' has become of much less value over recent years when attempting to describe family structures. In some areas nearly half of all births are to parents who are not formally married, although the proportion born to women unsupported by a partner is much smaller. It has been known for many years that mortality rates among babies of very young (teenage), and older mothers, is considerably higher than mothers in their twenties. However, young mothers are over-represented among more deprived social groups, while older mothers are frequently those who have delayed child-bearing in order to pursue a career or profession.

The estimated date of delivery (EDD) is normally calculated from the first day of the last menstrual period (Fig. 8.1), and the EDD is taken as 40 weeks (280 days) later. Routine ultrasound scans at 16–18 weeks of gestation are used to confirm the dates of the pregnancy by measurement of the biparietal diameter of the fetal skull, and to screen for major structural malformations such as spina bifida.

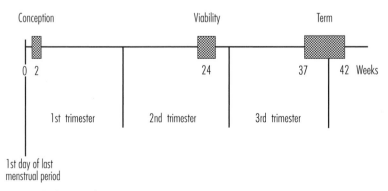

Fig. 8.1. The division of human pregnancy into trimesters.

The process of forming an attachment to the baby starts with the routine ultrasound scan. It is commonplace for both partners to be able to watch the scan and so get their first opportunity to 'see' their baby. Attachment is enhanced with the onset of the perception of fetal movements, is enormously advanced by the birth of the baby, and continues thereafter. The once fashionable concept of 'bonding', whereby mother and infant were believed to form a permanent attachment at a biologically critical time in the early postnatal period, has not been supported by empirical evidence in the human.

Fetal life is descriptively divided into three trimesters (see Fig. 8.1) but for the purposes of understanding the newborn baby the important division is between the first three months, and the rest. This is because the first trimester is the time of organogenesis, during which the embryo develops from a single fertilized ovum to a complete miniature fetus by completing all the major stages of organ differentiation and formation. Noxious influences at this stage of gestation may result in fetal loss (miscarriage), or malformation. For example, failure of closure of the embryonic neural tube at this stage gives rise to the anatomical malformation called spina bifida.

After the first trimester, the fetoplacental unit grows and the organs mature but no new organs are formed. Enormously rapid growth takes place which is normally accommodated by an expanding uterus and amniotic cavity, and is sustained by a growing placenta. This process can be disturbed by placental malfunction, leading to intra-uterine growth retardation, or physical constraints on fetal growth such as an inadequate volume of amniotic fluid. An effect of this may be to mould fetal growth, for example to produce the foot and ankle deformity of talipes equinovarus.

The uterus is generally a benign and protective environment for a fetus, but there are circumstances when this is not so.

- Alcohol abuse may globally damage the fetus, sometimes giving rise to a characteristic syndrome at birth.
- Cigarette smoking is a notorious fetal toxin, reducing birthweight and hampering subsequent neurodevelopment.
- Certain prescribed drugs, such as anticonvulsants (which cannot readily be stopped during pregnancy), may occasionally cause fetal harm, either by increasing the birth prevalence of a malformation (e.g. phenytoin and cleft palate), or causing a generalized effect (e.g. fetal valproate syndrome).
- Certain prescribed drugs, such as benzodiazepines, may be abused during pregnancy and give rise to a withdrawal state in the baby.
- Abuse of illegal drugs by the mother, particularly opiates or cocaine, can have a devastating effect on the well being of the neonate. Not only may there be a severe withdrawal state, but there is a very substantial subsequent infant mortality.
- Maternal diseases, such as diabetes, can give rise to malformations in the fetus, such as agenesis of the sacrum.
- Apparently minor viral illness in the mother, such as rubella and parvovirus B19 (which causes Fifth disease or 'slapped cheek' in children), may infect the fetus and cause severe damage.

Summary
Agents causing damage in utero

- alcohol abuse
- cigarette smoking
- anticonvulsants
- benzodiazepines
- heroin addiction
- maternal diabetes
- rubella, parvovirus

- Maternal protein–calorie malnutrition may reduce birthweight, but most of the variance in birthweight is accounted for by other factors unless maternal malnutrition is severe.

Because many of these noxious influences act on the embryo, they may have their effects even before a woman knows she is pregnant. Therefore, any public health strategy to minimize the chance of fetal harm can only be effective if it is implemented prior to conception. One model for this is the preconception clinic, which, if run as a primary care facility, has the opportunity to engage healthy women who may be planning a pregnancy.

Birth

Place of birth

Most babies are born in a maternity facility rather than at home, and most maternity facilities are in general hospitals. Around 0.6% of babies are born outside hospital, but only two-thirds of these are planned ('booked') home deliveries, since it has been health service policy for 30 years to encourage hospital birth. The concern about home delivery has been that of the safety of mother and baby, but even though current evidence suggests that planned home delivery is much safer than had been previously thought, there is widespread unwillingness among GPs and community midwives to return to the delivery of a larger proportion of babies at home.

In hospital, it is now routine practice for mothers to be supported throughout labour and delivery by their partner, a relative, or a friend, and this commonly extends to elective caesarean delivery. Such support is not merely socially pleasant: there are real benefits in shortening labour, diminishing pain, and reducing the likelihood of operative delivery.

When the baby is born

The vast majority of babies born vaginally breathe spontaneously and require no intervention at all. The presence of secretions, blood, or meconium in the mouth may justify suction of the mouth, nose, or oropharynx but is not usually required. Suction should be performed with a soft mucus extractor advanced not more than 5 cm through the lips.

The mother's own abdomen and chest are ideal sources of warmth. However, it is important to cover the baby with a blanket or towel (a sheet is not enough) to prevent excess heat loss. If the baby requires intervention, drying and wrapping the baby is essential (before active resuscitation) to prevent the addition of thermal stress to the baby's other problems. An overhead radiant heater is inadequate to compensate for the evaporative heat loss from a wet baby's skin.

The baby's condition at birth is best described in terms of the time to achieve a normal heart rate, time to first breath, and time to regular respiration. The

Apgar score (invented by Dr V. Apgar) is a shorthand way of conveying this information (Table 8.1), but it is seldom correctly ascertained and is poorly predictive of subsequent neurodevelopmental outcome.

Resuscitation of a newborn baby

Mrs Lyons is 25 years old, and is just delivering her baby after a 45-minute second stage of labour which has apparently been entirely normal. On complete delivery of the baby there is no attempt to gasp and the baby is floppy and blue. The midwife clamps and cuts the cord and passes the baby to you. What should you do?

Immediately, you dry the baby vigorously and leave him wrapped in a towel on a firm surface. It only takes you a few seconds and the stimulation might have been enough to encourage the baby to take a first breath. In fact he does not breathe, so you remember:

> A airway
>
> B breathing
>
> C circulation

Airway. You clear the mouth and nose with a mucus extractor, and keep the baby's head in the neutral position. Still no breath. You listen to the heart and the rate is slow.

Breathing. Simple measures have failed so you apply a neonatal bag and mask over the baby's nose and mouth and inflate the chest for 5 good breaths (Fig. 8.2).

Circulation. Listening to the heart again, it is speeding up.

Five more breaths from the bag and mask and the baby takes a gasp, then another and another. The baby becomes pink, and tone returns to the limbs; he is breathing regularly by 7 minutes. You watch him for another couple of minutes, then

Summary
Neonatal resuscitation

- Most occasions when a baby requires resuscitation at birth arise after a normal labour and delivery.

- A baby may not breathe because of maternal opiate analgesia. The ABC response is just the same, but you would also give naloxone intramuscularly to reverse the effects of the opiate.

- Resuscitating a baby in the presence of the parents is generally much less stressful for them than doing it outside the room.

Table 8.1.
The Apgar score

Attribute	Score: 0	Score: 1	Score: 2
Heart rate	Absent	< 100/min	> 100/min
Respiratory effort	Absent	Weak, irregular	Strong, regular
Muscle tone	Limp	Some flexion	Weak movement
Reflex response to stimulation of feet	None	Weak movement	Cry
Colour	Blue or pale	Pink body, blue extremities	Pink all over

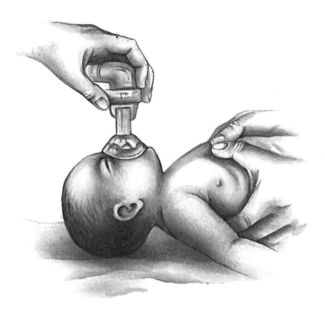

Fig. 8.2. Resuscitation procedure.

give him to his mother, explaining that in spite of his unexpected reluctance to breathe at delivery, he is now fine. You stay around for another 5 minutes, writing down what you did, and check that he is still pink before you leave.

Birth asphyxia

Janine is born by emergency caesarean section at term for fetal distress. At birth she is hypotonic, apnoeic, and the heart rate is around 40/minute. The pH of cord blood is 6.9. Full resuscitation, including cardiac massage, yields a heart rate > 100/min by 6 minutes. She is noted to gasp at 8 minutes and regular respiration is seen at 12 minutes, whereupon artificial ventilation is stopped. How might she be affected by this perinatal stress?

She may have received some hypoxic or ischaemic injury to her brain or other organs, in which case she might develop some degree of encephalopathy or related symptoms, or she may remain totally asymptomatic. Although she quite clearly had symptomatic asphyxia at birth, the long-term neurodevelopmental outlook is most closely related to the severity of any subsequent encephalopathy. Babies who do not establish regular respiration by 20–30 minutes postnatally almost invariably die, or are very severely disabled.

Cardiorespiratory adaptation

There are several stages of adaptation. The first breath initiates a cascade of changes (Fig. 8.3), but changes continue to take place over the first 24 hours.

During the first week, pulmonary arterial pressure continues to fall slowly, and the arterial duct becomes permanently shut by a process of fibrosis. Respiratory control changes over a similar time scale. In particular, the infant's oxygen sensor has been used to the oxygen tension present *in utero*—around 3 kPa. It now has to operate with oxygen tensions in excess of 10 kPa, and cannot immediately do so: it takes about a week to reset.

Adaptation continues long after the first postnatal week. *In utero* the right ventricle of the heart has been dominant; after birth the left ventricle assumes the work of pumping against systemic vascular resistance, while the right ventricle has the progressively less arduous task of pumping blood through the low resistance of the lungs. This reversal of dominance from right to left takes many months as the left ventricle progressively becomes more thick-walled and the right ventricle more thin-walled. The ribs gradually become more calcified and rigid, while the proportions of the thorax and abdomen alter with relatively greater thoracic growth; and the control of respiration becomes more robust.

Birth trauma

Serious birth trauma is rare, but occasionally a baby may fracture a clavicle at delivery, or suffer stretching of the nerves of the brachial plexus resulting in an Erb's palsy. Both of these injuries tend to heal well, although the outlook for a palsy where avulsion of the nerve roots has occurred is very poor.

Relatively common minor traumas include forceps marks on the face, which may result in localized fat necrosis with the formation of a small subcutaneous lump; and cephalhaematoma which is a subperiosteal collection of blood over one of the bones of the skull: this can look quite striking, and feels fluctuant on palpation.

Assessing gestation

Angela is 16 years old. She comes to the accident and emergency department with abdominal pain and examination reveals that she is in labour, at which point she admits to concealing her pregnancy. Her periods were always irregular and she has no idea of her dates. After transfer to delivery suite she gives birth to a baby boy weighing 1.8 kg, but the baby is not in good condition and spends the next three days on special care in up to 50% oxygen, though subsequently he does well. How can you begin to work out his approximate gestation?

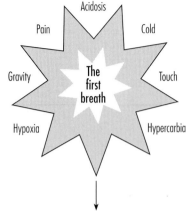

Pulmonary vascular resistance falls, so pulmonary blood flow increases

Large volumes of pulmonary venous blood enter the left atrium, causing left atrial pressure to rise and the foramen ovale to close functionally

The arterial duct closes functionally in response to rising arterial oxygen tensions

Systemic vascular resistance rises as the umbilical cord is severed

BUT

Pulmonary arterial pressure remains high initially

The arterial duct is not yet permanently closed

This potentially unstable ciculation is called transitional. It lasts for around 24 hours, by which time pulmonary pressure has fallen considerably and the risk of the baby flipping back into a " fetal " circulation, with persistent pulmonary hypertension, is much reduced.

Fig. 8.3. The mechanism of cardiorespiratory adaptation.

Normally, the length of gestation is known because the last menstrual period is a well-defined event and dates have been confirmed by ultrasound scan. However, the dating scan must be done early (≤ 18 weeks) to be valid, and where a woman has booked late, concealed her pregnancy, or been seriously uncertain of her history, antenatal dating becomes inexact. However, gestation may be estimated by assessing the maturity of a newborn baby. There are several methods for this, for example, using neurological features (Dubowitz), or external features (Parkin). In the above situation, the preferred system would be the Parkin score because only 4 external features are noted and minimal handling is needed to ascertain these.

Normal term babies

Around 95% of babies are born between 37 and 42 weeks of gestation (fullterm or term), and most of these are healthy. They need no interventions from any professionals and do not themselves need to be in hospital. Yet all of them are routinely examined. Why?

Additional Information
Ortolani test for congenital hip dislocation

- requires skilled examiner
- warm, contented infant
- supine on firm surface
- flex hips and knees
- thumb on inner thigh
- middle finger behind hip
- flex and adduct
- feel for 'clunk'

Babies are checked because a reasonably high proportion of them will have a relatively minor problem which, if dealt with appropriately, can have much less impact on the baby's subsequent growth and development. These problems can be sufficiently subtle not to be obvious without an examination to look for them. Examples are as follows:

- Cataract (p. 237)
- Cleft palate
- Inguinal hernia (p. 389)
- Undescended testes (p. 390)
- Hypospadias
- Dislocatable hips (p. 355)

The head circumference is also measured, not to diagnose hydrocephalus (you would do this with your eyes and fingertips), but to serve as a baseline for future measurements, just as for birthweight.

The timing of the examination is important. Too soon, and many babies will have a quiet systolic heart murmur ('flow' murmur) which is of no clinical significance; and the head will still be moulded from the passage through the birth canal, giving a falsely high value for the head circumference. Too late, and if a problem is identified, there may not be enough time to sort it out before the mother is due to leave, so that either it must be dealt with as an outpatient, or discharge is delayed, neither of which are desirable for the parents.

The ideal timing for neonatal examination is a full 24 hours after birth, although this is unattainable for mothers wishing for rapid (e.g. 6 hour) discharge. There is no evidence that two examinations are of more value than one.

Other minor congenital abnormalities are seen infrequently, for example, accessory digits, a two vessel cord (single umbilical artery), sacral dimple, mongolian blue spot, accessory auricles, single palmar crease, and capillary naevi or haemangiomas.

Neonatal screening

On day 6 babies have a blood sample from a heel prick taken on to a blood spot card (Guthrie test) for analysis to detect the inborn errors of metabolism *phenylketonuria* and *hypothyroidism*. The analytical methods vary, but the justification for the tests is as follows (see Wilson and Jugner criterion, p. 74):

- The conditions themselves, if not detected early and treated, will lead to permanent learning disability and the real possibility of never leading an independent life.
- With early detection and appropriate treatment, almost all affected children can expect normal neurodevelopment.
- By the time either condition presents clinically (i.e. without screening), substantial brain impairment has already occurred.
- Both tests are highly specific and very sensitive: that is, the tests are rarely positive in genuinely normal children, and virtually always positive in genuinely abnormal children.
- Phenylketonuria has a birth prevalence of around 1 per 6000 births, and hypothyroidism around 1 per 3000 births, so neither condition is excessively rare.
- The cost of screening, per case identified, is very much lower than the cost of caring for an undiagnosed or late-diagnosed child.

Jaundice

Amy is found to be jaundiced at 36 hours of age. Breast-feeding (2- to 3-hourly) is becoming well established and she sleeps between feeds but is alert when awake. She was considered normal at 24 hours on routine postnatal paediatric examination.

This is probably *'physiological' jaundice*, as the onset is on the second postnatal day, rather than the first. The hyperbilirubinaemia arises because there is an imbalance between the peripheral production of bilirubin from haem, and its disposal by conjugation in the liver. Production is increased because babies are born relatively polycythaemic, and they rapidly break down the excess red blood cells. Disposal

Additional Information
Colours of babies

- *Pale pink*. This is normal
- *Blue*. This is only normal when confined to the hands or feet, and within the first 48 hours postnatally. Central blueness is never normal and requires urgent investigation for cardiac or respiratory disease
- *Harlequin*. *Red* to the midline on one side, *white* on the other. This unusual and transient phenomenon has never been convincingly explained but is entirely benign
- *Very red*. This may occur with polycythaemia, and it is worth checking the baby's haematocrit. Generally the baby has received more than its fair share of blood from the placenta
- *White*. Anaemia, due to receiving too little blood at birth, occasionally because of a large feto-maternal leak, or placental abruption
- *Grey, or mottled*. This is a very sinister colour, often seen with septicaemia, and requires urgent intervention
- *Yellow/green*. This is the colour of conjugated jaundice.
- *Yellow/orange*. This is the colour of ordinary neonatal jaundice

Summary
Physiology of jaundice

- Unconjugated hyperbilirubinaemia is universal, and clinical jaundice common, in babies
- Breast-fed babies are more often affected
- Physiological jaundice is benign
- Only severe, early onset, and prolonged jaundice requires investigation

Additional Information
Rhesus disease

- The rhesus antigens are D, C, c, E, and e. Disease most commonly arises from antibodies to D.

- A Rhesus negative mother can make antibodies to the red cells of a Rhesus positive fetus as the result of a small feto-maternal leak.

- The maternal IgG anti-D antibodies then cross the placenta, causing destruction of the fetus's own red cells.

- The fetus then becomes anaemic, but excess bilirubin production is dealt with through the placenta until birth.

- After birth, an affected baby may need phototherapy or even exchange transfusion to treat hyper-bilirubinaemia. Seriously high bilirubin levels (>350 μmol/l) can cause bilirubin encephalopathy and sensori-neural deafness.

- Rhesus disease has now almost disappeared in the UK as a result of the use of prophylactic anti-D antibodies which are administered to Rhesus negative women after the delivery of a Rhesus positive baby, a miscarriage, or a termination of pregnancy. The exogenous antibodies coat any leaked fetal red cells, which are then destroyed in the mother's spleen before they can induce antibody formation.

is reduced because the hepatic conjugation process is immature and only becomes fully active in the second postnatal week.

Examination reveals a healthy girl. Estimation of the serum bilirubin from a heel prick sample reveals a concentration of 200 μmol/l; no action is required and the normality of the situation is explained to the mother.

Examination is important: a healthy infant with jaundice is common and normal, whereas an ill baby requires immediate investigation (particularly for infection). As babies can become ill quickly, the fact that she was well 12 hours ago is no help. The level of 200 μmol/l is unremarkable for the situation, and a more experienced doctor would probably have spared the baby the heel prick.

Jaundice can most usefully be considered potentially pathological if it occurs unusually early or is prolonged, compared with normal.

Early jaundice

In an otherwise identical situation, a bilirubin of 200 μmol/l is discovered at 18 hours postnatal age. Amy's mother is noted to be blood group O rhesus negative with no red cell antibodies detected during pregnancy.

This is unusual, because the jaundice has occurred early, and physiological jaundice cannot safely be assumed. After examination, and being satisfied of the baby's general good health, investigation is directed towards causes of excess bilirubin production. A haematocrit must be checked, and blood grouping and Coombs test may suggest the possibility of blood group incompatibility due to the presence of antibodies (which would be of maternal origin) directed against red cell antigens.

Haemolytic disease of the newborn used to be caused by rhesus antibodies. *Rhesus disease* is now rare, is unusual in a first pregnancy, and unlikely if the mother has previously tested negative for rhesus antibody. ABO incompatibility is more common in first than subsequent pregnancies, but may at best result in a weakly positive Coombs reaction, so it can be difficult to diagnose with certainty. Examination of the blood film is performed for the presence of spherocytosis as this and other abnormalities of red cell morphology result in short cell lives, but the diagnosis must be confirmed by a red cell osmotic fragility test. Glucose-6-phosphate-dehydrogenase deficiency is an important cause of neonatal jaundice world-wide, and must be considered in families of Mediterranean and Asian origin, but is rare in the UK.

The importance of early jaundice is that when caused by haemolysis, mild bilirubin toxicity leads to sensorineural hearing loss, while bilirubin encephalopathy (also called kernicterus) leads either to death or to severe athetoid cerebral palsy. Phototherapy is used to control jaundice non-invasively, and exchange transfusion if the jaundice reaches dangerous levels, or if there is severe haemolysis.

In fact the baby is also blood group O negative. All other investigations are negative, but over the next 3 days the bilirubin concentration rises to 350 μmol/l. The sister on the postnatal ward wonders whether the baby can safely go home with her mother.

Where haemolysis has been excluded, there is no evidence of any harm arising to the baby from levels of hyperbilirubinaemia even greater than this in healthy term babies. It is reasonable to ask the mother to return for a check of the bilirubin (or to organize for the community midwife to take the sample at home) to ensure the level has achieved a plateau within a further 48 hours, but thereafter no further action is required.

Prolonged jaundice

This baby's bilirubin may well remain elevated for many weeks while exclusive breast-feeding continues, although most breast-fed babies lose their jaundice within the first fortnight. The mechanism for prolonged breast-feeding jaundice is not entirely clear, but may relate to an idiosyncratic inhibition of bilirubin conjugation by some of the circulating fatty acids derived from breast milk.

Serious causes for *prolonged jaundice* are rare.

Prolonged conjugated hyperbilirubinaemia has serious implications. Urgent investigation is required, particularly to identify biliary atresia, since the results of surgery to correct this condition are much less good if assessment is delayed longer than 6 weeks postnatal age, (see also Ch. 14, p. 255).

Feeding

In the North of England and Scotland, only a minority of mothers initiate breast-feeding, and many of these give up after only a few days or weeks. Some people believe that breast-feeding adds nothing to health status in the developed world, and that formula-feeding is just as good. Is this true?

The neonatal period is not the time to discuss with a new mother the relative merits of breast- or formula-feeding: this should have been done antenatally. The role of staff is to support the mother in her chosen method of feeding her baby.

Additional Information
Investigation of prolonged jaundice > 2 weeks

- Serum conjugated ('direct') and unconjugated ('indirect') bilirubin
- Urine for bacterial culture, viral culture, bilirubin, reducing sugar (galactose), and amino acids (tyrosine).
- Thyroid function, alpha$_1$-antitrypsin
- Viral serology: rubella, cytomegalovirus, herpes simplex, hepatitis A, B, C
- Abdominal ultrasound for choledochal cyst

Summary
Benefits of breast-feeding

- Protection against infection
- Compensation for insensible fluid loss
- Provision of specialised nutrients
- Cost
- Convenience
- Loss of maternal fat from thighs and hips
- Probable protection against breast cancer

Breast-feeding has distinct benefits over artificial (formula) feeding, even in the developed world. The positive benefits can be summarized as follows:

- *Protection against infection.* This includes gastrointestinal infection, respiratory infections, and otitis media. Protection is conferred by the high iron binding of lactoferrin, which deprives Gram-negative organisms of this essential growth factor; promotion of an acid gut environment, colonized by lacto-bacilli, and inimical to Gram-negative bacilli; and secretion of maternal phagocytes and IgA immunoglobulins directed against pathogens such as rotavirus.
- *Automatic compensation for fluid loss.* In hot weather, a breast-fed baby will drink more milk but it is correspondingly more dilute, so the baby does not take excess calories while getting additional water.
- *Provision* of other specialized nutrients such as very long chain fatty acids which may specifically promote brain growth and development.
- *Cost.* Breast milk is free.
- *Convenience.* Breast milk is always there unless the health of the mother is severely compromised.
- For the mother, breast-feeding is the only known way (other than gross malnutrition) of *losing fat* from thighs and hips; and it may confer some protection against breast cancer.

World-wide, the lack of availability of clean water makes formula feeding a major cause of death. The relative health benefits of breast-feeding extend far beyond the nutritional and protective effects on the infant and include the lactational amenorrhoea which ensures wider spacing between consecutive children and hence more resources for the other children.

Artificial ('formula') feeds

If a baby is not being breast-fed, the only suitable food is highly modified cow's milk (or soya-based milk), often referred to as 'infant formula' feeds. Cow's milk, whether skimmed, semi-skimmed, or whole, is unsuitable for many reasons, not least the high phosphate load which can give rise to symptomatic hypocalcaemia if fed to very young human babies. Commercial dried milks, evaporated milk, and the so-called 'follow-on' formula milks are equally inappropriate for initiating artificial-feeding.

Care should be taken in *making up a formula feed*. A baby normally takes around 20 minutes to consume a formula feed, irrespective of his or her size. This is also true of the majority of breast-fed babies, after the first couple of weeks postnatally. If a baby takes longer, certainly over half an hour, there is almost by definition a feeding problem.

There are two kinds of infant formula based on modified cow's milk: those in which the protein is predominantly modified from the whey fraction, and those which are predominantly casein. Each manufacturer markets both types, with different and sometimes confusing brand names. *Whey-based* formulas have a

Summary
Making up a formula feed

- Bottles, teats, and utensils must be well-cleaned and sterilized

- Water for the feed should be boiled then allowed to cool

- The standard scoop provided with the formula should be used, 1 scoop to each fluid ounce (30 ml) of water. Granules may be shaken to level off; powdered milk should be levelled off with the back of a knife

- The capped bottle should be shaken well to mix; after putting on the teat, a few drops of milk shaken onto the wrist should feel warm but not hot

- Feeds can be made up in batches and stored in a fridge for up to 24 hours. They can be warmed up in warm water, but not heated in a microwave oven because the centre of the feed can become scaldingly hot, while the outside may feel only comfortably warm

protein composition which approximates to that of 'average' breast milk. *Casein-based* formulas have slightly higher concentrations of protein, carbohydrate and electrolytes.

Establishing feeding

Mothers who want to breast-feed are encouraged to put the baby to the breast very soon after birth (Fig. 8.4). Successful breast-feeding is made much easier for a new mother if she is motivated; she has consistent help, advice, and support from midwives; she has privacy if she wishes; and her baby is always kept close by her. In addition, the baby should not be offered any fluid other than breast milk, and should be fed on demand: babies form their own routine with their mother and do not conform to imposed expectations.

Difficulties with breast-feeding most commonly arise from poor positioning of the baby's mouth with respect to the areola of the breast, leading to inadequate milk delivery, prolonged feeds, an unsatisfied child, and sore breasts (Fig. 8.5). Some mothers find feeding easy while others need much help and support; some babies feed very frequently and others sleep for many hours between feeds. There is no such thing as a 'normal' pattern for breast-feeding.

Breast-feeding is a two-way process, with the baby exerting considerable influence over the mother by means of stimulating milk production in the act of suckling (the breast–pituitary–prolactin arc) and in provoking let-down (mediated by oxytocin). The content of human milk varies between different feeds, and within each feed: most of the water and protein is transferred in the first five minutes of the feed in the foremilk, but most of the fat (and hence the energy) comes across in the last minutes of the feed, in the hind milk. Breast-feeding is a

Additional Information
Curds and whey

- Casein (from Latin *caseus* = cheese) is that fraction of protein in raw milk which is precipitated at pH 4.6 at 20° C

- Whey is the rest (~20% of protein in cow's milk and ~80% in human milk)

- Modified formulas are often described in terms of the whey: casein ratio

- Infant formulas designed for initiating formula feeding are whey-dominant

Fig. 8.4. Breast-feeding.

Fig. 8.5. Breast-feeding: (a) correct latching-on: wait until the mouth is wide open, then the baby takes the whole areola into the mouth not just the nipple; (b) incorrect latching-on: the baby sucks on the nipple, as if the breast were a bottle.

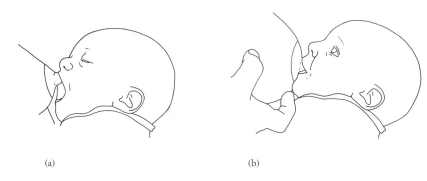

(a)   (b)

Additional Information
Ten steps to successful breast feeding (UNICEF)

- have a written breast feeding policy
- train health staff in breastfeeding support
- inform all pregnant mothers of the benefits of breastfeeding
- help mothers initiate breastfeeding within half an hour of birth
- show mothers how to breastfeed even if separated from infants
- give no food or drink other than breastmilk unless medically indicated
- room-in baby with mother 24 hr/day
- encourage breastfeeding on demand
- avoid dummies for breastfeeding infant
- foster breastfeeding support groups

A hospital which completes the ten steps will be designated as 'Baby friendly'.

most intimate process facilitating mother–infant attachment: it is far more than the mere passage of nutrients from a mother to her baby.

Hospitals can do much to promote breast feeding, whilst unsatisfactory practices can hinder it. *The UNICEF 'Ten steps to successful breast feeding'* are designed to encourage good feeding.

Bottle-fed babies require, on average, fewer feeds in the day than breast-fed babies because gastric emptying time is greater with formula feeds. Irrespective of the method of feeding, the milk intake increases over the first few days. There is no fixed amount that any healthy full-term baby should receive on any day; on average, many will be taking in excess of 100 ml/kg/day by day 3. Babies almost always lose weight over the first few days, except those with intra-uterine growth retardation, who sometimes do not; this loss may be up to 10% of their birth weight. The birthweight is usually regained between one and two weeks after birth and if feeding is progressing normally the baby will grow at a rate of around 7–10 g/kg/day (or about 200 g/week for an average-sized baby).

Immunization in newborn babies

Ashleigh is a first child, and her parents are on leave from their work in China. They will be returning to China when Ashleigh is 3 months old.

Ashleigh should receive BCG immunization because tuberculosis (TB) is endemic in China. Department of Health guidelines recommend BCG under the following circumstances:

- if the mother is from an immigrant family (particularly those from Asia or Hong Kong);
- where there is a definite history of TB in the parents (or in the household where the child is going to live) in the last 5 years; and
- if the infant will travel to any area where the risk of tuberculosis is high.

For babies whose mothers carry hepatitis B (and are positive for the e-antigen, a marker for high infectivity), an injection of anti-hepatitis B gamma-globulin followed by a course of active immunization, is recommended.

All babies, regardless of their gestation at delivery, should receive standard triple vaccine and *Haemophilus influenzae* type B (Hib) vaccine from 8 weeks of postnatal age (see Chapter 4, p. 72).

Congenital abnormalities
Cleft lip

Mrs Armstrong is about to deliver her second baby and her partner is with her. At birth it is immediately obvious that the baby girl has a left sided cleft lip, but she cries, turns pink, and is given to her parents. What should be done next?

- You keep the baby with the parents. Overt rejection at this stage is rare, however much grief and disappointment there may be.
- You explain gently and patiently what the obvious abnormalities are; don't try to hide anything. While avoiding any attempt to prognosticate, you emphasize that expert advice is available and will be sought. Do not be afraid to say that you have not got all the answers immediately.

Behind the cleft lip there may be a cleft palate, but in any case there will be difficulties in feeding the baby. The results of plastic surgery for the defect are extremely impressive, and the parents need to hear this, and see some 'before and after' pictures as soon as possible from the surgeon.

The community midwife will be a key figure in supporting the family after discharge home (up to 10 days postnatally); thereafter the health visitor will take on this role together with the GP. Once successful surgery (either one or two operations, depending on severity) has taken place, the baby will be normal.

Almost any imaginable congenital abnormality has been described, but the important ones are those most often seen, most preventable, and most treatable. Congenital abnormality can usefully be considered to be either of *genetic origin*, predating conception; a *malformation*, originating in the first trimester; or a *deformation*, arising at a later stage of pregnancy. Some, of course, are lethal in early life. Among surviving babies, many have special needs long after discharge from their primary hospital management, and close liaison between hospital staff, the primary care team and the local special needs team needs to be maintained.

Trisomy

This is the single most common group of chromosomal abnormalities, (see also Ch. 15, p. 264). Most frequent is trisomy 21 (Down syndrome, see Chapter 7), followed by 18 (Edward syndrome), and 13 (Patau syndrome). The last two are virtually always lethal within days of birth; survival for months is exceptional. Trisomy is less rare as the age of the mother increases, and mothers over 35 years are commonly offered early antenatal screening.

Neural tube defects

These are anencephaly and spina bifida. The defects are due to the failure of closure of the neural tube in early embryonic life (around 30 days). Anencephaly is always lethal, but spina bifida commonly is not and has profound repercussions for affected children (see Chapter 7, p. 138). Both may be screened for by measurement of maternal serum alpha-fetoprotein, detected by antenatal ultrasound, and prevented (in high-risk families with a previous affected child) by periconceptional folic acid supplements.

Oesophageal atresia and tracheo-oesophageal fistula

There are a number of variants of this condition, all of which result from maldevelopment of the foregut and lung bud (see Fig. 8.6).

The inability of the fetus to swallow usually gives rise to polyhydramnios complicating the pregnancy. In any baby born to a mother with this condition, and without any other explanation for it, it is worth attempting to pass a mucus extractor down into the stomach to prove oesophageal patency (passing a suction catheter or feeding tube is not diagnostic as it curls up in the oesophageal pouch). Depending on the anatomy of the lesion, it is not always possible to achieve an early primary surgical repair, so the establishment of a gastrostomy to enable enteral feeding is often the priority.

Other gut atresias: duodenal, anal

Duodenal atresia can be diagnosed antenatally by the 'double bubble' appearance below the liver, and is associated with Down syndrome, which again may have been diagnosed antenatally by fetal blood sample. Surgical repair is straightforward.

Anal atresia is usually diagnosed shortly after birth by the midwife who is unable to measure a rectal temperature. In girls, there may be a fistula to the

Fig. 8.6. The three predominant forms of tracheo-oesophageal fistulae.

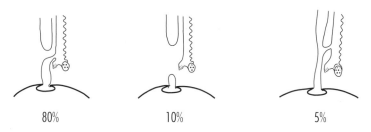

80%   10%   5%

vagina, but complete inability to pass meconium means that an urgent defunctioning colostomy is needed, with full corrective surgery when the infant is rather older. The aim is to allow the child to achieve full bowel control whenever possible.

Abdominal wall defects

On antenatal ultrasound at 18 weeks, the fetus is seen to have protrusion of abdominal contents out of the abdomen. It may be a gastroschisis, or an exomphalos. Why should the distinction matter to the parents?

In *gastroschisis*, the defect is separate from the umbilicus and loops of fetal bowel float outside the fetal abdomen in the amniotic fluid. There are not usually any other associated abnormalities, and the outlook for the baby is good with early neonatal surgery.

In *exomphalos*, the bowel protrudes in a sac through the umbilicus, and so is not in contact with the amniotic fluid; however, it may be part of an underlying chromosomal defect, part of another syndrome, or have an associated abnormality such as congenital heart disease. The surgical management is relatively straightforward, but the overall outcome is dependent on whatever other anomalies are present.

Explaining the further investigations and counselling the parents requires the distinction between the two conditions to be made.

Diaphragmatic hernia

A malformed diaphragm with a hole in it can allow mobile abdominal contents to herniate up into the chest, more often on the left than the right because of the site of the liver. The lung on the affected side is compressed, and may be hypoplastic. If the mediastinum is pushed across, the other lung will be compressed too. Some of these fetuses are picked up *in utero* on ultrasound scanning, but some are missed and may present after birth with cyanosis, either with obvious respiratory distress, or with persistent pulmonary hypertension. The treatment is surgical after stabilization, but the mortality is still high (around 40%). Survivors can expect a normal and disability-free life.

Congenital heart disease

A detailed discussion of the many varieties and their management are beyond the scope of this chapter (but see Chapter 10). Those considered here will be arranged according to their neonatal presentation. Serious congenital heart disease presents with either cyanosis or cardiac failure (frequently without a murmur), but in the neonate the process is often very rapid.

Cardiac murmur

Jonathan is found to have a significant systolic murmur on routine neonatal examination.

These are commonly found to be small ventricular septal defects. Most close by themselves over the next few months. The most life-threatening cause of such a murmur is aortic valvular stenosis, which may well be treatable. Sometimes the murmur of pulmonary stenosis is heard (as in the tetralogy of Fallot, see below).

Cyanosis

Leanne is found to be dusky at the age of 10 hours. Her measured oxygen saturation in air by pulse oximetry is only 75%, but she has no respiratory distress, and her saturation is unaffected by increasing her inspired oxygen to > 80%.

A blue baby whose oxygenation does not improve with high inspired oxygen levels and who has no respiratory distress must be considered to have cyanotic congenital heart disease. The most common diagnosis, presenting in the first 24 hours, is transposition of the great arteries, (see Ch. 10, p. 204, 207).

Simon is noticed to become rather blue with each feed, and when he is crying, but without any respiratory symptoms.

Babies with Fallot's tetralogy (ventricular septal defect, pulmonary stenosis, over-riding aorta, and right ventricular hypertrophy) often present in this way but they are generally pink at birth. They may also present with a pulmonary flow murmur (see above), (see Ch. 10, p. 204).

Cardiac failure

Jake is rushed to the special care unit on day 4 having collapsed after a feed. He requires vigorous resuscitation.

The collapse of a baby, previously thought to be healthy, occurs with such conditions as hypoplastic left heart, interrupted aortic arch, and coarctation of the aorta. Such babies have been dependant for their systemic output on the effort of the right ventricle, with systemic blood flow through the arterial duct, and often have a ventricular septal defect which allows the pulmonary venous return to be handled by the right ventricle. The crisis is brought on by closure of the duct. An important, and potentially life-saving treatment following resuscitation, is an intravenous infusion of a prostaglandin to re-open the duct. However, a hypoplastic left heart is nearly always lethal.

Small for gestational age

Lee is delivered at 38 weeks of gestation after induction of labour because of maternal pregnancy-induced hypertension. He weighs 2000 g, but is vigorous and lively.

Lee's weight is on the 1st centile, so he is probably pathologically small for his gestational age and it is likely that he suffered intra-uterine growth retardation (IUGR) as a result of his mother's hypertension. Measurement of his head circumference reveals that it is 33.4 cm, not far below the 50th centile. This 'asymmetry' of the centiles is common, and reflects relative sparing of brain growth in the face of borderline intra-uterine nutrition.

Recognizing his small size and thermal vulnerability, he is kept well wrapped in the delivery room and fed there. However a blood glucose measurement is made 3 hours later, as he is very sleepy when about to be fed again; this reveals a value of 1.0 mmol/l.

He is hypoglycaemic. A diminished level of consciousness is often the only symptom of this (the other important symptom is a seizure). He and his mother are taken to the transitional care ward and he is given intravenous glucose by infusion, as well as continuing to receive milk feeds. Over the next 24 hours the infusion is gradually reduced and Lee is shown to maintain a satisfactory blood glucose on 3-hourly milk feeds. He goes home with his mother once her hypertension has settled.

By one month, Lee's growth rate has been such that his weight is now on the 5th centile, and his head circumference is on the 50th, demonstrating very rapid catch-up growth.

The near term infant

Jodie is born by normal delivery at 34 weeks of gestation following a small antepartum haemorrhage, and is her mother's first baby. Jodie is in good condition at birth and weighs 2.2 kg. Should she stay with her mother or go to special care?

Summary
Babies with growth retardation

- Are vulnerable to hypothermia
- Need early and frequent feeding
- May develop hypoglycaemia
- Frequently do not lose weight after birth
- Often show rapid postnatal catch-up growth

Summary
Near term babies (32–36 weeks)

- Are smaller, need more thermal care, and may not feed effectively in the first days or weeks compared to full-term babies;
- If they develop respiratory distress, it is often due to incomplete clearance of the fetal lung fluid ('wet lung');
- May get respiratory distress due to surfactant deficiency, particularly if born by elective caesarian section (without labour);
- Do very well and are not at particularly high risk.

There is no good reason to separate baby from mother. Many hospitals now have an 'integrated' or 'transitional' care area for babies such as Jodie. This area needs to be kept warmer than ordinary postnatal wards in recognition of the thermal fragility of smaller babies, but at 2.2 kg she should not need an incubator.

An attempted breast-feed is a failure. Jodie is unable to co-ordinate sucking and swallowing.

Gestational age of about 34 weeks is when babies may start to be able to feed orally. Jodie will clearly require gastric tube feeds for a few days until she is ready to try breast-feeding again.

Her mother is initially afraid to hold or touch her, but gains confidence over 3 or 4 days.

Summary
Risk factors in pregnancy and at delivery

- Past history of adverse events
- Smoking
- Drug abuse
- Multiple pregnancy
- Maternal diabetes
- Antepartum haemorrhage
- Fetal growth retardation
- Cervical incompetence
- Heavy alcohol consumption
- Pre-eclampsia
- Age over 40 or under 16
- Other maternal disease
- Malpresentation
- Shoulder dystocia

This often happens. Although babies born mildly preterm seldom have much trouble, she arrived fully 6 weeks before the estimated delivery date (EDD), long before her parents were prepared for her, and may seem to them fragile and particularly vulnerable.

Serious prematurity (< 32 weeks)

Babies born more than 8 weeks early are often described as 'high risk' by virtue of their gestation. Similarly, a pregnancy is described as 'high risk' if there is a potential hazard to the fetus or neonate (or the mother) at any gestation. The term

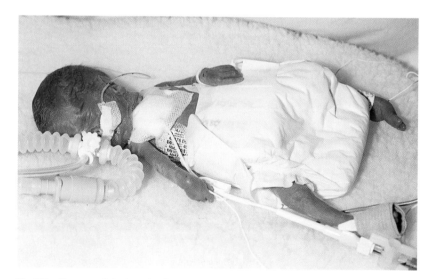

Fig. 8.8. Premature baby in an incubator.

'high risk' is really a shorthand for 'high risk of death or disability'. There are a number of important causes for being at 'high risk'.

Babies are born prematurely either because the obstetrician has identified a situation in which the mother is at risk from continuing the pregnancy (e.g. severe pre-eclampsia), or the fetus is at higher risk from remaining *in utero* than being delivered (e.g. severe growth retardation); or else some factor has stimulated preterm labour (e.g. infective chorioamnionitis). Sometimes the factors causing preterm labour require delivery to be hastened (e.g. antepartum haemorrhage with fetal compromise). Whatever the underlying reason, many premature deliveries are responses to an emergency situation, rather than being planned.

Only about 1% of all births occur before 32 weeks of gestation, and rather less than 2% of all babies require intensive care. Babies born at a unit not offering intensive care must be transferred to one that does, but the transport of an ill neonate may be hazardous unless undertaken by an expert team. If at all possible, a high-risk mother, particularly if she needs delivery before 30–32 weeks of gestation, should be transferred prior to delivery. This also helps to avoid unnecessary separation of mother and baby after delivery.

At 28 weeks of gestation, Susan Renwick notices that she is losing fluid from her vagina. In hospital it is confirmed that her membranes have ruptured, and she is told that she may go into labour and deliver prematurely. As a paediatrician you are asked to meet her and answer her many questions. The first question is usually, what are my baby's chances?

The answer depends more on the gestation than on the birthweight. Other things being equal, by 26 weeks of gestation, a live born baby has a greater than 50% chance of survival, and by 28 weeks this is up to around 85%. The proportion of survivors, and the rate of serious disability (compromising the ability to communicate or to lead an independent life) in survivors, is shown in Fig. 8.8.

The second question may be, can anything be done for my baby now?

Mothers can almost always be treated with a short course of corticosteroids, such as betamethasone or dexamethasone, prior to delivering a premature baby, provided that the attending obstetricians get some warning of the likelihood of delivery. This helps to mature the fetal lung, and reduces the severity of the respiratory distress syndrome in the baby (see below). In preterm, high-risk babies the reduction in mortality has been shown to approach 40%.

The third question is, what will happen to my baby?

Fig. 8.8. The number of survivors, by gestational age at birth, assessed at 2 years of age and their outcome. (From 'Report on perinatal and late neonatal death in the northern region' 1992.)

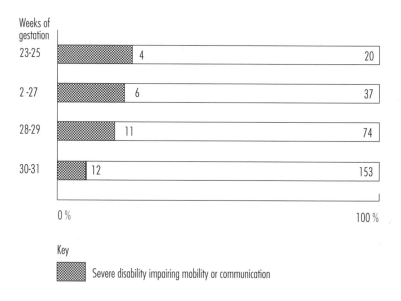

Weeks of gestation

23-25	4	20
2 -27	6	37
28-29	11	74
30-31	12	153

0 %   100 %

Key

Severe disability impairing mobility or communication

Summary
Respiratory distress syndrome

- Tachypnoea
- Recession between the ribs
- Bobbing of the head
- Flaring of the nostrils
- Audible grunt, or moan, with each expiration

Summary
Major complications of prematurity

- Acute respiratory disease
- Chronic lung disease
- Malnutrition
- Neurological damage
- Infection
- Necrotizing enterocolitis
- Retinopathy of prematurity

At birth, the paediatric team will be present, and the baby will be resuscitated if necessary. Not all babies, even when born 12 weeks early, require artificial ventilation from birth. All require to be transferred to the special care unit, but there is virtually never a situation when a mother (so long as she is not under an anaesthetic) is unable to see her baby before transfer to special care.

The problems a baby faces thereafter can best be considered functionally.

Breathing may be a problem due to inadequate production of surfactant molecules within the alveoli of the lung, giving rise to collapse of the small airways. A baby in this situation becomes tachypnoeic, has visible recession between the ribs, bobbing of the head, flaring of the nostrils, and produces an audible grunt, or moan, with each expiration. Without added oxygen to breathe the baby becomes blue, and may progress to respiratory failure, which is treated with mechanical ventilation and administration of exogenous surfactant.

This disease is called the *respiratory distress syndrome* (RDS) of the newborn, and is also known as hyaline membrane disease (HMD) from the histological appearances of the lungs of babies who perish from it. It becomes progressively more likely with *increasing prematurity*; its most important complications are as follows. First, the cardiovascular instability caused by RDS may lead to neurological damage (see below); second, the process of neonatal adaptation is disrupted, which may lead to failure of the arterial duct to shut, and this may require medical or surgical treatment; third, the lungs themselves may be harmed by mechanical ventilation, leading to prolonged oxygen dependency (bronchopulmonary dysplasia, BPD).

Nutrition is another major issue. The gut of premature babies is immature, and where there has been growth retardation the gut may be markedly hypotrophic. The goal is to establish enteral milk feeding as soon as possible, but this is commonly delayed in the presence of RDS, and a period of intravenous feeding is often needed in very premature babies. The nutritional requirements of premature babies are different from those of full-term babies: per unit bodyweight they need more energy and more nitrogen for growth, more calcium and phosphate for bone mineralization, and more sodium to compensate for the relatively high mandatory renal losses of this electrolyte. The composition of artificial milks for premature babies is designed to reflect these and other requirements. Where breast milk is available from the mother, use of this may enhance future neurodevelopment, but usually it cannot by itself meet the physical growth needs of premature babies.

Neurological damage is the biggest single hazard to the premature baby, and is made more likely by the presence of RDS and its complications. However, it is now known that sometimes the events giving rise to premature delivery can themselves cause hypoxic or ischaemic damage to the fetal brain before birth, so not all of this hazard is postnatal. Evidence for such damage is sought by performing ultrasound scans on the brains of babies at high risk of neurological damage, and haemorrhages into the ventricles (intraventricular haemorrhage, IVH) and brain substance are easily seen with this technique. Unfortunately, ultrasound is not good at picking up ischaemic damage to the brain substance, yet this is more important than the presence of IVH. Consequently, the ultrasound appearances are only poorly predictive of subsequent neurodevelopmental outcome.

Infection is much more common in the preterm baby than in any other groups of patients except those who are immunosuppressed. Preterm babies are poorly immunocompetent for a number of reasons, but the most striking is the fact that those born before the third trimester have not received maternal IgG antibodies, which are only transferred to the fetus in the last 3 months of pregnancy. Where infection was part of the process giving rise to preterm delivery the infant may be septicaemic at birth, and congenital bacterial infection has a high mortality, particularly when the causative organism is the group B beta-haemolytic streptococcus. Infection is an important component in the pathogenesis of *necrotizing enterocolitis*. This much feared complication of prematurity is occasionally seen in full-term babies. They present with an acute abdomen, feed intolerance, and the passage of blood in the stool. Breast milk protects against necrotizing enterocolitis.

Retinopathy of prematurity is uncommon but serious. Originally thought to be caused exclusively by an inappropriately high arterial oxygen tension, it is now known that this is at best a partial explanation and that many other factors are important. Low gestational age at birth remains the single most important risk factor. Regular examination of the eyes, by an ophthalmologist with appropriate expertise, from 34 weeks postmenstrual age, using indirect ophthalmoscopy to give a good view of the peripheral retina, is needed to detect serious retinopathy. Treatment (the success of which cannot be guaranteed) is with cryotherapy or laser photocoagulation, (see Ch. 13, p. 238).

The problems faced by the family are considerable.

- *Uncertainty*. Will my baby live? Will my baby be damaged? However careful the communication between staff and parents, the uncertainties are real and cannot be dismissed. Neurological outcome can only be truly determined in follow-up.

Additional Information
Some important neonatal bacterial pathogens

- Group B beta-haemolytic streptococcus: also called *Strep. agalactiae*
- Staphylococci (coagulase negative): also called *Staph. albus* or *Staph. epidermidis*
- *Staphylococcus aureus*
- Gram-negative organisms (e.g. *E. coli, Enterobacter, Pseudomonas*)

- *Cost.* In the process of supporting each other and being with the baby, jobs are lost, travel expenses mount up, and a high emotional cost is borne by close relatives.
- *Special needs.* A significant number of 'high-risk' babies (Fig. 8.8) will have some degree of long-term impairment, and a few of these will be severely disabled. These families then have to cope with a substantial and continuing economic, social, and emotional burden.

Multiple pregnancy

The rate of twinning, and higher order multiple pregnancy, is increasing with the impact of reproductive medicine. When the total mass of fetus outgrows the capacity of the placentation to support it, the fetuses become growth-retarded. Monozygotic twins may share a placenta, giving rise to the risk of twin-to-twin transfusion. Multiple pregnancies commonly deliver prematurely: not only is the process of delivery more hazardous than for singletons, but the babies are also subject to the consequences of prematurity as detailed above. The economic and social consequences of twins and triplets is considerable.

Teenage parents

Young teenagers sometimes become pregnant because of teenage risk-taking behaviour, sometimes out of ignorance, and sometimes because of rape or sexual abuse. Pregnancy is sometimes not noticed, and sometimes deliberately concealed; either way, pregnant children frequently miss out on antenatal care. After delivery at full-term, the majority of families look after the baby within the extended family, and placing a child for adoption in these circumstances is now a rare event.

Problems mainly arise when a baby is delivered preterm, since the emotionally immature adolescent girl is then faced with an emotional stress that can be overwhelming. The paediatrician is in effect dealing with two child patients simultaneously, together with a family which may be supportive, but may equally be blaming the teenager and wanting to take control of the situation. The putative father may or may not be involved, and may occasionally be blamed for the pregnancy when in fact a male relative is the real father.

Finally, the mother herself is disadvantaged by having her own secondary education interrupted. In some areas, special school-age mothers' facilities have been set up to enable mothers to continue academic education as well as learning parent craft skills.

Further reading

Roberton, N.R.C. (ed.) (1992). *Textbook of neonatology.* Churchill Livingstone, Edinburgh.

Tacush, H.W., Ballard, R.A., Avery, M.E. in Schaffer and Avery, (1991) *Diseases of the newborn.* W.B. Saunders, london.

Johnston, P.G. in Vulliamy (1994) *The newborn child.* 7th Edn. Churchill Livingstone, Edinburgh.

9

Accidents and poisoning

- Epidemiology of paediatric trauma

- Anatomy, physiology, and psychology relevant to trauma

- Initial assessment and treatment

- Case studies
  road traffic accident (head injury)
  scalds
  near drowning
  poisoning

Epidemiology of paediatric trauma
Accidents will happen

It is impossible to avoid accidents totally in children, nor is it desirable. Children's desire to explore their environment will inevitably expose them to hazards. The best experience is learned experience: by finding out first hand the unpleasant results of these hazards, the child learns to avoid them. The majority of accidents have no long-lasting effects, (e.g. scraped knees, bumped heads, etc.). Unfortunately, children may be exposed to accidents which result in severe morbidity or even death. Three to four children die each day from trauma and approximately 10 000 children are disabled permanently every year in the UK.

Trauma is the commonest cause of death from the age of 1 year throughout childhood.

It is a common perception that traumatic death is unavoidable. A true accident is an occurrence which appears to have no obvious cause. The majority of accidents which occur are not Acts of God, but have a definable cause. Many factors predispose children to suffering accidents, and these may be anticipated. The injury resulting from the accident may be mitigated. Injury prevention strategies are an essential part of paediatric accident and emergency medicine (see Chapter 2).

Anatomy, physiology, and psychology relevant to trauma

Children are not simply small adults. There are differences which lead to a different effect of injuries.

The small child represents a 'small target' when struck by an object such as a car. The energy transferred to the child dissipates over the smaller mass of the child, resulting in a higher concentration of force per metre squared, as compared to the same collision in an adult (see Table 9.1). The child's body has less elastic tissue and body fat to absorb this more intense energy which combined with closer proximity of organs, results in a high frequency of multiple organ injuries.

As a general rule with increasing age, bones become more brittle. For example, an elderly woman falling on an outstretched hand usually breaks her wrist, as the force applied rapidly overcomes the inherent strength of the bone. The wrist breaks absorbing the force. In the young adult, the bone is more resilient and so transfers the force. Therefore, the same mechanism of injury more commonly results in shoulder dislocation or clavicular fracture. In children, the skeleton is incompletely calcified and so transfers force more effectively, making fracture less likely.

In adults the bony skeleton usually fractures resulting in dissipation of the energy applied and this protects the underlying tissues and organs. Therefore, it is unusual in adults to have underlying brain or lung damage in the absence of skull or rib fractures. This is not the case in children. The growing bone may

Table 9.1.
Effects of injury: differences between child and adult

Injury	Adult	Child
Heavy impact	Energy dissipates Low frequency of organ injuries	Less mass for energy to dissipate High frequency of multiple organ injuries
Fall	Bone less resilient Force not transferred hence peripheral fracture	Bone more resilient Force is transferred through bone Fracture less likely (e.g. shoulder dislocation)
Major trauma	Bony skeleton fractures Energy dissipates Underlying tissues protected	Growing bone transfers force to underlying tissues Injury common in absence of fracture

easily transfer force to the underlying tissues and not fracture, (e.g. rib fractures are rare in children, but pulmonary contusion is relatively common). Therefore, the unwary may underestimate the severity of a child's injury in the absence of traditional markers of severity in adults such as skull fracture, etc.

The child's body proportions vary with age. For example, the relative contributions to the total body surface area change with age. At birth, the head accounts for approximately 20% falling to 10% by the age of 15 years. Body surface area to weight ratio increases with age and so young children lose heat much more rapidly which may compound the effect of the initial injury.

The young child's relatively large head, especially the occiput, results in a tendency to flex the neck when laid flat, resulting in an increased risk of a compromised airway in an unconscious child.

Young children lose heat very rapidly.

The infant has a higher metabolic rate and oxygen consumption resulting in the respiratory rate being highest at birth. Therefore, airway obstruction results in severe hypoxia more rapidly in children leading to metabolic acidosis.

Children develop hypoxia more rapidly than adults.

Stroke volume (SV) is very small at birth, but the high metabolic rate demands a relatively high cardiac output ($CO = HR \times SV$). Therefore, heart rate (HR) is high at birth, but decreases as the heart grows and stroke volume becomes a more significant variable factor. Systemic vascular resistance is low at birth and rises, resulting in a low systolic pressure which increases with age.

Physiological parameters must be interpreted according to the age of the child. A systolic blood pressure of 90 mmHg, respiratory rate of 30, and pulse of

120 beats per minute may indicate severe physiological compromise in an adult, but may be perfectly normal for a 4-year-old child.

Pulse, blood pressure, and respiratory rate must be interpreted in relation to the child's age.

The ability to compensate for fluid loss is more efficient in children. A child may lose up to 25% of circulating volume before measurable change occurs in physiological parameters compared to 10% in an adult.

Capillary refill time is a useful way of measuring peripheral vasoconstriction. The skin of the hand or foot is squeezed firmly for 5 seconds. The resultant blanching normally clears in less than 2 seconds. If prolonged, this may indicate decreased peripheral perfusion. Anatomical injury should be used to judge initial fluid resuscitation, for example, a child with bilateral femoral fractures will have significant fluid loss even if blood pressure and pulse are normal. *Fluid resuscitation must not be delayed.* Once the child becomes hypotensive, death may be imminent (Table 9.2).

Hypotension in a child may be a pre-terminal sign.

The smaller size of the child means that drug dosages, fluid replacement, etc, must be scaled down appropriately as should resuscitation equipment (e.g. intravenous lines, endotracheal tubes, etc.). There are various charts which may help in this assessment using the child's age or height where actual weighing of the child is impractical due to their injuries (Fig. 9.1).

Children react differently from adults to injury. Many children are frightened simply by the hospital environment. Parents must be present when the child is assessed. Children look to their parents for reassurance. The doctor must appear confident and in control, which will reassure both parents and child. Remember fear may alter a child's physiological parameters causing increased respiratory rate, blood pressure, and heart rate. However, *never assume such alterations are due to fear* until organic causes have been excluded.

Parents should accompany children at all times.

Table 9.2.
Clinical signs of blood loss in children

Blood loss (%)	< 25	25–40	> 40
Heart rate	Mild tachycardia	Marked tachycardia	Tachycardia or bradycardia
BP (systolic)	Normal	Normal or falling	Marked drop
Pulse volume	Normal/Reduced	Mildly reduced	Markedly reduced
Capillary refill time	Normal/increased	Mildly increased	Markedly increased
Skin	Cool, pale	Cold, mottled	Cold, deathly pale
Respiratory rate	Mild tachypnoea	Marked tachypnoea	Sighing respiration
Mental state	Normal/Agitated	Lethargic	Reacts to pain or unresponsive

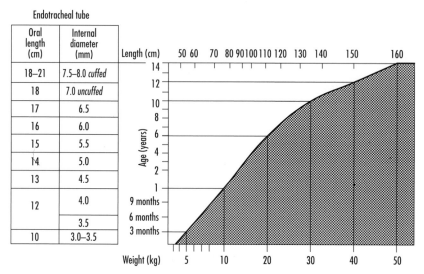

Endotracheal tube

Oral length (cm)	Internal diameter (mm)
18–21	7.5–8.0 *cuffed*
18	7.0 *uncuffed*
17	6.5
16	6.0
15	5.5
14	5.0
13	4.5
12	4.0
	3.5
10	3.0–3.5

Fig. 9.1. Paediatric resuscitation chart.

Summary
Assessment of severely injured child

1. Primary survey
2. Resuscitation
3. Secondary survey
4. Definitive treatment

Summary
Primary survey

Airway with cervical spine control

Breathing with ventilation if required

Circulation with haemorrhage control

Disability

Exposure

Initial assessment and treatment

The traditional method of assessment of patients beginning with history-taking, then examination, followed by investigation is not employed in the severely injured or unwell child. The general approach to the child in extremis has four phases. (see Summary)

Primary survey

The *primary survey* is a rapid assessment of the patient's overall state. Its purpose is to identify rapidly any life-threatening conditions and immediately treat them.

Airway with cervical spine control

The cervical spine is immobilized while the airway is assessed by talking to the patient or using direct visualization. The 'trauma handshake' is used: while the head is immobilized the attendant asks 'How are you?' to determine the reaction to verbal stimulus and hence conscious level.

Breathing with ventilation if required

A rapid assessment of the respiratory system to identify life-threatening conditions such as tension pneumothorax, open chest wound, etc. If the patient is not breathing, assisted ventilation may be required.

Circulation with haemorrhage control

The aim during the primary survey is to assess the circulation rapidly and preserve the remaining circulating volume until resuscitation is under way.

Disability

This involves a rapid mini-neurological assessment comprising of assessment of pupillary response and conscious level. The pupils are rapidly inspected for size, equality, and responsiveness to light.

The Glasgow coma scale is commonly used to assess conscious level, but takes too much time in the initial phase. Therefore, the *AVPU system* is used for rapid assessment.

Exposure

To complete the primary survey, the child is undressed completely to ensure no injuries, rashes, etc, are missed. Do not leave the child exposed for prolonged periods of time thus avoiding hypothermia and also embarrassment.

Resuscitation

The resuscitation phase may follow the primary survey, but is often performed simultaneously. A high concentration of oxygen is administered, intravenous access is achieved to obtain blood for cross match, and other appropriate tests and fluid resuscitation is commenced.

Blood pressure, pulse, temperature, and respiratory rate may be formally measured and recorded with various monitors being applied. Initial X-rays may be undertaken in the resuscitation area.

Splinting of fractures, administration of antibiotics and tetanus prophylaxis should occur now if required. As soon as respiratory and circulatory functions are secure, pain relief must be considered and administered safely. Intramuscular, oral, and rectal routes may be ineffective if fluid loss is present as absorption will be minimal in the presence of compensatory vasoconstriction. Analgesia should only be administered intravenously or intra-osseously in small doses which are titrated against the patient's response, (e.g. give small amounts and assess effect before proceeding to give more).

Secondary survey

A *secondary survey* is the taking of a full history and the performance of a top-to-toe examination. This is performed once resuscitation is established and ongoing. If the child has a life-threatening condition requiring immediate intervention (e.g. surgery), the secondary survey may have to be postponed but must not be forgotten.

A more complete history is obtained either from the patient or the accompanying parents or adults. A useful *aide memoire* is *AMPLE*. A thorough examination is then carried out including examination of the back. If a spinal injury is suspected, the patient must be turned in a rigid position which avoids any twisting of the patient's spine and is called the 'log roll' (Fig. 9.2). All orifices must be inspected, (eg ears, rectum). Once the examination is complete, all findings must be documented. Further investigation indicated by the examination are carried

Additional Information
AVPU system

Alert

Vocal: responds to vocal stimulus

Pain: response to painful stimulus

Unconscious

Additional Information
Secondary survey: AMPLE

Allergies

Medications

Previous illness/injury

Last ate or drank

Environment in which the injury occurred or illness developed

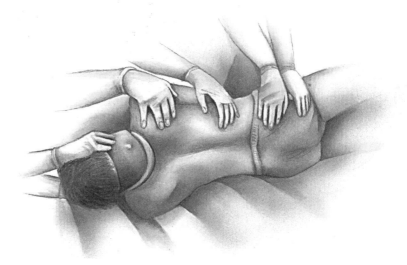

Fig. 9.2. The log-rolling procedure.

out at this time. Once these are complete, a definitive treatment plan is devised for the patient.

Definitive treatment

Definitive care is where the patient undergoes treatment for their illness/injury. For example, surgery for fractured limbs, renal dialysis for failure, etc.

The shorter the time from insult to definitive care, the better the outcome.

This above system should be employed in all patients presenting with emergency conditions and accidents. The majority will not require immediate intervention but this system ensures that those who do are not overlooked.

Case studies
Road traffic accident (head injury)

It is the summer holidays at about 8 p.m. and Paul (age 6) and his older brother are returning home on their bicycles. Neither is wearing a helmet. While crossing the road to reach their house, Paul is struck by a motor vehicle travelling at 45 miles per hour in a built up area. Paul's father witnesses the accident and immediately bundles Paul up, enlisting the aid of a neighbour to drive him to the accident and emergency department which is only half a mile away. Father runs into reception carrying Paul.

This is a relatively common presentation in this age group. Preschool age children are more likely to be the victim of a road traffic accident (RTA) as a passenger. As childrens' mobility increases, the incidence of pedestrian and bicycle accidents increases. Many children are unwilling to wear cycle helmets and as a boy, Paul is more likely to be involved in trauma than a girl of his age.

The accident occurred at twilight. At this time children are beginning to tire and their judgement begins to deteriorate. This is combined with failing light which makes judgement of both speed and distance more difficult for both Paul and the car driver. Drivers commonly exceed the speed limit in town unless specific physical speed-reducing measures such as speed bumps are present.

Most seriously injured RTA victims are transported to hospital by ambulance with initial first aid administered at the scene. However, anxious parents may react by taking children to hospital by whatever means of transport is immediately available. Small children are physically easy to carry compared to adults and it is difficult for parents to wait until help arrives and may panic.

Paul is rushed into the resuscitation area accompanied by his father and a rapid primary survey performed. The 'trauma handshake' is performed. Paul is moaning incoherently and does not respond to verbal stimulation. An obvious head wound is noted with blood around the right ear.

The airway is inspected and found to be clear. There is no evidence of airway obstruction. The respiratory rate seems very rapid. The trachea is central and there is obvious bruising to the left side of the chest and abdomen. Surgical emphysema is felt on the left side and breath sounds are decreased on the left. His pulse is rapid and of low volume. He appears pale and has a capillary refill time of 6 seconds. He responds to painful stimulation of his sternum by grasping your hand with his right hand. Pupils are equal but sluggish. The nursing staff begin to remove his clothing.

The *primary survey* aims to identify life-threatening injuries requiring immediate treatment. An orientated response to 'How are you' implies there is no immediate life-threatening problem. Frightened children in pain may be reluctant to answer and common sense must be used in assessing responsiveness. Paul is unresponsive to initial contact and so the primary survey continues. His airway appears clear.

Rapid examination of the *respiratory system* involves exposure of the chest to identify wounds. The work of breathing is assessed. The position of the trachea is determined to exclude tension pneumothorax. With gloved hands the whole of the chest is palpated for tenderness, crepitus, and surgical emphysema and blood on the examining glove may reveal posterior chest wounds which are not immediately visible. Cyanosis is rare in trauma patients and is a late and preterminal sign of hypoxia. Paul's respiratory rate is increased with signs suggestive of a left-sided pneumothorax. As there is no suspicion of a tension pneumothorax, intervention may be delayed for a few minutes until completion of the primary survey.

Cyanosis is a late and pre-terminal sign of hypoxia.

Circulation is assessed using the child's colour, capillary refill time, and rate and volume of a central pulse. Blood pressure is not measured at this stage. Any obvious external bleeding is treated with direct pressure and elevation only. Paul's circulation is obviously compromised. His rapid pulse may be due to pain but hypovolaemia must be assumed present until proven otherwise. It is a warm summer's night and Paul is unlikely to be hypothermic. Therefore, his prolonged capillary refill time is highly suggestive of peripheral vasoconstriction as a result of significant hypovolaemia.

Paul has a reduced conscious level. This may be secondary to hypoxia and/or hypovolaemia from other injuries causing decreased perfusion of the brain and not as a direct result of a head injury. Only once hypoxia/hypovolaemia has been excluded or corrected may the cause of a decreased conscious level be attributed solely to the head injury. Pupillary response to light may be sluggish in the presence of hypovolaemia.

Resuscitation is commenced by immobilizing the neck with a cervical collar, tape and sandbags, and administering oxygen. Intravenous access is obtained with difficulty and blood taken for cross-matching. Paul's weight is estimated by measuring his length with a tape measure which allows rapid conversion to weight using standard tables. Charts on the wall of the resuscitation room allow rapid calculation of initial fluid volumes to be infused. Monitors are used to measure heart rate, blood pressure, and oxygen saturation. Paul's pulse is 140 beats per minute, blood pressure is unrecordable, and an oxygen saturation monitor unable to detect a signal peripherally. Initial X-rays are taken in the resuscitation room of cross-table cervical spine, chest and pelvis. The chest X-ray shows no rib fractures but a pneumothorax on the left. Therefore a chest drain is inserted.

The aim of this phase is to restore tissue oxygenation as rapidly and completely as possible. If airway intervention (e.g. suction, intubation) is required, care must be taken not to manipulate the cervical spine. Oxygen must be delivered in high concentration and the lungs capable of effecting gas exchange with sufficient blood in the intravascular compartment to allow oxygen delivery to *all* parts of the body and not just the central circulation. Paul has both respiratory and circulatory components to his traumatized state which must be corrected. As the chest injury is not immediately life-threatening the presence of a pneumothorax is confirmed by X-ray before intervention. If Paul's condition had deteriorated, immediate chest intubation would be undertaken based on clinical findings.

Pelvic X-ray should always be undertaken as pelvic fracture may result in tor-
rential bleeding. (Clinical signs of pelvic fracture are notoriously unreliable.)

The phenomenon of SCIWORA (spinal cord injury without radiological abnormality) occurs in children. The cervical spine may be rendered inherently unstable in the accident. However, no fracture may result due to the relative elasticity
of a child's vertebral column.

While fluid is given intravenously, the *secondary survey* begins. Details of the injury
are obtained from the father. Paul was struck on his left hand side by the car and
thrown approximately 7 metres (20 feet) before landing on the road. Examination
reveals a scalp laceration to the right side of the head with blood coming from the
ear. There is bruising along the whole of the left side of his torso and a deformed left
thigh. On 'log-rolling' the patient, no injury is found. The first bolus of fluid is completed and the primary survey is repeated. Airway is still clear, respiratory rate is
decreasing to 35 per minute with equal air entry, the saturation monitor is now
reading 96%, and the systolic blood pressure is detected at 60 mmHg with a capillary refill time of 5 seconds. The fluid bolus is repeated.

While this is being performed, local anaesthesia is instilled around the left
femoral nerve and the left leg splinted.

The aim of this phase is to continue resuscitation whilst identifying and listing
all injuries. Obtaining a history is vital and must include a description of the
mechanism of injury if available. A child may be too young to describe the incident, may be too frightened or may be comatose. Initial information usually
comes from witnesses.

Mechanism of injury is important as it allows prediction of patterns of injuries.
The astute doctor seeks out anticipated injuries. This is particularly vital in children
as they are less likely to volunteer information about sites of pain. For example,
Waddell's Triad describes injuries sustained by a child pedestrian when struck by a
car. A child is small and so the car bumper strikes the femur. The child's thorax on
the same side strikes the bonnet. The kinetic energy delivered to the child's small
target catapults them down the road landing on the contralateral side of the skull.
Obviously, if the mechanism is known these injuries may be sought out directly.
There is no need to memorise various injury patterns as a good description of the
mechanism combined with the application of common sense is all that is required.

A full head to toe examination is performed and all evidence of injury recorded.
Further intervention such as splintage, tetanus prophylaxis, catheter insertion etc
may be undertaken but attention must be paid to adequate analgesia. After any
intervention is performed, or if any change in the patient's condition is detected,
the primary survey must be repeated.

If any change occurs, repeat the primary survey.

Further x-rays are undertaken and confirm the lung has re-expanded and a femoral shaft fracture. Paul's blood pressure, oxygenation, respiratory and pulse rates, and capillary refill times are now returning to normal and he is coherent; a blood transfusion is commenced. Paul undergoes an ultrasound scan of his abdomen in the resuscitation room which shows fluid in his abdomen possibly from a tear of his spleen. A CT (computerized tomography) scan of his abdomen is arranged.

Despite improvement in the haemodynamic condition, with his blood pressure returning to the normal range, the child is still hypovolaemic. It is physiological compensation which maintains blood pressure even in the presence of ongoing blood loss. Therefore, fluid resuscitation should be based both on physiological measurements and anatomical injuries. If a child has fractured long bones and intra-abdominal bleeding, fluid replacement must be given even in the presence of essentially normal blood pressure.

Urgent surgical opinion should be sought as ongoing blood loss may require immediate surgery. However, as a general rule in children, if fluid replacement restores good peripheral circulation, conservative management is undertaken with further imaging of the abdomen to determine exact pathology.

A CT scan confirms a splenic tear but Paul's condition remains stable, hence surgery is not required. He has good perfusion and urinary output and is admitted to the general surgical ward and his femoral fracture treated by splintage. He makes a slow but steady recovery. He is finally discharged from follow-up 3 months following the accident with full mobility. However, his father reports that he is behind in his school work and has become increasingly dependent on his mother. The driver of the vehicle has been charged with reckless driving.

The long-term outcome is dependant on rapid, effective, and complete resuscitation with early institution of definitive management. However, in children it is not simply a case of achieving physical recovery but also ensuring growth and psychological development continue normally. Frequently following traumatic episodes children revert to an earlier developmental stage (i.e. a 5-year-old may act like a 3-year-old). A head injury may have a prolonged effect on learning abilities. Follow-up by a paediatrician with involvement of the school health service will be required and extra help provided in school if necessary. The lasting emotional trauma to the parents should also be recognized and managed with skilled counselling.

Scalds

Ailsa (18 months old) is an inquisitive toddler and an only child. Her mother is boiling water in an electric kettle but does not appreciate the flex is hanging over the edge of the work surface. She leaves the cooking area briefly to answer the telephone. She hears Ailsa scream and returns to find she has pulled the kettle full of boiling water over herself. There is a large laceration on her forehead. Mother rapidly removes all Ailsa's clothes and puts her under the cold shower. She then wraps her in a blanket and drives directly to the accident and emergency department.

Burns and scalds are the second commonest cause of accidental death in children after road accidents. The majority of deaths occur outwith hospitals and are usually the result of house fires. The cause of death is usually asphyxiation combined with carbon monoxide or other toxic gas poisonings.

Fig. 9.3. Child pulling flex with risk of boiling water pouring on to head.

Approximately 60 000 to 70 000 attendances per year to accident and emergency departments in the UK are due to burns and scalds in children with approximately 10–15% requiring admission: 70% occur in preschool children with the most common age being between 1 and 2 years.

A full history is essential. The important factors in determining the severity of the burn, are the temperature of the stimulus and the duration of contact. It is important to have an estimate of the temperature (e.g. if a hot drink had milk in it this would reduce the temperature). If a child's clothes are soaked in hot water, the clothes hold the heat against the child increasing the duration of contact. Those children unable to remove their own clothing will sustain a more severe burn.

First aid involves all clothing being removed and the affected area cooled rapidly under cold water for 5 minutes. The burnt area should have nothing applied to it. There are many 'old wives tales' about applying butter and other substances but these have no benefit and may be detrimental. The child should be wrapped in clean linen such as a pillow case or tea towel.

The child should *not* be transported with cold soaks in place over the burn as this will lead to rapid heat loss from the child. Children with scalds often arrive in accident and emergency department's mildly hypothermic as a result of unwitting over-intervention.

Ailsa is immediately taken to the resuscitation area and a primary survey performed. She is screaming hysterically. A cervical collar is applied and oxygen given. The airway is clear, respiratory rate is 45 per minute, pressure is applied to the wound and the pulse estimated at 160 per minute with capillary refill less than 2 seconds.

It is important to assess fully any burnt or scalded child to exclude coexisting injuries such as head or neck injury. In house fires, injuries may be sustained from blasts or during attempts at escape. If the child has been exposed to flames or a superheated environment, signs of upper airway damage may be present due to inhalation of hot air. If there are signs suggestive of an upper airway burn, elective intubation to secure the airway before obstruction occurs may be required.

Fluid loss from burns takes time to occur so unless attendance at the accident and emergency department has been delayed, hypovolemia on presentation is likely to be due to coexisting injury rather than to the effects of the burn.

Ailsa has intravenous access established rapidly and blood sent for various investigations. Intravenous morphine is given at a dose initially of 0.1 mg/kg. Tetanus prophylaxis is not felt necessary. Initial fluid resuscitation is commenced with crystalloid solution. Blood pressure, pulse, respiratory rate, and oxygen saturation are all monitored.

If there is no evidence of an immediate airway problem or coexisting severe injury, the emphasis moves to early pain relief. Oral administration may be appropriate for very minor burns but in severe burns the analgesia should be delivered directly to the circulation in small aliquots. Morphine is the drug of choice but mandates close monitoring of the child's vital signs.

Blood is sent for haemoglobin and haematocrit, urea and electrolytes, and cross-matching. These initial values will serve as a baseline to monitor the ongoing resuscitation of the child.

The burnt area will rapidly develop oedema and fluid will also be lost from the circulation directly via the surface of the burn. This takes time to develop but should be anticipated. Replacement is commenced prior to the extent of the burn being estimated with the flow rate being adjusted once the expected requirement is calculated.

Ailsa is comfortable and sleeping with mum holding her hand. Blood pressure, pulse, respiratory rate, and oxygen saturation are stable. The area of the burn is mainly over her face, neck, chest and arms. The percentage burn is estimated at approximately 15% combined partial and full thickness burns. The fluid replacement required is calculated and the flow rate adjusted accordingly. The burns are now covered in cling film and a clean warm blanket placed over Ailsa. She is then transferred to the nearest burns unit after consultation with the duty plastic surgeon.

Assessment of the burn involves several factors. First, the surface area is estimated. Children's relative proportions vary with age. Wallace's rule of 9s used in adults does not apply in children. Specialized charts for children of varying ages exist (see Fig. 9.4) and should be used to estimate the burnt area. If none are available, an estimate may be derived by using the patient's palm and adducted fingers to represent 1% of the body area (Fig. 9.5) Secondly, the depth of the wound must be assessed. Burns are either superficial, partial, or full thickness (Fig. 9.6).

A *superficial burn* affects the epidermis only resulting in erythema and no blister formation. This simple erythema should not be included in the surface area assessment.

A *partial thickness burn* results in some damage to the dermis. Blistering occurs but sensation is usually retained.

A *full thickness burn* damages epidermis, dermis and perhaps even deeper structures. The skin looks white, charred and is painless. If such burns are circumferential, tissue swelling may cause constriction compromising circulation or the ability to breathe.

Once the combined surface area of the partial and full thickness burn is known, the fluid replacement for the first 4 hours is calculated using the following formula:

$$\text{Plasma amount (in ml)} = \frac{\% \text{ burn} \times \text{weight (kg)}}{2}$$

Fig. 9.4. (a) and (b) Burn/plastic surgery charts. (Figure 9.4.(b) on page 187)

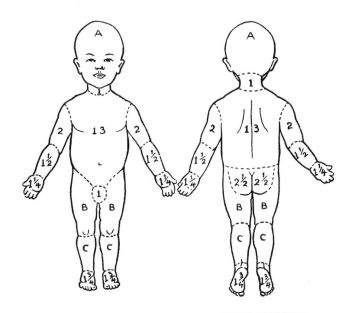

Grampian Health Board

Plastic and Reconstructive Surgery

BURN RECORD, CHILD		SURNAME	UNIT NO.
		FIRST NAMES	
Royal Aberdeen Children's Hospital		ADDRESS	MARITAL STATUS
			DATE OF BIRTH
CONSULTANT			RELIGION
		POST CODE	
WARD/DEPT	AGE	BAR CODE	PLACE LABEL WITHIN BRACKETS
Ages: Birth to 7½ Years		G.P.	DATE OF OBSERVATION

RELATIVE PERCENTAGES OF AREAS AFFECTED BY GROWTH

Area	Age	0	1	5
A = ½ of Head		9½	8½	6½
B = ½ of One Thigh		2¾	3¼	4
C = ½ of One Leg		2½	2½	2¾

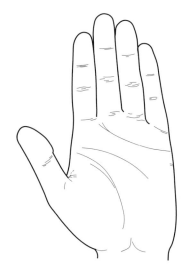

Fig. 9.5. Palm represents 1% of body area.

As a general rule, full thickness burns in excess of 5% or combined burns in excess of 10% should be referred to a burns unit. Smaller burns in significant areas such as the face, sole of foot, or over the joints, etc should also be referred.

The burns must be dressed to reduce pain and the likelihood of infection. Once dressed the burn should not be re-exposed until arrival in the burns unit. Repeated needless examinations often occur in the resuscitation room. These may be reduced by taking Polaroid pictures prior to the dressings being applied. The burnt child will lose heat rapidly and exposure should be minimized.

Grampian Health Board

Plastic and Reconstructive Surgery

Fig. 9.4.(b)

BURN RECORD, ADULT		SURNAME		UNIT NO.

FIRST NAMES

HOSPITAL — ADDRESS — MARITAL STATUS — DATE OF BIRTH

CONSULTANT — RELIGION

POST CODE — OCCUPATION

WARD/DEPT. — AGE

BAR CODE — PLACE LABEL WITHIN BRACKETS

Ages: 7½ Years to Adult — G.P. — Date of Observation

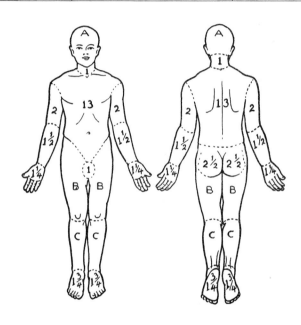

RELATIVE PERCENTAGES OF AREAS AFFECTED BY GROWTH

Area	Age	10	15	Adult
A = ½ of Head		5½	4½	3½
B = ½ of One Thigh		4¼	4½	4¾
C = ½ of One Leg		3	3¼	3½

Ailsa does well in hospital but requires skin grafting for full thickness burns of her face and upper chest. This involves multiple operations over several years.

Although a major cause of mortality, burns also result in significant cosmetic morbidity. This may well result in future psychological morbidity.

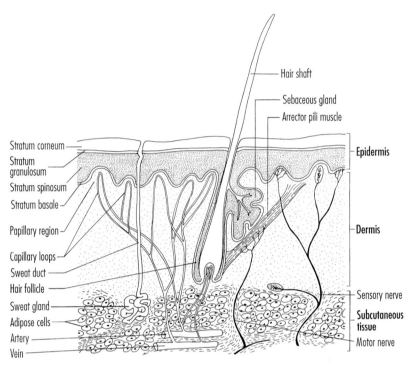

Hair shaft
Sebaceous gland
Arrector pili muscle

Stratum corneum
Stratum granulosum
Stratum spinosum
Stratum basale
Papillary region
Capillary loops
Sweat duct
Hair follicle
Sweat gland
Adipose cells
Artery
Vein

Epidermis

Dermis

Sensory nerve
Subcutaneous tissue
Motor nerve

Fig. 9.6. Skin layers.

Near drowning

Rajinder (age $3\frac{1}{2}$ years) is the second son in a family of four of a wealthy business man. The family have a large estate just outside of town with its own outdoor swimming pool. It is October and someone has inadvertently left the gate to the swimming area open. When Rajinder's mother calls the children in for lunch Rajinder does not appear. Mother finds Rajinder face down floating in the swimming pool. She is unable to swim, so dials 999 and calls her neighbour for help. Her neighbour pulls Rajinder from the pool but finds fluid in the airway which he drains by turning Rajinder on to his side. He is apnoeic so the rescuer gives 5 rescue breaths. There is no evidence of a major pulse so 5 compressions are given with the heel of one hand, 1 finger breadth above the xiphisternum. He continues basic life support at a ratio of 1 ventilation to 5 compressions until help arrives in the form of a Paramedic closely followed by an ambulance crew.

Drowning is death from asphyxia associated with submersion in a fluid.

Near drowning occurs if any recovery (however transient) occurs following a submerging incident.

Drowning is the third commonest cause of accidental death in children in the UK following road accidents and burns. The highest incidence occurs in the under-fives in private swimming pools, baths, ponds, and in land waterways. Coexisting disease such as epilepsy, diabetes, and ingestion of alcohol increases the risk of drowning.

The mechanism of drowning progresses in sequence and outcome is dependent on when rescue occurs. Initially, the child breath holds and the diving reflex slows the heart rate. As apnoea continues, the child becomes progressively hypoxic and acidotic and the heart rate begins to rise. Hypoxia potentiates vagal activity. Eventually, the child is unable to continue breath-holding and takes an involuntary breath. The inhaled water hits the glottis or vocal cords which are already sensitized due to hypoxia. Profound vagal outflow occurs causing laryngeal spasm and extreme bradycardia or even cardiac standstill. If rescued at this time and resuscitation attempts are unsuccessful, the post-mortem shows no fluid in the lungs.

This is 'dry lung drowning' which occurs in 15% of cases. If rescue does not occur, eventually the laryngeal spasm breaks and involuntary breaths lead to aspiration of water with its associated debris and then death.

First aid involves removing the child from the water rapidly. The airway is assessed and cleared of all material taking care not to manipulate the neck excessively (see below). Breathing is then checked for 5–10 seconds. If absent, 5 rescue breaths are immediately given. A major pulse is then sought and, if absent, external cardiac massage commenced: 20 cycles of 1 ventilation to 5 compressions are given per minute. In this situation, basic life support should not be interrupted and must *never* be abandoned before reaching hospital as the prognosis, even in seemingly desperate cases, may be good (see below and Fig. 9.7).

Never abandon pre-hospital resuscitation in a drowned child.

The ambulance crew continue basic life support and the paramedic intubates the child ensuring the neck is not manipulated. Once intubation is complete, the neck is immobilized with a cervical collar, sandbags, and tape. Intravenous access is impossible. A cardiac monitor is applied and shows possible ventricular fibrillation but the ambulance crews' defibrillator does not have low enough settings for a $3\frac{1}{2}$-year-old child. Basic life support is continued and Rajinder is transferred to the nearest accident and emergency department.

Cervical spine injury should always be assumed in a drowning accident. The child may have dived into shallow water and struck his head on the bottom. Therefore, care is taken to avoid undue movement of the neck but *must not* delay measures to secure the airway and deliver high concentration oxygen.

Always assume a cervical spine injury in near drowning.

Paediatric Basic Life Support

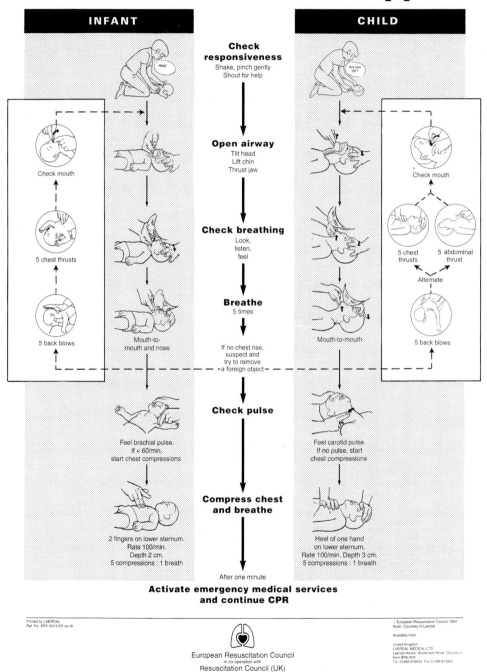

Fig. 9.7. Basic Life Support chart. Reproduced by kind permission of European Resuscitation Council. Available from Laerdal Medical Ltd, Orpington, Kent BR6 0HX.

The child will swallow large amounts of water during the drowning episode. There is a high risk of regurgitation resulting in secondary aspiration particularly if external massage is in progress. Therefore, the airway should be secured by endotracheal intubation as soon as practical. Thereafter, the stomach should be decompressed by a large oral/gastric tube.

The presenting rhythm of a cardiac arrest following drowning may be asystole due to vagal outflow or prolonged time to resuscitation. However, if hypothermia has occurred, it may often resemble ventricular fibrillation. Ventricular fibrillation is extremely rare in children and so most ambulance defibrillators do not have the capability of defibrillating children.

On arrival in accident and emergency a primary survey is performed again. This child is being ventilated with a secure cervical spine. There is still no pulse so external cardiac massage is continued. The monitor trace continues to appear like ventricular fibrillation. The child's weight is estimated using an Oakley chart. Defibrillation is attempted immediately. After delivery of the first 3 shocks the child is further assessed. The child feels cold and is peripherally vasoconstricted. Therefore intravenous access is impossible and an intra-osseous needle is inserted. Blood is drawn and sent for full blood count, urea and electrolytes, and a blood glucose stick performed. Adrenaline is given in a dose of 10 μg/kg initially. The patient continues to be in ventricular fibrillation and therefore a further 3 shocks are given and the adrenaline repeated. Cardiopulmonary resuscitation continues at all times except to allow defibrillation. Following defibrillation a rectal temperature is obtained with a low reading thermometer and is found to be 29 °C. Attempts are made to rewarm Rajinder. All his wet clothing is removed, warm blankets applied as well as an overhead heater. Using a blood warmer, intravenous fluids are given at 37 °C and the ventilator gas is heated to 42 °C.

Advanced cardiac life support (ACLS) measures should be undertaken. Adrenaline has been shown to improve cerebral and coronary artery blood flow during external cardiac massage. Defibrillation is unlikely to be effective at temperatures below 30 °C. Resuscitation measures must be continued until the core temperature is at least 32 °C.

Methods of external re-warming include removing wet cold clothing, warm blankets, heating, blankets, as well as infrared radiant lamps.

Core rewarming may be performed by a variety of methods. The simplest of these is heating the child via normal resuscitation lines. Therefore, intravenous fluids and gases may be warmed and gastric and urinary catheters may be lavaged using fluids warmed to 40 °C. If more aggressive re-warming is required, body cavities may be entered and lavaged.

After 20 minutes of resuscitation and warming Rajinder spontaneously regains cardiac output but continues to require ventilation. Primary survey is repeated and a secondary survey commenced. Temperature is now 32 °C. Intravenous access is now possible. Blood gases and arterial blood gases are drawn. In addition, X-rays of cervical spine, chest, and pelvis are taken. A 12-lead ECG (electrocardiogram) is also performed. Rajinder is transferred to the intensive care unit for further monitoring and care. Rajinder's mother wishes to know the prognosis.

Once cardiac output is restored, the child requires standard post-cardiac arrest care. Blood tests are repeated. Arterial blood gas machines assume a normal temperature for the sample so the laboratory must know the patient's actual temperature if allowances are to be made and accurate results obtained. A thorough examination is performed to exclude coexistent injury.

There are various factors that affect prognosis. The duration of immersion is important with immersion times in excess of 3–8 minutes carrying a *poor prognosis* The time to first gasp is a good prognostic factor. If this occurs within 3 minutes of initiating basic life support, the prognosis is good. If there is no gasp within 40 minutes (in the absence of hypothermia) prognosis is very poor.

Children cool very rapidly due to their relatively large surface area and this may have a protective effect. Therefore, resuscitation should be aggressive until a core temperature of at least 32 °C is achieved. Whether the water was salt or fresh has no bearing on prognosis. Resuscitation should always be continued until the child is warm.

Up to 70% of children survive near drowning if basic life support is provided at the scene. Of those who survive having required cardiopulmonary resuscitation approximately 70% make a full recovery and 25% have only a mild neurological deficit. The remaining 5% will be severely disabled or in a permanent vegetative state.

Obviously those children who do not suffer a cardiac arrest have a better prognosis. However, they may have aspirated during the episode but have no signs or symptoms on initial assessment. As a general rule, all victims of near drowning should be admitted to hospital for 24-hours observation.

Prevention of drowning requires that all children should be taught to swim at an early age, all private swimming pools should be securely fenced and covered when not in use, and public facilities should be supervised by qualified instructors.

Poisoning

Daniel (2 years old) is staying with his grandparents whilst his mother is in hospital after the birth of his younger brother. His grandparents find him in the bathroom

Summary
Poor prognostic factors in near drowning

- Time to first gasp > 40 min
- Immersion Time > 8 min
- Absence of hypothermia
- Persistant coma (no sedation)
- pH < 7.0, PO_2 < 8.0 kPa (despite treatment)

before his bed time with an open bottle of his grandmother's tablets used for her depression. The tablets look like Smarties and are across the floor. Grandmother has no idea how many tablets were in the bottle. They immediately take him to the local accident and emergency department.

Suspected poisoning is an extremely common presentation in children accounting for in excess of 50 000 presentations to accident and emergency departments per year, with 50% being admitted. Deaths are uncommon and usually due to tricyclic antidepressants or household products. The commonest poisoning resulting in death is carbon monoxide in house fires. The introduction of child-resistant containers has significantly reduced mortality from accidental poisoning but 20% of children under 5 years are capable of opening them.

The vast majority of poisonings in children are accidental in nature most commonly occurring in the 2–3 year age group in whom curiosity leads to ingestion. In addition, many modern drugs resemble sweets. Accidental ingestion occurs more frequently when there is disruption in the household such as moving house, a new baby, etc. Deliberate overdose does occur but usually in teenagers. These patients will require psychiatric referral as well as medical attention for the overdose. Recreational drug abuse is an increasing problem amongst teenagers with the biggest culprit being alcohol. Very rarely, the poisoning may be iatrogenic or may be deliberate on the part of the parent, either as part of a non-accidental injury pattern or to gain access to hospital in the case of Munchausen by proxy syndrome (see Chapter 5, p. 87).

On arrival at hospital, a primary survey is normal and no immediate resuscitation is required. The tablet involved is identified as a tricyclic antidepressant and Daniel had last been seen 20 minutes before being found. It is assumed that if any ingestion occurred it did so within one hour. Physical examination was entirely unremarkable.

The majority of children will attend with no symptoms or physical signs. However, in more severe poisonings, the child may present in a drowsy or frankly collapsed state. The general principles for management using the primary survey apply here. In any child who has a decreased conscious level of unexplained origin, poisoning must be considered.

Identification of the poison is paramount if the child is not unwell. Wherever possible remains of the poison should be obtained. Obtaining the drug bottle will allow identification of the substance. In the case of household products, the bottle should have a list of the ingredients. If it is a plant ingestion, another plant of the same type or the remains of the ingested plant may well assist identification. Tablets often have identifying marks on them and so may be reliably identified using lists or via a *poisons information centre*. This is particularly important if pills have not been kept in their original container. Remember, there are an infinite amount of poisons and tests do not screen for all known substances. The laboratory relies on clinical information and only serves to confirm your suspicions.

There are no blanket toxicological screens.

An estimate of potential quantity may be made by counting remaining tablets. The largest potential overdose should always be assumed to have been taken. The time from ingestion should be assessed if possible.

There is no objective evidence of ingestion but in view of the potential toxicity of tricyclic antidepressants, Daniel is given 15 ml of paediatric syrup of ipecacuanha with diluted orange juice to follow. He vomits several times but no tablet remnants are identified in the vomitus. He is admitted for cardiac monitoring and observation.

Summary
Treatment of acute poisoning

- General supportive measures
- Reduction of exposure to the drug
- Neutralization of the drug
- Increased elimination of the drug

Summary
Drug removal

- Gastric lavage
- Induction of emesis

Treatment for acute poisoning may involve various measures. If the drug is felt to be toxic enough to require removal and the time since ingestion is relatively short, then *removal* may be indicated.

Gastric lavage involves passing an oral–gastric tube and washing out the stomach with fluid to try and recover the toxin. This is obviously distressing in conscious children and of dubious effectiveness as the tube used in children is often too small to allow removal of tablets. Recent evidence has cast considerable doubt on the therapeutic effectiveness of gastric lavage especially if not used in the first hour or so after ingestion. Its only regular use is in the unconscious intubated patient where lavage of the gastric contents may be combined with installation of a neutralizing agent. If gastric lavage is used, a sample of the gastric lavage should be sent to the laboratory to try and identify the drug ingested.

Induction of emesis is the traditional method of emptying the stomach in children, usually using syrup of ipecacuanha. Use only in a fully conscious child and avoid in children who have ingested tablets which may make them drowsy in the near future. It should never be given if corrosives or petroleum could have been ingested. Recent evidence has also cast considerable doubt on the therapeutic effect of this if emetics *are not given within 1 hour* of ingestion.

If there is any doubt about the substance ingested, gastric contents from lavage or emesis should be sent for analysis.

Neutralization of the toxin may be carried out by giving an agent to bind the toxin in the gut or by administration of a specific antidote intravenously. Activated charcoal acts by binding with the toxin in the gut and is effective on a wide range of drugs including tricyclic antidepressants and digoxin. If gastric lavage has been employed, it may be instilled down the tube at the end of lavage.

Other specific antidotes may be given intravenously to neutralize the effects of the drug once absorbed. Examples are *N*-acetylcysteine in paracetamol overdose, ethyl alcohol in antifreeze ingestion, and naloxone in opiate overdose. Some of these act by being competitive inhibitors of the toxic substances and others by binding the toxic by-products of the toxin.

Increased elimination of the drug is achieved generally by maintaining a good diuresis. In severe cases or if renal and/or liver failure intervene, elimination of the drug may be achieved by means of dialysis, charcoal haemoperfusion, etc. However, these techniques are only very rarely required and specialist advice should be sought.

> The following day Daniel is well and discharged home. Written advice is given with regard to safe storage of drugs.

It is important to use contact with patients and their relatives to increase awareness of drug safety. Much has already been done to increase safety through child-resistant containers and packaging paediatric paracetamol in amounts that are unlikely to be toxic. In addition, some manufacturers ensure their substances taste unpleasant to prevent ingestion. The majority of household products tend to be distasteful and so discourage young children from ingesting large quantities.

Products or drugs which are likely to be harmful to children are listed in the Additional Information box.

There are several national *poisons information centres* in the UK, which may be consulted with regard to the management of various individual substances. It is impossible to remember the management of all poisons, and the centres maintain large databases on various substances including household products and plants. However, it is important to be conversant with substances that are commonly ingested.

Further reading

BMA (British Medical Association) (1993). *Advanced paediatric life support*. The practical approach. BMJ London.

Mills, K., Morton, R., and Page, J.G. (1995). *A colour atlas and text of emergencies*, (2nd edn). Mosby-Wolfe, London.

Muir, I.F.K., Barclay, T.L., and Settle, J.A.D. (1987). *Burns and their treatment*, (3rd edn). Butterworth, Sevenoaks, Kent.

Morton, R.J. and Phillips, B.M. (1992). *Accidents and emergencies in children*. Oxford University Press.

Additional Information
Substances harmful to children

Drugs
- Digoxin
- Diphenoxylate (Lomotil)
- Iron
- Paracetamol
- Salicylates
- Tricylic antidepressants

Chemicals etc.
- Alcohol
- Bleach
- Paraffin

Additional Information
Poisons information centres

Belfast	01232 240503
Cardiff	01222 709901
Dublin	00353 1 8379964
Edinburgh	0131 536 2300
Leeds	0113 243 0715
London	0171 635 9191
Newcastle	0191 232 5131
West Midlands	0121 554 3801

10

Heart disease

Almost all heart disease in children is congenital in origin, with a birth prevalence of 8 per 1000 live births. Heart defects account for over 30% of all congenital anomalies. The mortality rate for children with congenital heart lesions has fallen over the last two decades, mainly due to earlier recognition of symptoms, the skill of cardiac surgeons, and the significant improvement in cardiopulmonary bypass techniques. There are eight commonest structural *congenital heart defects* (accounting for over 80% of the total), and it should be remembered that rhythm disturbances such as heart block and supraventricular tachycardia can also present as congenital lesions.

Acquired heart disease still presents a few problems, but the incidence of rheumatic fever in children has now fallen from 12 per 100 000 in the 1960s, to 0.2 per 100 000 in the late 1980s. However, myocarditis, infective endocarditis and Kawasaki's disease (mucocutaneous lymph node syndrome) are still a cause of death and disability in children. The heart can also be involved in a number of systemic diseases such as Marfan's syndrome, muscular dystrophy, and the connective tissue disorders.

Children with actual or suspected heart disease require specialist evaluation, and resources and expertise are concentrated in tertiary centres in the UK. However, GPs and paediatricians in general hospitals will see most children with their initial presentation of cardiac disease, and have to decide when it is appropriate to seek the services of the specialists.

As with other surgical conditions of the newborn, many congenital cardiac anomalies are now diagnosed antenatally on routine ultrasound, enabling management plans to be formulated prior to the delivery of the baby.

Aetiology

Both genetic and environmental factors have been implicated as a cause of congenital heart disease (CHD). Multifactorial inheritance is the most likely genetic base for most CHD, with the abnormal genes only expressing themselves in the presence of external environmental stimuli. Over a third of children with chromosomal problems will have CHD, and just under a tenth of all children with CHD will have a chromosomal anomaly. A number of children with a q11 deletion on chromosome 22 have abnormalities of the aortic arch and ventricular septum. Other organs are also involved and the association is known as *Catch* 22.

A number of syndromes also have *associated cardiovascular anomalies*. Some of these syndromes may have autosomal dominant or recessive inheritance and some may be associated with chromosomal aberrations.

In embryological terms, the heart initially consists of a single tube with a number of dilatations along its length. These dilatations are the precursors of the systemic and pulmonary veins, the great vessels, the atria and ventricles, and the heart valves. The heart develops between 20 and 60 days of gestation and if *external factors* are a cause of CDH, they must exert their effect during this 5-week period.

Summary
The eight commonest congenital heart defects

1. Ventricular septal defect
2. Patent ductus arteriosus
3. Atrial septal defect
4. Tetralogy of Fallot[*]
5. Pulmonary stenosis
6. Coarctation of the aorta
7. Aortic stenosis
8. Transposition of the great arteries[*]

[*] Cyanotic heart lesions

Summary
Acquired heart disease

Rheumatic fever

Myocarditis (viral)

Infective endocarditis (bacterial)

Kawasaki disease (mucocutaneous lymph node syndrome)

Additional Information
'Catch 22'

- Cardiac defects
- Abnormal facies
- Thymic hypoplasia
- Cleft palate
- Hypocalcaemia
- Chromosome 22 q11 deletion

Additional Information
Syndromes with associated heart disease

Down syndrome (*trisomy* 21): atrioventricular septal defect

Turner syndrome: coarctation of the aorta

Noonan syndrome: pulmonary stenosis

William syndrome (D): aortic stenosis

Marfan syndrome (D): dilatation and aneurysm of the aorta

Holt–Oram syndrome (D): atrial and ventricular septal defects

Ellis–van Crefeld syndrome (R): atrial septal defect

Friedreich's ataxia (R): cardiomyopathy

R, autosomal recessive;
D, autosomal dominant

Summary
Environmental factors and CDH

Maternal infection
Toxoplasma, rubella, cytomegalovirus

Maternal disease
Diabetes mellitus (TGA*, transient cardiomyopathy)
Systemic lupus erythematosus (complete heart block)

Drugs
Alcohol: TGA*
Warfarin: ventricular septal defect
Phenytoin: pulmonary/aortic stenosis
Amphetamines: atrial and ventricular septal defects

TGA, transposition of the great arteries (see below)

As well as, 'Why did it happen?' one the of the commonest questions asked by parents is, 'Will it happen again?' If normal parents have had a child with a congenital heart defect, the risk of the next child being affected is increased threefold. If one of the parents has a heart defect, the risk of them having an affected child is increased fourfold. Subsequent pregnancies can be monitored using fetal echocardiography or, in the case of chromosomal problems, amniocentesis.

Case studies
Heart murmur

Jack, aged 5, is the second child of unrelated parents. Father is away from home every alternate 2 weeks as he works offshore on the oil rigs, mother works part-time in the local superstore. He was born at term by spontaneous vaginal delivery and has had all his immunizations. Developmental assessment at 6 weeks, 18 months, and 3 years was age-appropriate. He developed a harsh barking cough followed by loss of appetite and vomiting. After 24 hours, he was noted to have a fever and his mother took him to his GP. A diagnosis of viral gastroenteritis was made but clinical examination revealed a soft mid systolic heart murmur localized at the apex.

The majority of children with heart disease present in one of three ways:

1. Detection of a heart murmur
2. Cyanosis
3. Signs of heart failure

Other presenting symptoms, especially in older children, are syncope and dizziness, chest pain, poor exercise tolerance, and palpitations.

On examination Jack has no signs of respiratory distress (intercostal and subcostal indrawing). There is no tracheal tug. He is pink and has no finger clubbing. His apex beat is not displaced. There is no evidence of a forceful right or left ventricle and no palpable precordial or suprasternal thrill. In his brachial pulse the rate is 80/minute, and the volume, character, and regularity are unremarkable. He has normal femoral pulses and no delay between the upper and lower limb pulses (and therefore no clinical evidence of aortic coarctation). He is not visibly anaemic. He does not have hepatomegaly or peripheral oedema. His blood pressure in the right arm is 90/45, making both aortic stenosis and coarctation of the aorta unlikely since both may cause hypertension. The blood pressure in all 4 limbs must be recorded if you suspect coarctation of the aorta as lower limb pressure may be reduced.

Jack made a good recovery from his gastroenteritis but on subsequent review, 2 weeks later, the systolic murmur was still present. He was referred for further investigations because of parental anxiety and desire for a second opinion.

Electrocardiography (ECG) and a chest X-ray (CXR) should be carried out on all children with suspected heart disease. The ECG provides information on chamber size, evidence of hypertrophy, and conduction disturbances. There are several *features of cardiovascular disease* on a CXR. Echocardiography confirms most anatomical abnormalities of the heart and can provide functional and haemodynamic information. Magnetic resonance imaging (MRI) is also a useful non-invasive investigation of cardiac and great vessel anatomy, but most young children need to have this investigation under light general anaesthesia. Cardiac catheterization and angiography provide more detailed functional and anatomical information especially in complex cardiac cases.

Normal heart sounds are described in the summary box. Abnormal heart sounds include ejection clicks found in aortic and pulmonary stenosis and occasionally in abnormalities of the mitral valve. A loud second heart sound is heard in the presence of pulmonary hypertension. The time between the closure of the aortic and pulmonary valve is increased during inspiration and the second heart sound is said to be 'split'. In cases of atrial septal defect, an excess of blood in the right ventricle delays pulmonary valve closure and the splitting does not vary with respiration. This is 'fixed splitting'.

Heart murmurs are classified as innocent or organic and can be heard in either systole or diastole, or both (continuous murmurs); see Table 10.1. The intensity of the murmur is usually graded 1 to 6 and all murmurs 4 to 6 or greater are associated with a palpable thrill. Murmurs may be located to one site on the praecordium but some may radiate throughout the praecordium, to the head and neck, and through to the back. In cases of aortic coarctation, listen for the continuous murmur of collateral blood vessels over the scapula.

Jack's CXR and ECG are normal, so he almost certainly has an innocent murmur.

Over 50% of children will be heard to have a murmur at some stage. An *innocent murmur* is usually detected at a routine clinical examination or when a child presents with an intercurrent febrile illness. These murmurs, caused by turbulence in a normal heart or great vessels, are also accentuated by exercise. The murmurs can present at any time in childhood, but to be deemed innocent they have to satisfy

Additional Information
Cardiac disease and the chest X-ray

- Increased cardiac size and cardiothoracic ratio
- Altered cardiac contour
  - absence or dilatation of pulmonary artery
- Lung fields
  - plethora
  - oligaemia
  - oedema
- Rib notching (coarctation of aorta)

Summary
Normal heart sounds

First. Closure of the mitral and tricuspid valves

Second. Closure of the aortic and pulmonary valves

Third. Rapid ventricular filling in diastole

Fourth. Increased flow into left ventricle caused by atrial contraction

Additional Information
Common innocent murmurs

- *Vibratory murmur* (Still's murmur) 'Buzzing' in quality, mid systolic in timing, varies with position, age group 3–7 years, best heard between the apex & lower left sternal border
- *Pulmonary flow* (ejection) murmur Soft and 'blowing' in quality, early systolic ejection murmur in timing, age group 8–14 years, best heard at upper left sternal border
- *Venous hum* Caused by blood cascading into great veins, continuous murmur, age group 3–6 years, best heard above and below clavicles, can be diminished by Valsalva manoeuvre or turning neck

Summary
Characteristics of an innocent murmur

- Asymptomatic
- Systolic (except venous hum)
- Accentuated by fever and exercise
- Varies with position
- Does not radiate
- No accompanying cyanosis
- Normal pulses
- Normal heart sounds
- No associated thrills
- Normal ECG and CXR

Table 10.1.
Heart murmurs

Murmur	Defect
Ejection systolic	Aortic stenosis
	Pulmonary stenosis
	Coarctation of the aorta
Pansystolic	Ventricular septal defect
	Mitral incompetence
	Tricuspid incompetence
Early diastolic	Aortic incompetence
	Pulmonary incompetence
Mid diastolic	Mitral stenosis
	Tricuspid stenosis
Continuous	Patent ductus arteriosus
	Venous hum
	Systemic collaterals

a number of *criteria*. Most innocent murmurs are diagnosed and dealt with by GPs. Children are referred for further investigation if there is:

(1) any doubt over the diagnosis; and
(2) parental anxiety and a request for a second opinion, which was the reason for Jack's referral.

Murmurs are frequently heard in the neonatal period and may be due to a closing ductus arteriosus, mild tricuspid regurgitation or flow through the branch pulmonary arteries. These murmurs usually disappear after the first few weeks of life, (see Ch. 8, p. 163). Once the diagnosis of an innocent murmur has been made, the parents and the child should be reassured and no follow-up is necessary.

However, some organic conditions may be asymptomatic and the murmur may be picked up on routine examination. The differential diagnosis includes: small ventricular septal defect, atrial septal defect, coarctation of the aorta, and mild pulmonary or aortic stenosis. These murmurs are usually louder than innocent murmurs and may radiate through to the back.

The murmur of a ventricular septal defect has a maximum intensity at the lower left sternal border; over 60% of small intramuscular ventricular septal defects will close spontaneously. The aortic stenosis murmur may radiate to the neck and the pulmonary stenosis murmur is best heard at the upper left sternal border. In coarctation of the aorta, there may be a collateral circulation over the scapula and there will be weakness and delay of the femoral pulses. The classical finding of an atrial septal defect is 'fixed splitting' of the second heart sound.

Cyanosis

Hannah is 9 months old. She was born at 36 weeks gestation by emergency caesarean section for breech presentation. She spent one week in special care because of feeding problems. She was breast-fed for the first 4 months of life. Solids were introduced at 6 months of age. She was fully immunized. At her 8-month assessment her mother commented that Hannah had developed blue lips and face whilst crawling and moving in her baby walker.

Intermittent episodes of cyanosis during the first year of life are the usual presentation of *tetralogy of Fallot*, which is the most common cyanotic congenital heart defect. It may present at birth with cyanosis or be picked up on routine clinical examination if a murmur has been heard. Cyanotic episodes may occur early in the morning, with activity or with intercurrent illness. During the cyanotic episodes the child may be breathless, irritable, and crying.

The two main components are the right ventricular outflow tract obstruction and the ventricular septal defect. As the obstruction increases, the right ventricular pressure rises, becomes suprasystemic and blood flows from right to left across the ventricular septal defect, causing cyanosis. When flow to the lungs improves, the right ventricular pressure falls and the infant becomes pink once more.

Hannah was still feeding well and thriving. On clinical examination, she was pink but had a harsh, pansystolic murmur best heard at the lower left sternal border radiating through to the back. There was an associated precordial thrill. In view of the history and clinical findings (which were not suggestive of an innocent murmur) Hannah was referred for a cardiac opinion.

On further assessment Hannah was found to have a right ventricular 'tap' and systolic thrill with an associated 4–6 ejection systolic murmur at the upper left sternal border. There was evidence of right axis deviation and right ventricular hypertrophy on the ECG. Chest X-ray revealed oligaemic lung fields, a 'concave' pulmonary artery bay, and a right-sided aortic arch ('boot-shaped' heart Fig. 10.1). The diagnosis of *tetralogy of Fallot* was confirmed on echocardiography.

Hannah started to walk unaided at 10 months of age. By the age of 1 year, her mother noted Hannah had a tendency to squat after she had been walking. She was again cyanosed and slightly breathless. On one occasion, she developed deep cyanosis, lost consciousness for a few seconds, and became extremely pale.

Summary
Tetralogy of Fallot

- Ventricular septal defect
- Pulmonary or infundibular stenosis
- Overriding aorta
- Right ventricular hypertrophy

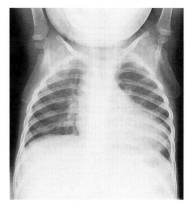

Fig. 10.1. Chest X-ray showing a typical boot-shaped heart and poor pulmonary vascular pattern seen in tetralogy of Fallot.

Fig. 10.2. A 1-year-old child, breathless and squatting during a cyanotic episode. Squatting improves blood flow to the lungs.

Additional Information
Complications of tetralogy of Fallot

- Cyanotic episodes ('spells')
- Polycythaemia/anaemia
- Exercise limitation
- Bleeding tendency
- Cerebral thrombosis
- Cerebral abscess
- Bacterial endocarditis
- Cerebral anoxia/death

Summary
Transposition of the great arteries (TGA)

- Commonest type of cyanotic congenital heart disease
- Presents with neonatal cyanosis when the arterial duct shuts
- Immediate treatment is prostaglandin infusion to open the duct
- Definitive corrective treatment is a 'switch' procedure

These 'spells' or hypoxic episodes are due to insufficient blood flow to the lungs. During these spells, the heart murmurs may be reduced or inaudible. Squatting delays venous return, increases systemic vascular resistance, and improves blood flow to the lungs. There are a number of *complications of tetralogy of Fallot*. The cyanotic episodes are caused by infundibular spasms and the spasm can be relieved by squatting, placing young infants in the knee-elbow position, or by treating with propranolol.

Cyanotic heart disease occurs when there is a right to left shunt of blood, when oxygenated and de-oxygenated blood mix together (common atrium, single ventricle) or when the infant has *transposition of the great arteries (TGA)*. TGA is the commonest cyanotic heart defect presenting in the first postnatal days. The aorta arises from the right ventricle and the pulmonary artery from the left ventricle. This arrangement creates two parallel circulations. If the ductus arteriosus is patent or if there is an accompanying ventricular septal defect, the two circulations can mix and some oxygenated blood reaches the aorta. If mixing does not occur, the infant will become acidotic and develop heart failure soon after birth. Emergency management consists of the infusion of prostaglandins E_1 or E_2, to open the duct and allow time for the baby to be transported to the intensive care unit. This may be followed by the 'Rashkind's procedure', where an artificial atrial septal defect is created, allowing mixing of oxygenated and deoxygenated blood. A definitive repair involves transposing the aorta on to the pulmonary root and the pulmonary artery on to the aortic root in the first or second week of life (the 'switch' operation).

Cardiac failure

Lee, aged 10 weeks, was brought to the surgery with a 2-day history of cough and fever. His mother reported that Lee was sweating a lot and his pillow was wet every morning. He was breathless when feeding and was now only taking 30 ml (1 oz) of milk every feed. His mother said he normally took 6 feeds of 120 ml (4 oz) of formula milk daily. His weight at 6 weeks of age was 4.35 kg, but was now 4.5 kg. He had had one respiratory infection. He had his first immunizations at 8 weeks. His mother thought that he had become more breathless over the past 3 weeks.

Lee is developing *classical symptoms of heart failure*. He is dyspnoeic and tachypnoeic with a poor weight gain over a 4-week period. He is also sweating a lot, a cardinal sign of heart failure in the first few months of life.

On clinical examination, Lee was restless and apprehensive. His respiratory rate was 80 per minute and he had marked intercostal and subcostal recession. His heart rate was 165 beats per minute and he had a gallop rhythm and palpable thrill. He had a 4–6 pansystolic murmur loudest at the lower left sternal border radiating to the back. His liver was palpable 3 cm below the costal margin. Chest X-ray showed cardiomegaly and pulmonary plethora (increased vascular markings). Again, the heart murmur did not sound innocent and was associated with clinical symptoms and an abnormal CXR. The GP arranged for Lee to be admitted to hospital.

The most likely cause of heart failure at 3 months of age is a large left to right shunt through either a ventricular septal defect, a patent ductus arteriosus or an atrioventricular septal defect. Although these defects are present at birth, they do not become symptomatic until the pulmonary vascular resistance falls, thus creating a pressure difference between the left and right side of the heart. The additional blood flow to the right heart may cause hypertrophy and dilatation of the right ventricle. The extra blood flows through the lungs leading to pulmonary plethora (Fig. 10.4). Failure of the right ventricle leads to oedema, weight gain, and hepatomegaly. The extra blood flow eventually compromises the left ventricle which leads to pulmonary venous congestion and the symptoms of tachypnoea, dyspnoea, feeding difficulties, and poor weight gain.

Echocardiography confirmed the diagnosis of a large ventricular septal defect.

If Lee had a patent ductus arteriosus he would have 'bounding' or 'full' pulses with a wide pulse pressure and he would have a continuous murmur heard at the upper left sternal border below the clavicle. If the diagnosis was an atrioventricular septal defect then a mid diastolic murmur may be audible (due to a relative narrowing of the mitral/tricuspid valve) and the ECG will show a 'superior' axis (i.e. the QRS axis between −10° and −150°). Thirty per cent of atrioventricular septal defects occur in children with Down syndrome. Other causes of heart failure in the first weeks and months are shown in the further information boxes.

Lee's medical management will aim to decrease the systemic and pulmonary congestion, improve cardiac performance and consequently peripheral perfusion. He should be given oxygen to keep his oxygen saturation above 95%. Factors such as acidosis, anaemia, infection, and hypocalcaemia, which impair cardiac function, should be corrected.

The mainstay of treatment will be a combination of inotropic drugs (digoxin, dobutamine) to support cardiac contraction, diuretics to decrease preload, and vasodilators to decrease the afterload on the left ventricle.

Summary
Signs and symptoms of heart failure

- Restlessness
- Excessive weight gain (acutely)
- Poor weight gain (chronically)
- Tachypnoea
- Tachycardia
- Sweating
- Gallop rhythm
- Cyanosis
- Hepatomegaly
- Oedema

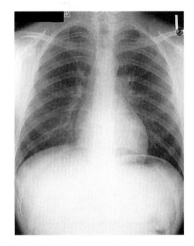

Fig. 10.3. Chest x-ray showing a normal heart.

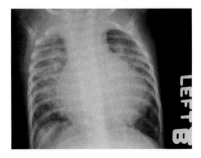

Fig. 10.4. Chest x-ray showing an enlarged heart with pulmonary plethora.

Additional Information
Causes of heart failure in the first weeks

- Coarctation of the aorta
- Hypoplastic left heart syndrome
- Critical aortic valve stenosis
- Myocardial ischaemia (secondary to birth asphyxia)
- Supraventricular tachycardia
- Severe cardiomyopathy/myocarditis

Additional Information
Causes of heart failure at 3 months

- Ventricular septal defect
- Patent ductus arteriosus
- Large ostium primum atrial septal defect
- Atrioventricular septal defect
- Anaemia
- Supraventricular tachycardia
- Arteriovenous fistula

Summary
Infective endocarditis

Symptoms
Fever
Malaise/lethargy
Loss of appetite
Pallor/Anaemia

Signs
Loud or new heart murmur
Splenomegaly
Splinter haemorrhages
Skin/Conjunctival petechiae

Investigations
White cell count
Erythrocyte sedimentation rate (ESR)
Positive blood cultures
Microscopic haematuria

If medical management is not successful, blood flow to the lungs can be decreased by placing a band around the pulmonary artery or the defect can be closed or corrected surgically. If the excess pulmonary blood flow is not decreased, the pulmonary arterioles react by contracting. This increases the pulmonary vascular resistance which if unabated, can cause irreversible pulmonary hypertension. As pulmonary pressure increases to a value greater than that in the left ventricle, deoxygenated blood flows across the VSD (a right to left shunt) causing permanent cyanosis. This is called Eisenmenger's syndrome.

Pyrexia and congenital heart disease

Fiona (aged 7) is known to have a moderate sized ventricular septal defect. While on holiday, she developed severe toothache and a carious tooth was extracted at the local hospital. On returning home, she developed a fever, had lost her appetite, and was extremely lethargic. On clinical examination, her temperature was 39.5 °C, she had an easily palpable spleen, and had 'splinter' haemorrhages in her nail beds. She had petechiae on her skin and conjunctivae. She had developed a palpable thrill and a 5–6 heart murmur. Blood tests revealed anaemia and a raised white cell count and *Streptococcus viridans* was isolated from blood cultures. Urinalysis revealed microscopic haematuria.

Children with congenital heart disease are at great risk of *infective endocarditis* (sometimes called subacute bacterial endocarditis). Turbulent blood flow across the heart defect leads to platelet adhesion to the endocardium. Organisms in the bloodstream are caught within the platelet clot, subsequently multiply, and create vegetations. Pieces of these vegetations may slough and create septic emboli.

Fiona demonstrated most of the classical symptoms and signs. Other clinical findings can include hemiplegia, meningitis, or breathlessness due to pulmonary emboli.

Management consists of a 6-week course of the appropriate intravenous antibiotic(s). The disease can be prevented by scrupulous dental hygiene and prophylactic antibiotics prior to and after dental or surgical procedures. It is imperative that children and parents are educated with regard to the prevention of infective endocarditis.

Congenital heart-disease in the neonate

In the first week of life, infants with congenital heart disease may present with signs of cardiac failure, cyanosis, or a mixture of the two (see Chapter 8). Most of these defects do not manifest themselves *in utero* because of the *fetal circulation*. However, at birth the *circulation changes*, (see Ch. 8, p. 153).

Presentation of congenital heart disease in the first few weeks of life is commonly related to the closure of the ductus arteriosus. When systemic blood flow is dependent on the arterial duct, the infant presents with heart failure. A common cause of duct-dependent systemic blood flow is critical coarctation of the aorta. If the coarctation is very narrow, blood can only reach the descending aorta from the pulmonary artery via the ductus arteriosus. When the ductus closes, the lower half of the body becomes poorly perfused and acidosis and heart failure ensue.

When pulmonary blood flow is dependent on the arterial duct the infant presents with cyanosis: in lesions such as atresia of the pulmonary or tricuspid valves, blood flow to the lungs is dependent on an open duct, since blood can only reach the pulmonary artery via the duct from the descending aorta. When the ductus closes, severe *cyanosis ensues*. The immediate treatment for any lesion, which is dependent on the duct being open, is the infusion of prostaglandins E_1 or E_2.

Further reading

'Moss' heart disease in infants, children and adolescents, (4th edn). (1989) Ed. Forrest Adams. Williams & Wilkins, Baltimore, MD

Anderson, R.H., Macartney, F.J., Shinebourne, E.A., and Tynan, M.J. (1987) *Paediatric cardiology*. Churchill Livingstone, Edinburgh.

Jordan, S.C. and Scott, O. (1989). *Heart disease in paediatrics*, (3rd edn). Butterworth Heinemann, Sevenoaks, Kent.

Summary
Circulation changes at birth

- Pulmonary vascular resistance falls
- Pulmonary blood flow increases
- Systemic vascular resistance rises
- Ductus arteriosus closes
- Foramen ovale closes
- Ductus venosus closes

Additional Information
Causes of cyanosis in the first postnatal week

- Transposition of the great arteries
- Pulmonary atresia
- Severe tetralogy of Fallot
- Tricuspid atresia
- Ebstein's anomaly of the tricuspid valve
- Total anomalous pulmonary venous drainage

11

Skin disorders

First impressions do count. The skin has a social function as well as its importance as a vital organ of the body. Thus children present because of illness, an uncomfortable symptom such as itch, unhappiness at perceived disfigurement, or because of the reactions of others. Primary skin conditions are a common and important presentation in primary and secondary care, and the skin may reflect systemic conditions specific to childhood. At one end of the spectrum this may be a non-specific viral rash; less commonly a systemic illness such as Henoch–Schönlein purpura; and rarely, but devastatingly, meningococcal septicaemia.

Patterns of skin disease are changing. There is evidence that atopic dermatitis is becoming even more common. Malignant melanoma, although rare, is increasing in incidence in adults, giving rise to a greater consciousness of the importance of sun protection for children; campaigns in the media have led to an increase in parental concern about benign pigmented naevi in children. Teenage acne, on the other hand, is a declining problem.

Major functions of the skin

The dermis is derived from mesoderm and at the third week of fetal life a single layer of undifferentiated cells precedes the formation of the periderm. The melanocytes migrate from the neural crest. As the true epidermis keratinizes beneath it, the periderm is ultimately lost *in utero* contributing, with sebum, to the vernix caseosa. Keratinization is progressive, so that by about 32 weeks of gestation the keratin forms a fairly effective barrier to water loss. The skin of the full-term term baby is still a delicate structure which is easily subject to irritation or mechanical damage. The skin of young children remains particularly susceptible to damage from sunlight, and considerable adult vigilance is required to protect babies and children from sunburn.

As the quality of the skin changes with age, so do its disorders. For example, the onset of atopic dermatitis, one of the most prevalent skin disorders in childhood, is generally in infancy: fortunately, with treatment and time most children have progressively less discomfort. But it is wise to avoid making predictions since children rightly resent hearing 'they will grow out of it'— especially when their experience proves otherwise. Another chronic condition with an unpredictable course is psoriasis. This disorder may suddenly erupt in the early teenage years and can significantly impair a young person's social development.

The issues of water loss, temperature regulation, and infection are especially important for neonates (Chapter 8), and that of vitamin D synthesis for children with heavily pigmented skin. Poor nutrition increases susceptibility to skin infection, and cutaneous infection in older children may be the initial presentation of immune deficiency or the side-effect of immunosuppressive drug therapy.

Summary
Major functions of the skin

- Protection from:
  - chemicals
  - infections
  - ultraviolet radiation
- Barrier to water loss
- Regulation of temperature
- Sensation
- Synthesis of vitamin D
- Immunological organ

Case studies
Atopic dermatitis (eczema)

For some weeks Alexander, aged 2 years, has been very itchy. His mother tells the GP that he wakes every night scratching, and his sheets are bloodstained. The 1% hydrocortisone cream which was initially prescribed has been helpful but his mother is reluctant to continue using it daily since she read in a magazine that steroid creams have side-effects. His mother is exhausted.

Unfortunately, parents are often so concerned about scares of steroid side-effects in the media that undertreatment is common, and each flare up is inadequately treated (usually by stopping using the treatment too soon). The continual demands of an itching, fretful child who sleeps poorly, and scratches at night are very exhausting for parents.

Alexander has had dry, itchy skin since he was three months of age. Initially there was scaling of his scalp which the health visitor described as 'cradle cap'. She recommended that warm olive oil be massaged in gently each night for 20 minutes until there was an improvement, to gently rub aqueous cream onto his skin each evening and she also suggested that Alexander's mother was careful to avoid dressing him with wool next to his skin.

Topical emollients such as aqueous cream are useful in preventing recurrence of symptoms. Systemic antihistamines given 1 hour before bedtime are effective in reducing night-time scratching, mainly through their sedative effect. It is important to specify sucrose-free syrups to reduce the risk of dental caries.

His two elder brothers also had atopic dermatitis in infancy but have not had troublesome skin since they were 5 years old. One brother has asthma. His mother, aged 31 years, is now living apart from the children's father. The boys visit their father every weekend—where they are fond of playing with his pet cat.

The clinical history and family history are important. It is crucial to establish when and where the skin trouble first developed and then to define the course and response to treatment. The history may suggest allergic eczema, or there may be emotional factors which can also affect eczema.

On examination, Alexander is fretful but co-operative. At any opportunity he scratches his wrists and when undressed he scratches his legs. His skin is generally dry and there are painful fissures on his right thumb. On his knees and legs there are pustules. At his elbows and wrists, popliteal fossae and ankles there is dermatitis. At his ankles there are weeping, crusted areas. There are no scabies burrows on his hands, wrists, feet, and genitalia. The lymphatic glands in both groins are enlarged.

It is necessary to examine the nails and hair as well as the skin, and it is usually advisable to complete the physical *examination*. Although he clearly has *atopic dermatitis*, it is important to establish if there could be any other reason for his itch. Papular urticaria from cat fleas must be considered, even if no other family members are affected. These lesions tend to occur on the legs and can either be a papule with a central punctum or a small blister. *Scabies* must always be sought, by examining for the burrows especially on the hands, wrists and feet (and the groins of infants). Children may acquire this infestation outside the family from neighbouring children or directly from older relatives. (Figs 11.1–11.3). It is essential in the *treatment* of scabies to treat the whole family.

A swab for bacteriological culture was taken. Acute flares, including pustule formation, may be caused by group A alpha haemolytic streptococci, so Alexander

Summary
Examining the skin

- Examine all the skin
- Nappy area in infants
- Flexures
- Hair and scalp
- Nails
- Mucous membranes of the mouth (and tonsils)

Additional Information
Atopic dermatitis

An itchy skin condition (or parental report of scratching or rubbing by the child) plus at least three of the following

- A history of flexural involvement (elbows, knees, fronts of ankles, or around the neck; or cheeks in children under 4 years)
- A personal history of asthma or hay fever (or history of atopic disease in a first-degree relative in children under 4 years)
- A history of generally dry skin in the last year
- Visible flexural dermatitis (or dermatitis involving the cheeks/ forehead or outer limbs in children under 4 years)
- Onset under the age of 2 years

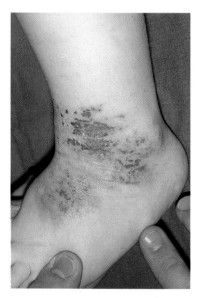

Fig. 11.1. *Atopic dermatitis* characteristically affects the flexures in children. There are itchy, inflammatory papules and crusting. In the more chronic lesions there is lichenification.

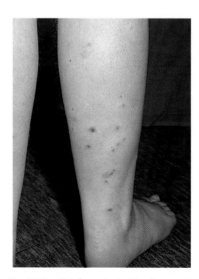

Fig. 11.2. *Papular urticaria.* The characteristic lesion is an urticarial papule with a central punctum. Blisters may be seen particularly on the legs. Present-day houses appear to be a favourable habitat for cat fleas, the usual cause of this condition.

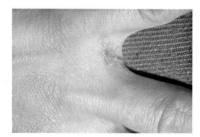

Fig. 11.3. *Scabies.* The characteristic lesion in scabies is the burrow. In infants, the soles of the feet or the groins are commonly affected. In older children with a persisting infestation, the generalized truncal rash may be difficult to distinguish from atopic dermatitis.

Additional Information
Treatment of scabies

- Record everyone in close personal contact with the child

- Ascertain how these people are to be informed of their need for treatment

- Explain that treatment has to be synchronized within a family group

- Advise on appropriate products and the method of use (e.g. if permethrin cream is used there is a maximum dose depending on the child's age)

- Arrange any required practical help from community nurses

- Include a warning that repeated unnecessary treatments cause irritation

Summary
Management of atopic dermatitis

- Identify avoidable/treatable factors:
  - overbathing
  - pets
  - food
  - allergy to any medication
  - infection by bacteria, virus, or fungus
  - infestation by scabies or lice

- Explain the natural history

- Consider prescribing:
  - a soap substitute
  - an emollient (moisturising cream)
  - topical corticosteroid
  - oral antihistamine especially at night
  - oral antibiotic

- Offer a review appointment

was given erythromycin pending the result of culture. He was also prescribed a topical antibiotic for the crusted lesions.

Children with atopic dermatitis are often colonized by *Staphylococcus aureus* and the inappropriate use of antibiotics, either topically or systemically, can allow the emergence of resistant strains.

Many factors have to be taken into account in the *management of atopic dermatitis* but the mainstay of treatment is a *topical steroid* ointment or cream. The golden rule is 'the least amount of the weakest preparation which gives relief'. Combination products, such as antibiotic with corticosteroid, can stain clothes and carry a greater risk of inducing allergic contact dermatitis, so they are better avoided. How a preparation is used is as important as what is used.

> Three weeks later Alexander's mother brought him back for review. She reported that he appeared to be scratching more on two occasions after eating a particular brand of potato crisps.

For many children review appointments are required to optimize therapy and to assess the importance of social, psychological, and other factors. It is always relevant to review a child's diet since food allergy can both cause urticaria and worsen atopic dermatitis. Generally, the clinical history and evaluation of dietary changes is more informative than skin testing or blood tests such as serum IgE. In Alexander's case it was straightforward to request the ingredients of that particular brand of crisps and to recommend that his diet exclude that colouring. Advice from a dietician is generally required if more complicated alterations are needed.

It is important to offer assessment for every exacerbation since occasionally herpes simplex may be responsible. In this situation steroid cream is contra-indicated, and systemic therapy with acyclovir is required. In the hospital setting it may be possible to obtain direct examination of vesicle fluid by electron microscopy, which is a rapid method of confirming herpetic infection. Alternatively, the fluid may be swabbed and then the swab sent to the laboratory in virus transport medium.

Additional Information
Topical corticosteroids

- Prescribe the least dose of the weakest preparation that works
- Sometimes needs a short course of strong steroid to get control
- Ointment may work better than cream but children like it less
- Should be lightly spread, not rubbed in vigorously
- Should be applied once or twice daily: there is no advantage in more frequent use
- Should be continued for 4 or 5 days after the surface of the skin has healed, to deal with the deeper inflammation

Psoriasis

Elaine, aged 11 years, is distraught. In the past 16 days, itchy red spots have appeared on her body, arms, and now her neck. She is frightened that they will extend on to her face. Additionally, her scalp is scaly and she has dandruff. Three weeks before she had had a severe sore throat. She has been rubbing in a variety of hand creams and body lotions. Her mother is anxious since she recalls an uncle who had severe psoriasis and was very disabled by arthritis. Looking at her medical records, her GP notes that she had a very florid nappy rash as a baby.

Nappy rashes may occur from irritation or from infection by *Candida*. Children with a genetic predisposition to psoriasis can present with a sudden severe nappy rash, often with a more generalized rash.

Elaine has enlarged tonsils and associated lymphadenopathy. In her scalp there are hard, scaly areas but no loss of hair. Diffusely over the trunk there are oval lesions that are red and superficially scaling. Where she has scratched, tiny circular lesions are developing.

Elaine has the features of *guttate psoriasis* (Fig. 11.4). In older children with the genetic predisposition it is not uncommon for guttate psoriasis to develop after tonsillitis due to streptococcal infection. It is relevant to take a throat swab since if group A alpha-haemolytic streptococci are isolated, penicillin should be prescribed. If episodes of tonsillitis are frequent then referral for consideration of tonsillectomy is appropriate.

The response to scratching is called the 'Koebner phenomenon'. This term is used in several conditions for the isomorphic response, for example, after trauma, the Koebner phenomenon can develop in the damaged areas, and may also occur in surgical scars if the psoriasis is in an active phase.

Both the patient and her parents require an explanation of the natural history of psoriasis. They need to know that childhood psoriasis often presents acutely in the guttate form and resolves in a few weeks. Treatment is required if the itch is severe and an emollient cream can minimize the scaling. Shampoos containing prepared coal tar can be beneficial. It is important to apply the therapies gently to avoid further trauma to the already irritated skin.

The practice nurse has a video of the treatments available for psoriasis. She has also a leaflet illustrating how to get the most benefit from the prepared tar shampoo prescribed for Elaine. The nurse suggests that there is no need for the

Summary
Guttate psoriasis

- Often follows streptococcal sore throat
- Presents acutely in the guttate form and resolves in a few weeks
- Treatment is required if the itch is severe
- An emollient cream can minimize the scaling
- Shampoos containing prepared coal tar can help
- Avoid further trauma to the already irritated skin

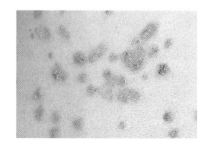

Fig. 11.4. *Guttate psoriasis* tends to erupt. The erythematous scaling lesions are widely distributed over the trunk but may also affect the scalp.

family to cancel their holiday since she expects Elaine's skin to improve and points out that natural sunshine in moderation will help.

Pigmented skin lesions

It is September and Barry (aged 13 years) and his family have recently returned from a holiday in southern Spain. His father brings him to the GP stating that two moles on Barry's back have changed significantly in the past year. Barry has made no complaint but admits that both moles have been itchy at times. His father is requesting that these moles are removed. Barry is soon to enroll at a boarding school and his parents will fly to the Far East for an extended business trip of six months. An urgent dermatology consultation is arranged.

When parents present expressing concern that their child may have cancer it is particularly important to establish all the reasons for their anxiety. In Barry's case his father had read a newspaper report about the dangers of melanoma.

On examination Barry's skin is tanned. On his back there is one prominent dark pigmented naevus which has the features of a compound naevus (Fig. 11.5). There is also a pre-existing circular pigmented naevus which has a halo of non-pigmented skin. This has the characteristics of a halo naevus (Sutton's naevus) (Fig. 11.6).

The newspaper article had not explained that moles may suddenly develop during adolescence and that existing lesions may alter either by increasing in thickness or by darkening in colour. At this age, some compound naevi also involute giving the characteristic halo naevus.

The slight risk from a general anaesthetic and the associated experience of surgery must be weighed against the reassurance to be obtained by removal and a subsequent histopathology report of a benign lesion. Surgery inevitably results in scarring and keloid scars can be painful as well as disfiguring.

After discussion it was agreed that good quality clinical photographs be taken. Copies would be sent to the school doctor with a request that he re-examine the lesions.

Although rare, melanoma may occur in childhood. There is no formula which may be applied at this age to aid the decision as to whether to excise a pigmented lesion and it remains a difficult area of clinical judgement. Several factors may influence the decision to *remove pigmented lesions*.

Summary
Pigmented skin lesions

- Freckle
  - tiny areas of increased pigmentation after sun exposure
- Naevus (plural naevi)
  - an imprecise term applied to a variety of hamartomas in the skin
- *Café au lait* lesion
  - irregular areas of even pigmentation which are not raised; if numerous they may suggest neurofibromatosis
- Compound naevus
  - the most frequently seen lesions; the centre tends to be darker and more raised
- Spitz naevus
  - spindle cell tumour which histologically resembles melanoma.

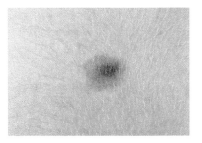

Fig. 11.5. *Compound naevus.* Adolescents develop such lesions which often have an irregular border and a darker raised centre.

Additional Information
Factors influencing the need to remove pigmented lesions

- First degree relative treated for melanoma
- Multiple atypical lesions
- Spitz naevus (Fig. 11.7)

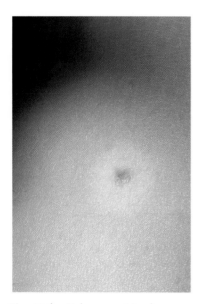

Fig. 11.6. *Halo naevus* (also known as Sutton's naevus) is a pigmented lesion which is spontaneously regressing and has a rim of depigmentation.

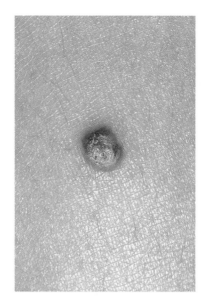

Fig. 11.7. *Spitz naevus.* Characteristically a single red or red-brown nodule suddenly develops on sun-exposed skin. After a period of rapid growth it stabilizes and persists.

Summary
Sunburn

- Risk factor for malignant melanoma

- Avoid sun exposure between 11 a.m. and 3 p.m.

- Clothing and a hat are useful barriers

- Additional ultraviolet is reflected from water

Summary
Acne

- Is becoming less common

- Is a cause of undivulged teenage unhappiness

- May be helped by topical benzoyl peroxide

- May need systemic tetracycline or erythromycin

- In severe cases may require isotretinoin

The opportunity was taken to discuss sun protection. This remains a controversial subject since it has been argued that the introduction of sun blocks which reduce the erythema from short wavelength ultraviolet rays (UVB: 295–320 nm) result in longer overall sun exposure. Thus sunscreens now generally also contain protection from longer wavelength ultraviolet (UVA: 320–400 nm). Dermatologists recognize that sunburn in childhood is a risk factor for the subsequent development of multiple naevi and of malignant melanoma. There is no controversy that it is wise to avoid exposure to the sun between 11 a.m. and 3 p.m. Clothing and an appropriate hat may also be useful barriers to ultraviolet radiation. Thus, wearing a cotton T-shirt makes sense especially for activities such as water sports when there is additional ultraviolet reflection from the water.

Acne

It is also evident that Barry has acne. There are several comedones and pustules on his forehead and nose.

With acne, teenagers are usually very self-conscious and may be embarrassed to ask for help (Fig. 11.8). It is worth suggesting a preparation that contains benzoyl peroxide taking care to explain that it is likely to cause some dryness of the skin

Fig. 11.8. *Acne.* The epitome of teenage years. The hallmark is the comedone but the more obvious lesions are pustules and cysts. Treatment aims to lessen the severity and to limit the extent of long-term scarring.

and that it will not bring about instant resolution of the problem. This is another situation when an explanatory leaflet, which the youngster can read in his own time, may be a useful means of providing information. A thoughtfully written publication may help a teenager realize that a number of treatment approaches are available, (e.g. systemic antibiotics), or in severe cases isotretinoin.

Viral warts and molluscum contagiosum

Mary, who is 5 years old, is due to attend for her school medical examination in three weeks time. Her mother is aware that Mary has had warts on her fingers for three months. However in the past two weeks she has begun to develop warty lesions on her arm.

The viruses that cause *warts and verruca* are papilloma viruses of which many types are now recognized (Fig. 11.9).

The school doctor confirms that Mary has indeed got viral warts on her fingers. The new lesions are more pearly and some have an umbilicated appearance. They are in groups at the axilla and in the flexor aspect of her forearm. The surrounding skin is dry and irritated. These lesions are called molluscum contagiosum.

Molluscum contagiosum is also a cutaneous viral infection (Fig. 11.10). It has been usual not to treat either warts or molluscum with cryotherapy (liquid nitrogen) because it is painful. Unfortunately, molluscum contagiosum may become very widespread and the lesions do not always resolve without scarring. The introduction of effective local anaesthetic cream means that children are often able to tolerate cryotherapy if this cream has first been applied under occlusion for one hour. However, young children who are untroubled by their warts need not be treated, and older children may respond to topical salicylic acid or podophyllin. Molluscum contagiosum is generally more troublesome in children who have atopic dermatitis. It is important to employ emollients and to avoid topical corticosteroids at the affected sites.

Mary's mother decides that she would like to discuss the options with Mary's father. She has a recollection that some of his relatives suffer from atopic dermatitis and accepts a prescription for aqueous cream to counteract the associated dermatitis reaction. After discussion, Mary's parents decide to adopt a wait-and-see policy.

Summary
Viral warts

- Are common
- Are caused by *papilloma virus*
- Usually resolve spontaneously after a few months
- May be treated with topical keratolytics
- Plantar warts (verruca) are often painful and need treatment
- Respond to cryotherapy (with topical anaesthetic cream)

Summary
Molluscum contagiosum

- Is caused by a pox virus
- Is often associated with dermatitis
- May spread widely
- May persist for years
- Responds to cryotherapy (with topical anaesthetic cream)

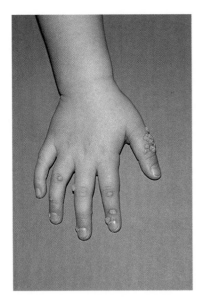

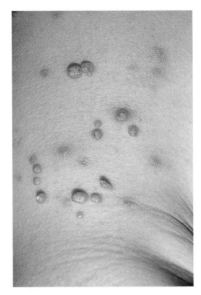

Fig. 11.9. *Viral wart.* The correct term to describe this familiar viral infection is 'verruca vulgaris' on the hand, and 'verruca plantaris' on the foot. However, 'wart' is acceptable as a description of a benign virus-induced tumour. The pain from a verruca on the sole of the foot may be lessened by using a ringed dressing to reduce pressure. Whichever treatment is adopted, persistence is required.

Fig. 11.10. *Molluscum contagiosum.* Early lesions are skin-coloured papules. More chronic lesions have an umbilicated appearance or become very inflamed. In some children, pustular areas develop. The cause is a pox virus and untreated the condition may persist for months or years.

The school doctor faces a dilemma. Should a child with molluscum contagiosum and warts be banned from the swimming pool?

At present this is unlikely to be suggested. There is no proof that molluscum is contracted from swimming pools but the method of infection is still unknown.

Nappy rash

Jane (3 months) presented at the practice baby clinic for weighing and immunization. The health visitor noticed that she had a bad nappy rash which extended over the whole buttocks and down the legs. There were also some red satellite lesions extending onto the abdomen. The mother was quite upset about it, saying that she had tried several creams but it was getting worse. She had also noticed that Jane had

Summary
Causes of nappy rash

- ammoniacal dermatitis
- Candida (thrush)
- seborrhoeic dermatitis
- napkin psoriasis

Summary
Treatment of nappy rash

- Ammoniacal dermatitis: exposure, good hygiene, protective cream, topical corticosteroids
- Candida: nystatin orally and topically, exposure, topical steroid
- Seborrhoeic dermatitis: scalp—clean with olive oil apply sulphur, and salicyclic acid ointment if severe .
  bottom: exposure, topical steroid possibly with antifungal.
- Napkin psoriasis: weak topical steroid.

some white spots in her mouth and wasn't feeding so well. The health visitor asked her to see the doctor.

It is extremely common for a baby of this age to have a rash in the buttock area. Often it doesn't upset the baby at all but the parents become quite worried, partly because there is a feeling of blame—nappy rash tends to be associated with the idea of poor hygiene and infrequent nappy changes. It is certainly true that close contact with decomposing urine will cause a rash (ammoniacal dermatitis), but nappy rash can also be caused by the fungus thrush or Candida, and by seborrhoeic dermatitis which is a skin disorder associated with cradle cap (heaped-up yellow scales on the scalp), and also by napkin psoriasis.

The GP examined Jane carefully and noticed that the rash covered the skin creases at the hips and there were satellite lesions on the abdomen; it appeared very red, with sharply demarcated edges. She also had small white spots on her tongue and on the gums which looked like thrush. These are all signs of Candida infections, so he treated her with oral and topical Nystatin an antifungal agent and advised the parents to expose the bottom after changing the nappy. A week later the rash had nearly disappeared.

Ammoniacal dermatitis tends to spare the flexus whereas Candida infection covers them, and has noticeable satellite lesions. The fungus is not always seen in the mouth but tends to commence there and work its way down, so it is usually best to give a course of oral treatment. Seborrhoeic dermatitis is diagnosed by the presence of cradle cap and also the rash tends to be in all the flexures: neck, axillae, and also behind the ears. Napkin psoriasis resembles true psoriasis with well-defined, slightly raised scaly plaques in the napkin area, which may spread up the trunk. The flexures are usually involved.

Resistant cases are best treated by exposure and with a topical corticosteroid, mixed with an antibacterial and antifungal agent if appropriate.

Further reading

Champion R.H., Burton J.L., and Ebling, F.J.G. (ed.) (1992). *Textbook of dermatology*, (5th edn, 4 vols). Blackwell, Oxford.

Harper, J. (1990). *Hand book of paediatric dermatology*, (2nd edn). Butterworth Heinemann, Sevenoaks, Kent.

12

Ear, nose, and throat problems

Childhood ear, nose, and throat problems are extremely common and make up a significant proportion of the workload of many GPs. Most of these problems are caused by the many upper respiratory infections to which young children are exposed. Existing as a recluse on a remote mountain top would protect against many of these ills which can be seen as a consequence of normal social interaction, exacerbated by airborne allergens in some and probably also by passive smoking and environmental air pollution.

Most ear disease in children affects the middle ear, so that earache and deafness are the common symptoms. Otitis externa is also seen in children and to the young child, the external ear canal provides a tempting receptacle for a variety of foreign bodies. This temptation to insert objects into the ear canal is not limited to young children, however; the delicate nature of the deep meatal skin, eardrum, and ossicles means that the well-known advice to insert nothing smaller than one's elbow is generally sensible.

Sensorineural deafness is dealt with elsewhere in this book (Chapter 7, p. 132) and so will not be considered here.

A blocked and runny nose is a normal part of growing up although in most children these symptoms are relatively infrequent and short-lived. A significant proportion have more persistent trouble and are often labelled 'catarrhal'. Most of this group have no serious pathology and the symptoms reflect a transient tendency to upper respiratory infections (often associated with starting to attend playgroups, nursery, or school), a degree of nasal allergy, or prominence of the nasopharyngeal pad of lymphoid tissue (the adenoids). Many small children are unconcerned by the presence of excessive nasal mucus and show no desire and often little ability to 'blow' their noses unless prompted by adults. Some children have a habit of adopting an open-mouth posture even when the nasal airway is patent. Significant underlying pathology, such as mucosal cilial abnormalities (as in Kartagener's syndrome) or cystic fibrosis, is uncommon and is usually suggested by unusually troublesome or persistent symptoms and associated features such as pulmonary involvement or nasal polyps.

Physiology and development

The tonsils and adenoids make up part of a collection of lymphoid tissue surrounding the upper aerodigestive tract. This arrangement is completed by lymphoid tissue in the base of the tongue (the lingual tonsils) and is known as Waldeyer's ring. The tonsils and adenoids are not usually apparent at birth but enlarge through hyperplasia and multiplication of lymphoid follicles which in turn produce antibodies (immunoglobulins), IgA locally and IgG and IgM systemically.

A pad of adenoids which is small in relation to the size of the nasopharynx usually causes no problems. The size of the adenoids increases after birth usually reaching a maximum between the ages of 3 and 7 years. In the first few years of life the facial skeleton and the nasopharynx are relatively small in comparison with the cranium and the child as a whole and enlargement of the adenoids can obstruct the nasopharyngeal airway, particularly in response to recent or recurrent infection. In addition, because of the situation of the adenoids close to the nasopharyngeal openings of the Eustachian tubes, enlargement or inflammation of the adenoids is thought to be a factor in the development of chronic otitis media in some children. Growth of the nasopharynx and regression of the adenoids with increasing age ensure that problems due to the adenoids usually resolve before adolescence. The tonsils may also cause difficulties for the child because of relative size.

In young children, the tonsils tend to project into the oropharynx with only a small proportion buried between the anterior and posterior pillars of the fauces (the palatoglossal and palatopharyngeal folds), whereas in adults the opposite is more often the case. In addition, a child's tonsils are relatively much larger than they will be when the child reaches adult life. If the tonsils are particularly large (sometimes in a small child they may meet in the middle) then breathing may be impaired, particularly at night and when the adenoids are large, resulting in sleep apnoea. Swallowing may also be impaired although large tonsils are probably blamed for poor eating more often than is justified.

Case studies
Earache with fever: otitis media, glue ear

Jamie (18 months old), the second of two children in the family, is brought to the GP's surgery irritable and hot, having had a poor night's sleep, apparently crying with pain. The doctor sees from the baby's records 6 different episodes of acute otitis media over the past 10 months, each time resolving on amoxycillin. The mother reports an improvement in Jamie's condition the previous evening after paracetamol, with deterioration again overnight. Examination reveals Jamie to be pyrexial and after a struggle the doctor can see that the right eardrum is red and bulging. The left drum is obscured by soft wax. The child's mother reports that he and his older sister, Kathleen, have both been prone to colds since Kathleen started nursery school the preceding autumn. His mother feels that Jamie hears well although often ignores her. Speech development seems slow in that he has no definite words although he tries to say 'ma', 'da', and 'Jess' (the dog), and babbles with a variety of sounds. He had failed the first health visitor distraction test of hearing at 8 months, passing the re-test 6 weeks later.

Earache is probably a universal symptom in childhood, although very young children have difficulty localizing the pain and communicating the sensation to adults. Reports of children poking, pulling, or hitting their ears may indicate middle ear pathology and it is well known that in a sick child otoscopy is essential to assist in localizing the illness. The severity of earache is not necessarily a

Summary
Causes of earache in children

- Acute otitis media
  - viral
  - bacterial
- Foreign body in ear canal (not necessarily painful)
- Trauma
- Impaired Eustachian tube function (e.g. during aircraft flight)
- Referred pain, usually from dental causes or tonsillitis
- Furuncle of external ear canal

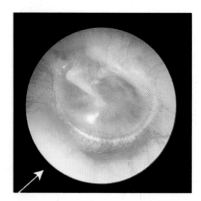

Fig. 12.1. A normal eardrum. Note the light reflex.

good guide to the underlying pathology as acute otitis media can be extremely painful yet resolve spontaneously, whereas the relatively uncommon form of chronic otitis media with destructive and potentially complicated cholesteatoma is usually painless.

A tendency to *acute otitis media* associated with upper respiratory infections is common in toddlers, possibly due to a sibling bringing home infections from play-group or school. The history often begins after 6 months of age when the effect of maternal immunity declines. The child may go through a spell of recurrent otitis media which resolves spontaneously with increasing age and development of immunity, but progress is unpredictable and varies greatly between individuals. The tendency to infection is often worse in winter months and may be linked to poor housing, nutrition, and parental smoking. Acute otitis media often begins with a viral infection and resolves on symptomatic treatment alone but bacterial infection may require antibiotics. A common complication of bacterial otitis media is rupture of the bulging tympanic membrane with resulting blood-stained discharge. The severe pain of the acute infection usually resolves with this rupture and as the condition resolves, the drum usually heals rapidly, although frequent infections may prevent healing and result in a chronic perforation. Febrile convulsions (see Ch. 20, p. 343) may accompany acute otitis media in young children.

At Jamie's consultation, the GP prescribes amoxycillin in view of his general upset and the red, bulging drum. However, Jamie is brought back to the surgery the following day, no better. His mother reports that the right ear seems to be sticking out and very sore. Jamie appears drowsy and is again pyrexial and the doctor sees that the right pinna is indeed more prominent than the left. Closer inspection reveals a tender, boggy red swelling filling in the post-auricular sulcus and displacing the pinna forwards. The GP telephones the local ear, nose, and throat department and arranges urgent admission to the paediatric ENT ward. On admission, Jamie is drowsy but rousable with no neck stiffness and although he is unco-operative, there are no obvious cranial nerve abnormalities. A diagnosis of acute mastoiditis is made and treatment with intravenous Co-amoxiclav is started, along with intravenous fluids as Jamie is reluctant to drink. Over the next 24 hours there is a marked improvement in his general condition and the mastoiditis is settling. Jamie rapidly returns to health and is allowed home. Follow-up in the outpatient clinic confirms complete resolution with normal hearing and over the next 3 months there is a marked increase in his speech and language development.

Acute mastoiditis, once a common emergency, is now relatively rare, the decrease in incidence due to antibiotics and also perhaps a reduction in the virulence of the infecting organisms, so that an individual approach is necessary when attempting to quantify the problem.

Summary
Management of acute otitis media

- Analgesia
- Antipyretics
- Antibiotics
- Advice to parents

Many episodes of acute otitis media will resolve without antibiotics, which should be avoided if possible

Additional Information
Complications of acute otitis media

- *Acute mastoiditis*
  Now an uncommon complication. May resolve with antibiotics or require surgical drainage

- *Perforation of tympanic membrane*
  Relatively common, usually heals with minimal after effects but may become chronic

- *Tympanosclerosis*
  Scarring of middle ear structures including tympanic membrane. Usually minimal and of little significance.

- *Intracranial complications*
  These are rare but can include meningitis and intracranial abscess

Summary
Chronic suppurative otitis media

1. *Persistent perforation of the tympanic membrane*
   This leads to hearing loss and a tendency to recurrent otorrhoea. Surgical repair is possible

2. *Cholesteatoma*
   This is a collection of keratin from the skin of the drum and deep external meatus which accumulates in the middle ear and mastoid. Chronic infection causes local damage and complications. Surgical eradication is vital

Summary
Glue ear (chronic otitis media with effusion)

- Persistent viscous mucus in middle ear

- Very common in preschool children

- May lead to hearing loss, poor concentration at school, and chronic middle ear damage

- Initial management: watchful waiting

- Surgery (grommet insertion) indicated for significant persistent deafness or recurrent infection

Additional Information
Grommet

A tiny plastic tube inserted through the tympanic membrane into the middle ear to allow drainage of fluid. This operation is performed frequently, and some think too often, for a condition which is self-limiting.

Hearing loss in acute otitis media is usually transient and of little significance. However, it is common in small children, particularly in the presence of recurrent upper respiratory tract infection, that the middle ear fluid is slow to disappear and an effusion persists (glue ear), creating an environment in which re-infection is more likely and a hearing impairment occurs. The hearing impairment varies in severity and may be of little significance. However, it may impair speech and language development or in older children hinder school progress and middle ear effusions are a frequent cause of failed hearing screening tests in young children. If the hearing loss or the tendency to *chronic infection* (Figs 12.2 and 12.3) persists then a prolonged course (4–6 weeks) of a broad spectrum antibiotic may bring about resolution. If this fails, surgical drainage of the middle ear with insertion of a ventilation tube in the eardrum (a grommet) may be indicated.

Otitis media with effusion (OME, *glue ear*) also occurs in the absence of acute infections. The underlying pathology in this situation is thought to be poor Eustachian tube ventilation of the middle ear, in certain children associated with enlarged adenoids or nasal allergy. Medical treatment of nasal allergy may cure the glue ear. Otherwise, medical treatment is disappointing. Surgical treatment of the glue ear with grommets (and adenoidectomy in selected children) should be reserved for the minority who suffer significant handicap from the condition.

The highly variable natural history of glue ear means that ideal management usually involves a period of 'watchful waiting' during which the child is monitored, observing hearing, speech and language development, school progress, behaviour, and tendency to infection, before surgical intervention is contemplated.

Surgical intervention for glue ear consists of myringotomy (incision of the eardrum) with drainage of the middle ear fluid through this incision and insertion of a tiny plastic tube (a *grommet*, See Fig. 12.4) into the incision. The grommet therefore ensures ventilation of the middle ear, usually for between 6 and 12 months, before the tube is extruded from the drum. If a grommet is not inserted, the incision heals rapidly and the middle ear effusion may recur.

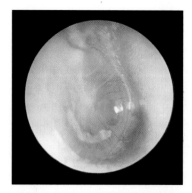

Fig. 12.2. A retracted eardrum with amber-coloured fluid and tympanosclerotic scarring. This change is seen in chronic otitis media.

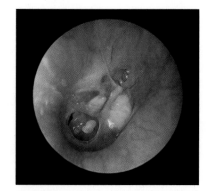

Fig. 12.3. Eardrum with dry central perforation and severe tympanosclerotic scarring.

Alternative or additional surgery is adenoidectomy, the rationale for which is the removal of a chronically enlarged or inflamed mass of adenoids from the nasopharynx to improve Eustachian tube function. However, adenoidal enlargement is not a factor in all children with glue ear and adenoidectomy is generally reserved for children whose adenoids are also causing troublesome nasopharyngeal obstruction or infection.

Surgery for glue ear has become controversial, particularly as these operations are very frequently performed in Europe and the USA for a condition which is often self-limiting. It has been estimated that at least 75% of children in the UK experience glue ear at some time but only 5% have persistent and bilateral middle ear effusions severe enough to warrant attention. Nevertheless, this 5% constitutes a large number of children who may need assessment, monitoring, and consideration of surgery in the absence of satisfactory alternative treatment. Reliable measurement of the outcome of surgical intervention in glue ear has proved difficult because of the variable and multifactorial nature of the condition.

A small minority of children with ear disease have persistent structural abnormalities of the middle ear. A chronic perforation of the eardrum may be asymptomatic or may predispose to infection, especially if water enters the middle ear through the hole. Repair of the defect may then be indicated. A relatively rare but important form of chronic otitis media involves a perforation or retraction pocket in the attic or posterosuperior portion of the drum, through which the keratin of the squamous epithelium covering the outer surface of the drum and deep meatus can accumulate in the middle ear and mastoid air cells. This is known as cholesteatoma and is associated with chronic, low grade infection, often with offensive, smelly otorrhoea. Destruction of middle ear structures occurs, causing deafness, and the condition may slowly progress to cause facial nerve palsy, dizziness, and occasionally intracranial complications. Surgery is usually required to eradicate the condition and referral to an ear, nose, and throat department is indicated.

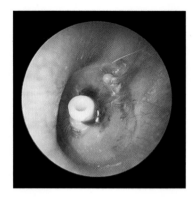

Fig. 12.4. An eardrum with a grommet *in situ.*

Sore throat: tonsillitis, sleep apnoea, glandular fever

A 5-year-old boy, Daniel, is brought to the GP's surgery soon after Easter with a 2-day history of worsening sore throat and pain on swallowing so that he is unable to eat and is drinking little. His mother reports that he is not his usual lively self, lacking energy and distressed. She also reports that his breath smells bad and he has swollen neck glands. The doctor has a look at Daniel, finding him hot and flushed with a rapid pulse. He has tender, enlarged glands below the angle of the lower jaw on both sides and an open-mouth posture with a tendency to drool saliva. On opening his mouth, the boy's tonsils are seen to be enlarged, inflamed, and with a white coating on the surface. The eardrums look slightly pink.

Summary
Causes of sore throat in children

- *Viral pharyngitis*
  Associated with upper respiratory infections. Mild, does not usually impair eating and drinking.

- *Acute tonsillitis*
  Tonsils inflamed. Usually systemic upset. Impairs eating and drinking.

- *Quinsy*
  Peritonsillar abscess. In addition to symptoms and signs of tonsillitis, drooling and trismus (inability to open the mouth widely) occur

- *Glandular fever*
  Similar symptoms to acute tonsillitis and also marked glandular enlargement

This is an extremely common presentation which is not easy to manage scientifically. Ideally, a bacteriological diagnosis should be made by taking a throat swab and sending it for culture to look for *Streptococcus haemolyticus*: type A is associated with serious complications in the form of scarlet fever, rheumatic fever (now rare), and acute glomerulonephritis. Most cases of tonsillitis are viral in origin and do not require antibiotics, and it is not certain whether an antibiotic will shorten the course even when a bacterium is the cause. However, throat swabbing is now rarely practised outside hospital and it is conventional to treat a case such as the above with penicillin to cover a possible streptococcal origin. Can a bacterial tonsillitis be diagnosed from the appearance? Unfortunately not: even large red tonsils with white spots can be due to a virus such as Epstein–Barr (EBV), the cause of glandular fever, well-known for its association with a widespread rash if ampicillin is used in treatment.

Summary
Glandular fever (infectious mononucleosis)

- *Cause:* Epstein Barr virus
- *Transmission:* intimate oral contact
- *Features:* fever, malaise, headache, anorexia, petechiae over soft palate, lymphadenopathy, splenomegaly. Occasional rubelliform rash
- *Diagnosis:* excess of mononuclear cells, positive Paul Bunnell test, positive Monospot test
- *Treatment:* symptomatic

The doctor confirms Daniel's mother's diagnosis of acute tonsillitis, and in view of the worsening clinical history over the past 48 hours, decides to prescribe antibiotics to add to the mother's management with paracetamol. The doctor notes from the child's record that Daniel came out in a rash on penicillin in the past and so writes out a prescription for erythromycin.

Erythromycin is the usual alternative to penicillin in patients who are possibly allergic. If there are many large glands present in the submandibular and tonsillar regions then *glandular fever* should be suspected. This can be a severe illness with a prolonged course and subsequent fatigue lasting several weeks, especially in older children and adolescents.

At this point the boy's mother complains that Daniel always seems to be getting tonsillitis, she feels he doesn't eat well and is not growing, and he has missed several weeks of school since he started last September. On further questioning from the GP, she reports that he snores extremely loudly at night and sleeps very poorly, especially when he has a sore throat. The doctor presses her on the point of the sleeping pattern and she admits to being very worried about Daniel's breathing at night, because in addition to the snoring, the child struggles to breathe at times and has spells when he stops breathing. Her own sleep is disturbed as a result and she feels something needs to be done to solve the problem. During the day the child has little difficulty with breathing, but always has his mouth open and seems unable to blow his nose. She also complains that one of the GP's partners refused to prescribe antibiotics for Daniel for a previous episode of tonsillitis.

This history of recurrent tonsillitis is very common in GPs surgeries and ear, nose, and throat clinics. Children are most often affected during their early school years, when they pick up infections from other children and the autumn and winter months are usually the worst time of the year. Most children who suffer tonsillitis have a few episodes and grow out of the tendency, but like acute otitis media and glue ear, the natural history is extremely variable and unpredictable. Management includes distinguishing acute tonsillitis from more minor pharyngitis associated with the common cold and, in teenagers, from glandular fever. Symptomatic treatment with oral fluids and paracetamol during acute attacks is indicated, and antibiotics are not always necessary, although antibiotics will shorten acute attacks and reduce the incidence of complications such as quinsy (peritonsillar abscess).

Tonsillitis can occur at any age, most frequently in children and adolescents, but is relatively uncommon in the first 2 or 3 years of life. The peak incidence is usually in a child's early school years and as genuine tonsillitis (as opposed to milder, general pharyngitis with common cold) is usually associated with systemic illness, loss of time from school is often a feature of the condition. The natural history in most children suffering recurrent attacks is of spontaneous resolution after a variable period of time and it is this variation along with the often seasonal pattern (tendency to be worse in the winter in the UK), which can make a decision about intervention difficult. In some cases the recurrent attacks do not resolve and in this situation the severity and frequency of the episodes of tonsillitis will need to be taken into account when deciding management.

The long-term *management of recurrent tonsillitis* largely depends on the frequency and severity of the individual episodes, the extent to which the child is disadvantaged by the condition (e.g. loss of time from school) and an estimate of the length of time the child will take to grow out of the condition.

It is this estimate of the duration which is the most difficult and imprecise element in the process of deciding whether to continue treating episodes as they arise or to intervene. Intervention usually means *tonsillectomy* although a prolonged course of antibiotics, as in recurrent otitis media, may be effective in halting a cycle of frequent, recurrent tonsillitis, and so avoid surgery. As in troublesome recurrent otitis media, therefore, a 'wait-and-see' policy, looking for signs of spontaneous improvement, is often the best approach. This will require careful discussion with parents, going over the advantages and disadvantages of antibiotics, the natural history of the condition, and the advisability or otherwise of surgical intervention.

In the case of Daniel, there are additional considerations. There may be some degree of 'Failure to thrive' because of repeated infections and possibly difficulty swallowing. Some children appear to improve dramatically in this respect after tonsillectomy although this is unpredictable. Daniel also has a clear history of upper airway obstruction at night, with snoring and sleep apnoea, and so is likely to have large

Summary
Management of recurrent tonsillitis

Treatment of acute episodes: supportive measures: analgesics and antipyretics; antibiotics if resolution slow or severe systemic upset

Management of recurrence: monitor progress, looking for spontaneous improvement with age. Prolonged antibiotics may curtail frequent recurrences.

Tonsillectomy if severe adverse effect on health or educational progress and no evidence of spontaneous improvement

Summary
Indications for tonsillectomy

- Recurrent tonsillitis with significant adverse effects on health or educational progress and no sign of spontaneous improvement with increasing age
- Quinsy, particularly if recurrent
- Obstructive sleep apnoea

Additional Information
Sleep apnoea

Episodes of cessation of respiration during sleep, which if significant, lead to hypoxia. In children usually obstructive sleep apnoea due to enlarged tonsils and adenoids, only rarely due to central neurological cause.

Diagnosis from the history from observers, usually parents.

Confirmed by monitoring during sleep, including pulse oximetry to assess severity, ECG to look for right heart strain.

adenoids in addition to his troublesome tonsils. If the tonsils are not large in between episodes of infection and yet the airway obstruction persists, then the adenoids can be assessed with a lateral X-ray of the nasopharynx. If there is nasal allergy, this should be treated medically, with antihistamines or topical steroids before surgery for nasal obstruction is contemplated.

Sleep apnoea is a condition which has been better recognized and documented over the past decade or so. In children, it is almost always due to enlarged tonsils and adenoids which obstruct the upper airway when the palatal and pharyngeal muscles relax in sleep, particularly in the supine position (Figs 12.5 and 12.6). Typically, the child has a restless sleep pattern with struggling to breathe. During periods of obstructive apnoea the arterial oxygen saturation falls, rousing the child, who will then usually change position and resume breathing until obstruction recurs. In severely affected individuals the quality of sleep is poor, resulting in daytime sleepiness, irritability, and possibly poor intellectual performance as a result. The condition can be documented in hospital, the most useful advance in this area in recent years being the pulse oximeter, a painless and accurate indicator of oxygen saturation. With this device the significance of the sleep apnoea as measured by the fall in oxygen saturation during apnoeic spells can be estimated.

In view of his mother's concern and the history of sleep apnoea, Daniel's GP decides to refer him to an ENT surgeon. Although Daniel has no further tonsillitis over the next few weeks, his tonsils remain very large. His sleeping remains restless but the apnoea seems less noticeable. There is no history of nasal allergy and the nasal

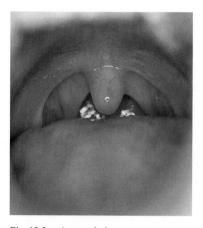

Fig. 12.5. A normal pharynx.

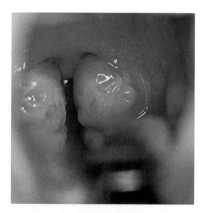

Fig. 12.6. Tonsillar hypertrophy in a child with obstructive sleep apnoea.

mucosa appears normal. The ENT surgeon and Daniel's mother agree that there has been some overall improvement since the winter and opt for conservative management with a review in the autumn, although tonsillectomy and adenoidectomy remain a possibility.

Rarely, sleep apnoea in children with extreme falls in oxygen saturation can lead to pulmonary hypertension and right heart strain. It is not clear how many children with a clear history of sleep apnoea suffer serious long-term consequences, however.

It seems likely that a significant number of children have a transient tendency to apnoeic episodes at times when the tonsils and adenoids are relatively large, but do not develop sequelae and surgery can be avoided.

Chronic nasal discharge

Summary
Nasal symptoms in children: pathology

- *Upper respiratory tract infections*
  - universal
- *Foreign body*
  - usually unilateral
  - foul discharge +/− bleeding
- *Chronic rhinitis*
  - allergic
  - vasomotor
- *Adenoidal hypertrophy*

Sarah is 3 years old and has been referred to her local ear, nose, and throat department because of her persistently runny nose. Her mother reports that Sarah's nose is rarely dry and that she tends to be a mouth breather with noisy breathing, especially at mealtimes and during sleep. After closer questioning of Sarah's mother by the ENT surgeon it appears that there is no history of sleep apnoea. The rhinorrhoea is usually clear mucus but Sarah seems prone to upper respiratory tract infections during which her nasal discharge turns thick and green, often persisting for weeks after the infection. Sarah's family doctor has been reluctant to prescribe antibiotics in this situation, although antibiotics have rapidly eradicated the green discharge when used in the past. Sarah seems otherwise very healthy although her brother has asthma and eczema and there is a family history of hay fever and Sarah's mother is allergic to cat fur. The only family pets are two goldfish.

The ENT surgeon examines Sarah and considers her to be generally very well but with a snuffly nose and somewhat hyponasal speech. Her nasal mucosa looks rather pale and purple rather than the usual pink and there is excessive mucus in the nose although it is clear and watery. Checking Sarah's nasal airway by holding a cold, shiny, metal spatula under her nose reveals a reasonable nasal airflow through both sides as shown by two mist patches on the spatula when she breathes out. Sarah's throat and tonsils look healthy and unremarkable and her ears look fine. Examination of her neck shows small, palpable and mobile glands on each side which are not tender.

The runny nose is due to chronic rhinitis, which in Sarah's case with the family history may have an allergic cause. There is a tendency for small children to put things such as peas, beans, small toys, and other objects up their noses: these normally give rise to a persistent foul-smelling discharge with total obstruction of the airway on that side, often with bleeding. The other common problem at this age is enlarged adenoids. These are easily demonstrated on a lateral soft tissue X-ray of the nasopharynx.

In view of the history and the appearance of Sarah's nasal mucosa, the surgeon opts for treatment with topical steroids in the form of betamethasone drops, one drop each side of the nose twice a day. He explains to Sarah's mother that this treatment is a steroid, but used in this way for up to a month at a time is quite safe, without risk of systemic side-effects. He advises the use of the drops for 2–4 weeks as there may not be immediate benefit and suggests an X-ray of the adenoids only if there is no improvement. He also reassures Sarah's mother that this is usually a passing condition which children grow out of.

Further reading

Birrell, J.E. (1986). *Paediatric otolaryngology*, (2nd edn), (revised by D.L. Cowan and A.I.G. Kerr). Wright, Bristol.

Browning, G.G. (1994). *Updated ENT*, (3rd edn). Butterworth Heinemann, Sevenoaks, Kent.

Scott-Brown's Otolaryngology 5th Edition, Editor A.G. Kerr Vol. 6. Paediatric Otolaryngology. Editor: John N.G. Evans, Butterworths 1987.

13

Eye disorders

Introduction

Vision has the largest representation on the brain cortex occupying $\frac{1}{3}-\frac{1}{2}$ of its total area. Visual maturity is reached at about 8 years old but the most critical period of visual development is from a few weeks until 2 to 3 years of age. Any block to clear vision during this time has a profound effect on vision in later life: the earlier the visual block occurs the greater the eventual deficit will be. Therefore early diagnosis is essential in the management and outcome of childhood visual problems.

Early detection of treatable disease may be best achieved by screening of either the whole population or certain at risk groups.

At birth the anteroposterior diameter of the eye is 17 mm (24 mm in the adult) with most postnatal growth of the eye complete by 3 years. On day 1 the newborn infant will follow a face picture in preference to a picture where the same facial features are jumbled.

Clinical assessment

Vision in infants is measured by preferential looking and/or visually evoked cortical potential (VECP). Preferential looking relies on the fact that an infant will look to a striped black and white pattern in preference to an even grey target of the same overall luminosity. The spatial frequency of the stripes is increased until the infant shows no preference for the striped over the grey target. The VECP are electrical responses measured from occipital skin electrodes to an alternating or on/off black and white checkerboard pattern on a TV monitor.

A baby is examined for response to a visual target. A face, the baby's parent's or the examining doctor's is usually preferred to a torch. Visual attention and the ability of the eyes to follow a moving target is noted. A blind baby will often follow a face when the examiner is talking. Speech or a noisy toy or rattle may be needed to engage visual attention but then visual fixation should be central, steady, and maintained in the absence of auditory stimuli. Smooth pursuit movements are looked for in response to a moving target. The examiner's hand or a sticky patch is used to occlude either eye when measuring monocular acuity. Abnormal eye movements (nystagmus) are looked for and noted. Pupillary constriction to light is tested with the child in a dim room, moving a pen torch from one eye to the other.

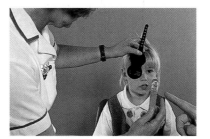

(a) (b)

Fig. 13.1 (a and b). *The Cover Test*: Jodie is asked to describe a small picture, the accommodative target, held about 0.3 of a metre in front of her. Any misalignment of the visual axes is noted. She has a right convergent squint. When the squinting eye is covered the fixing eye does not move. When the fixing eye is covered the squinting eye moves to take up fixation. The test is repeated with a larger target at 6 metres.

Summary
Amblyopia

- Reduced vision due to disuse of one or both eyes in infancy or childhood.

- Can occur in squint, refractive error, cataract, ptosis (drooping upper eyelid blocking vision) or corneal scar.

Fig. 13.2. *The Sheridan Gardiner Test*: The visual acuity of each eye is measured with the other eye occluded with a patch. Jodie identifies the letters, held at 3 metres or 6 metres pointing to the same letter on her key card.

Fig. 13.3. *The Snellen Test*: Jodie reads the letters with each eye from 6 metres. A minor degree of unilateral amblyopia may be detected by a reduced acuity in one eye by the Snellen test when the visual acuity is equal in each eye with a single letter test (Sheridan Gardiner test). The adult with normal vision is able to read the reference sized letter on the Snellen chart at 6 metres, ie 6/6.

At around 3 years old a child will match letters or pictures of reducing size held at 3 or 6 metres by the examiner. The child will hold a key card and pick out the letter or picture by pointing with a finger. The right and left eye are tested separately using a sticky patch to occlude the other eye. At 4–5 years old a child will read the letters on the Snellen test type held at 6 metres. Some normal children do not achieve adult visual acuity, 6/6, using the Snellen test until 6–7 years of age.

A family history of refractive error in childhood or squint in either parent or siblings should alert the family practitioner when seeing any child. Parental worries regarding vision should always be taken seriously. Vision testing in the non-verbal child is difficult and if any doubt exists about visual acuity referral to an experienced ophthalmologist is indicated.

Case studies
Squint

Scott was 2 years old when his mother first noticed his right eye turning in towards his nose. This squint was intermittent at first but had become constant in recent weeks which prompted his mother to take him to the family practitioner. On examination the general practitioner noted the reflections of a pen torch on his cornea while Scott looked at her. This showed quite definite asymmetry which was confirmed when the GP spoke to Scott and moved her head from side to side. This clearly showed the convergent right eye from the lateral position of the reflection of her pen torch on Scott's right cornea as against the central position of the reflection on his left cornea.

Squint is usually recognized by a parent but unilateral visual impairment due to refractive error can only be noticed when the child's vision is assessed monocularly. Childhood screening programmes would expect to pick up particularly refractive unilateral amblyopia at an earlier age and allow better treatment results, with better vision than an unscreened population although this has yet to be statistically proven.

Squint is a common problem occurring in approximately 4 per cent of all children. There is a strong familial incidence. It is usually noticed by parents and parental report of squint should always be taken seriously. It is detected by the corneal reflection test or by cover test. The cover test requires the child to fixate on an object while one eye and then the other is covered. Any movement of the uncovered eye is noted, see Fig. 13.1.

Scott was referred to an ophthalmologist. At $2\frac{1}{2}$ years his binocular visual acuity was 3/6 with the Kay picture test. He did not cooperate with monocular vision tests.

Fig. 13.4. *Examination of newborn for congenital cataract*: Normally a clear even red reflex is seen in each pupil when viewed through the ophthalmoscope. Cataract will show up as a black area in the red reflexes. Dim room lighting and a new battery in the ophthalmoscope is required!

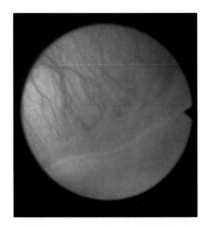

Fig. 13.5. *Retinopathy of prematurity*: At eight weeks of age this 800 gm birth weight premature infant has a ridge of neovascular tissue at the advancing edge of the retinal vessels with white avascular retina seen beyond the ridge. The ridge extends for 12 clock hours around the retinal fundus.

He had a constant right convergent squint and objected strongly to occlusion of his left eye. He fixed a small picture held at $\frac{1}{2}$ metre with his left eye but did not fix well with his right eye when his left eye was covered with an occluder. 1 per cent Cyclopentolate drops were instilled into both his eyes and after 30 minutes retinoscopy was performed. The retinoscope held at $\frac{2}{3}$ metre from each eye directed a slit of light from a linear bulb filament through a mirror into his pupil. The red reflex of the bulb filament image on the retina was seen by the ophthalmologist through a hole in the centre of the retinoscope mirror. The direction of movement of the red reflex in the pupil was noted as the light from the retinoscope was moved from side to side across each eye with correcting lenses held in front of Scott's eyes in a spectacle trial frame.

In a long sighted eye, the pupillary red reflex moves **with** the direction of movement of the retinoscope light across the patient's eye. In a myopic eye greater than 1 dioptre the pupillary red reflex moves **against** the direction of movement of the retinoscope light.

Scott was long sighted. His right eye had 4 dioptres of long sight and his left eye had 2 dioptres of long sight. (The power of a lens in dioptres is the reciprocal of its focal length measured in metres). Ophthalmoscopy was normal in both Scott's eyes.

Summary
Squint

- A misalignment of the visual axis of one eye.
- May be right, left or alternating.

Summary
Tests for squint

- corneal reflection
- cover test

He was given spectacles worn constantly to correct his long sight and his mother was instructed to patch his good left eye for 4 hours each day, ideally when he was looking at books and pictures or watching TV. When he returned to the eye clinic after 6 weeks he was tolerating his left eye patch for the full 4 hour period each day. Wearing his glasses his right convergent squint was noticeably less than without them. At 3 years of age his visual acuity was 6/6 in his right and left eye with the Kay picture test and his daily occlusion treatment was no longer required.

Double vision in the young squinting child does not occur due to **suppression** of the image of the squinting eye. If the squint is constant amblyopia rapidly ensues with reduced vision in the squinting eye. Surgery has a role in maintaining binocularity in an intermittent squint but is usually only cosmetic for a constant squint of long duration.

Retinopathy of prematurity

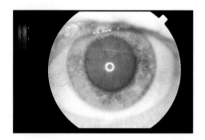

Fig. 13.6. A congenital cataract is seen as a black object blocking the pupillary red reflex.

Peter was born at 25 weeks gestation when he weighed 650 g. The pregnancy was complicated by recurrent vaginal bleeding. His mother went into labour spontaneously. She was given intramuscular Betamethasone to accelerate Peter's lung maturation (See Chapter 8, p. 167). He was born by spontaneous breech delivery and required intubation and was ventilated for six days. The highest oxygen concentration required was 0.5. His fundi were examined by binocular indirect ophthalmoscopy when he was 7 weeks old after dilating his pupils with $\frac{1}{2}$ per cent Cyclopentolate drops. At this time retinal vascularization had proceeded two-thirds of the distance from his optic discs to the ora serrata in each eye. Retinal vascular growth had stopped and a ridge of pink coloured neovascular tissue projecting slightly forward into the vitreous had formed at the front edge of the advancing retinal vessels for 360° around each eye (Fig. 13.5). The ocular media were clear and posterior pole retinal vessels appeared normal.

The developing retina first receives oxygen and nutrition from the choroidal vessels. As the retina develops in complexity, its metabolic requirement increases and blood vessels grow across the retina from the optic disc. This vascularization is not complete until full term. Extreme prematurity may interfere with normal vascularization of the retina resulting in retinopathy of prematurity (ROP) and sometimes severe visual impairment. All premature infants less than 32 weeks gestation or 1.5 kg weight at birth are screened for ROP using indirect ophthalmoscopy through dilated pupils.

The appearance was unchanged at each weekly retinal examination until he was 10 weeks of age when the height of the neovascular ridge was seen to be increasing in both eyes and traction on the ridge was seen temporally due to some anterior vitreous fibrosis in the left eye. There was some vitreous haemorrhage in the left eye and pre-retinal haemorrhage was seen in front of the ridge in each eye in the area of avascular retina. Two days later he was anaesthetized and the avascular retina between the neovascular ridge and the ora serrata of each eye was ablated with a diode laser using the binocular indirect opthalmoscope in an effort to reduce further neovascular proliferation and retinal traction from retrolental fibroplasia.

Laser treatment to the avascular peripheral retina in severely affected infants reduces the incidence of blinding traction retinal detachment by about 50 per cent. Most cases of ROP regress without treatment.

At one year of age he has good vision in his right eye but his left visual acuity is reduced. The retina is fully vascularized in both eyes but in the left eye there is a localized area of retinal detachment not reaching the macula but dragging it laterally. He may well develop a degree of myopia, worse in his left eye which will require spectacle correction. His mother is patching his right eye for 30 minutes each day in an effort to improve the left vision.

Poor vision in an infant

Kay was born by normal vaginal delivery after induction of labour at term plus two weeks. Her mother had been immunized against rubella in childhood, this was her first pregnancy. There was no family history of congenital or childhood cataract. Kay's pupillary red reflexes were not examined with the ophthalmoscope at birth. At 3 weeks her mother noticed that Kay did not look at her and that her eyes moved about aimlessly. She saw her family practitioner for a routine check at 8 weeks of age who found that Kay did not fix a face with her eyes. She did follow a light at 1.5 metres with both eyes open. She had a constant coarse roving bilateral nystagmus. When both eyes were examined with the ophthalmoscope a red reflex could be seen but no retinal details were visible. There appeared to be a central black area within the red reflex. The GP immediately telephoned the local ophthalmologist who saw her the same day.

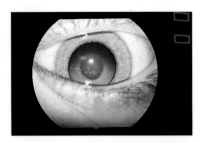

Fig. 13.7. *Retinoblastoma*: Louise's left eye before enucleation. A white tumour can be seen elevating the retina just behind the crystalline lens.

It is most important that every newborn is screened at birth for congenital cataract using the ophthalmoscope (Fig. 13.4). Congenital cataract is the commonest cause of treatable visual impairment or blindness occurring in infancy. The pupillary red reflex is seen through the ophthalmoscope held at about 30 cm from either eye of the baby after first focusing the examiner's free thumb held at a similar distance in front of the instrument. The ophthalmoscope is held close to the examiner's eye. Any congenital cataract will show up as a black area within the pupillary red reflex (Fig. 13.6). When diagnosis is delayed until the child's mother notices that her child does not see fully or its eyes develop irregular movement, or nystagmus, a normal adult level of vision will not be achieved. Causes of congenital and infant cataract are shown in Table 13.1.

Kay was admitted for a full neurological assessment. She appeared normal apart from the cataracts. Electroretinograms were performed with both eyes tested together, as was a visually evoked cortical potential (VECP) in response to a flash stimulus to both eyes. Both tests were normal. VECP to a pattern stimulus would have been reduced by the light diffusing effect of the cataract. Three days after her initial consultation she underwent left lensectomy and anterior vitrectomy under general anaesthesia. Her left eye was padded until right lensectomy was performed 6 days later. Next day she was later given 30 dioptre extended wear contact lenses for both eyes, focusing her eyes at about 1 metre.

At $4\frac{3}{4}$ years old her visual acuity was right 6/9 and left 6/36 with distance spectacle correction. Her right eye was patched 5 hours a day. At 5 years her visual acuity was right 6/9, left 6/12 wearing spectacles. She still has a slight fine irregular jerking of both eyes (nystagmus) and a small variable left squint. She attends a normal school.

Table 13.1.
Causes of congenital and infantile cataract

Isolated hereditary defect (autosomal dominant or autosomal recessive)
Down's syndrome
Metabolic: hypoglycaemia
hypocalcaemia
galactosaemia
galactokinase deficiency
Infection: rubella, mumps, measles (now rare with MMR immunization)
Prematurity
Ocular trauma
Uveitis

Squint and abnormal pupil in an infant

Louise was noted by a health visitor to have a left convergent squint at the age of 10 months. This was confirmed by the community orthoptist when her older sister was being seen for routine vision screening. An appointment was made for her to see the local ophthalmologist but before she could be seen her parents noticed what appeared to be blood vessels visible through the pupil at the back of her left eye. The family doctor was consulted and she arranged an urgent appointment with the ophthalmologist later that day.

On examination Louise was able to pick up tiny cake decorations with both eyes open but appeared to have no vision when her right eye was covered. The pupils only constricted when a pen torch was shone into the right eye not the left. A retinal detachment was visible in the left eye with her pupils dilated. Examination under anaesthetic the next day confirmed that the left eye had a total retinal detachment due to underlying tumour which obscured the optic disk from view. The right eye showed four retinal tumours not involving the macula, the largest measuring 5.7mm in diameter on ultrasonography which also demonstrated tumour calcification. A diagnosis of bilateral retinoblastoma was made. There was no family history and ophthalmoscopy of both parents was normal. Her older sister who also had a squint was examined regularly under anaesthetic until 5 years of age and no retinal tumours were detected up to that age. Her younger brother is also tumour free in both eyes at 16 months of age.

Louise was referred to a tertiary paediatric ophthalmology centre where her left eye was removed surgically and her right eye treated with lens sparing external beam radiotherapy and freezing treatment applied through the intact sclera. A further tumour in the right retina developed at 3 years and was treated by cryotherapy. Louise is now 6 and has 6/5 vision in the remaining right eye. She has a satisfactory left eye prosthesis.

Summary
Retinoblastoma

- may present as a squint, or a white 'cat's eye' pupil

- early diagnosis = effective treatment and care

- approximately half are hereditary (autosomal dominant inheritance)

- all children with a known family history require regular indirect ophthalmoscopy under general anaesthetic from birth, to 7–8 years of age

Retinoblastoma is a malignant ocular tumour of childhood occurring in approximately 1 : 20 000 live births. Forty per cent of cases are hereditary including those with a family history or where the tumour is bilateral. Louise is likely to have a germinal mutation of one allele at the retinoblastoma locus of chromosome 13. The analysis of DNA from a blood sample did not show the microdeletion on chromosome 13 seen in 10–15% of children with the hereditary condition. Louise carries a 40% risk of her children being affected with retinoblastoma.

Severe bilateral visual disability is rare in childhood (2–3/10 000 population) but early and accurate diagnosis is important to allow parental support, special educational needs, and correct genetic counselling. Most visually impaired children now attend normal local school but their visual impairment service teacher acts within the child's home and in conjunction with the class teacher. The visual impairment teacher advises and helps with stimulation of the impaired sight of the affected child and can help to arrange for enlargement of printed matter, use of ideal lighting, and low vision aids at home or at school. The visually impaired child is always encouraged to be as independent as is safely possible and to join in the same sports and pastimes as normally sighted peers and siblings, (see Ch. 7).

Further reading

Taylor, David. (1990). *Paediatric ophthalmology*. Blackwell Scientific.

Buckingham, T. (1993). *Visual problems in childhood*. Butterworth/Heinemann.

Forrester, J., Dick, A., McMenamin P., and Lee W. (1996). *The eye—Basic sciences in practice*. W. B. Saunders, London.

14

Gastrointestinal disease and nutrition

- The development and structure the gut

- Function of the gut

- Gastrointestinal symptoms

- Nutritional requirements in childhood

- Case studies
  acute infective diarrhoea
  chronic diarrhoea
  vomiting in infancy
  recurrent abdominal pain
  persistent jaundice
  failure to thrive
  faecal soiling (encopresis)

Gastrointestinal problems remain among the most common seen both in general practice and in hospital. Enteric infections are not only a major cause of morbidity in the UK but on a world-wide basis may result in approximately 2.5 million deaths in those under 5 years old each year. This enormous mortality is entirely preventable, caused as it is by a lack of access to clean potable water and sewage disposal by a large proportion of the world's population.

Problems such as recurrent abdominal pain, chronic diarrhoea and constipation, and failure to thrive although rarely life-threatening occur very commonly in children and form a major part of the workload of paediatricians.

Disease of the gastrointestinal tract and the ability of the child to digest and absorb nutrients and grow normally are very closely linked and therefore assessment of any child with gastrointestinal disease requires a knowledge of their nutritional intake and of their growth (see Fig. 14.1).

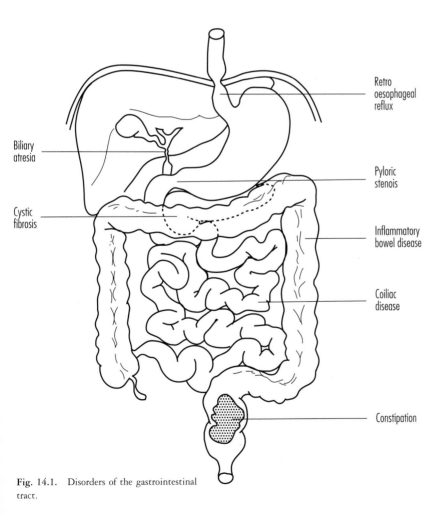

Fig. 14.1. Disorders of the gastrointestinal tract.

Development and structure of the gut

During the early stages of embryogenesis the mesodermal layers of the embryonic disc differentiate to form the muscle and connective tissue elements of the gastrointestinal tract, while the endodermal layer differentiates to form the epithelium of the intestinal tract and the parenchyma of the liver and the pancreas. With cranial and caudal extension of the embryonic disc the formation of the foregut gives rise to thyroid, respiratory tract, oesophagus, stomach and the upper duodenum, liver and pancreas with the mid gut extending from the lower duodenum to the distal third of the transverse colon, and the hind gut extending down to the anus. The primitive gut undergoes rapid elongation and extension through the umbilical orifice between 5–10 weeks and subsequently between 10–12 weeks the gut returns to the abdominal cavity and undergoes a counter-clockwise rotation of 270 degrees which leaves the small intestine positioned centrally with the caecum in the right iliac fossa and the colon lying in a lateral position. Failure of these developmental processes leads to major structural and functional abnormalities of the gut, many of which present in the neonatal period as surgical emergencies, (see Ch. 8, p. 162).

While the structure of the gut will be well advanced by the mid trimester of pregnancy, its digestive and motor function lags behind. This explains why in the very preterm infant enteral nutrition may be poorly tolerated.

Function of the gut

Digestive functions of the small intestine are carried out in two phases with an initial intraluminal phase where pancreatic enzymes digest large macromolecules into smaller units which can then be further digested by brush border enzymes into small molecules that can be directly absorbed. Most nutrients can be absorbed anywhere along the length of the small intestine although vitamin B_{12} and bile salts can only be absorbed from the terminal ilium. Food which is not digested, passes into the colon where luminal fluid is re-absorbed allowing the excretion of waste with the passage of a formed stool through the anus. Where digestion is disrupted the malabsorption of nutrients results in diarrhoea and poor growth.

Disturbances of the normal movement of food through the gastrointestinal tract may occur due to disturbances of the nervous and muscular control of intestinal function in the immature preterm infant or the child damaged by a severe brain insult. Problems with swallowing and regurgitation of food from the stomach are common in these children and where the excretion of waste by the colon is compromised symptoms of constipation are likely to develop.

The gastrointestinal tract also has a very important immune function. It protects the body from the passage of bacteria and viruses through the wall of the gut and develops tolerance to foreign proteins which are commonly found in ingested food. Breakdown of these mechanisms can lead to chronic intestinal infection or the development of food intolerance.

There is a link between 'psyche' and 'soma' in gut function and this is as true in children as it is in adults. The commonest manifestation is with recurrent abdominal pain, sometimes associated with vomiting ('periodic syndrome'), which occurs in school-age children in association with emotional triggers such as bullying, anxiety over parental stress, or exams.

In young children, control of food intake or output may be used as a means of asserting independance in the toddler who refuses food or is reluctant to undergo potty training

Gastrointestinal symptoms

A major part of the assessment of any child with gastrointestinal disease is an accurate history of their symptoms and in particular the timing of the symptoms relative to the ingestion of food. Common gastrointestinal symptoms are listed in Table 14.1 and it must always be remembered that the likely cause of each symptom will depend on the age of the child and that symptoms apparently related to the gastrointestinal tract may arise from disease elsewhere. It should be noted

Table 14.1.
Common gastrointestinal symptoms

Symptom	Cause	Age
Acute diarrhoea	Acute gastroenteritis	Any age
Chronic diarrhoea	Toddler's diarrhoea	6 months–5 years
	Food allergy	4 weeks–5 years
	Post gastroenteritis	Any age
	Coeliac disease	> 6 months
	Cystic fibrosis	Any age
Vomiting (effortless)	Gastro-oesophageal reflux	0–1 year (developmentally normal)
		Any age (developmentally abnormal)
Vomiting	Infection	Any age
(short onset)	Pyloric stenosis	1–4 months
	Poisoning	Any age
	Head injury	Any age
	Metabolic disease	Any age
Abdominal pain	'Irritable bowel syndrome'	2 years–adulthood
	Constipation	Any age
	Food allergy	Any age
	Urine infection	Any age
	Peptic ulcer	5 years–adulthood
	Inflammatory bowel disease	5 years–adulthood
Constipation	Idiopathic	1–12 years
	Hirschsprung's disease	0–6 months
Jaundice	Breast milk	0–6 months
(unconjugated)	Hypothyroidism	0–6 months
Jaundice (conjugated)	Biliary atresia	0–1 years
	Hepatitis syndrome	0–2 years
	Infectious hepatitis	Any age

that vomiting may be a symptom of infection, poisoning or head injury, and abdominal pain may occur in a child with pneumonia or diabetic ketoacidosis.

Physical examination and accurate assessment of the child's height and weight is important and with the aid of previous weights (if available) a centile chart should be used to assess whether the child is growing appropriately for his or her sex, ethnic origin, and parental size. The timing of growth failure may give important clues to the aetiology of any underlying pathology, as a failure from the time of birth may relate either to a feeding problem leading to reduced intake or alternatively may be seen in conditions such as cystic fibrosis. In coeliac disease, however, gastrointestinal symptoms do not develop until after gluten has been introduced into the diet and are thus very unlikely to develop before 6 months of age. A list of physical signs and their clinical significance is outlined in Table 14.2.

Nutritional requirements in childhood (see also Chapter 3, p. 48)

The early stages of life are an unprecedented period of growth and development and as a consequence, if the needs of the child are not met, failure in weight gain will occur very quickly and will be followed a few months later by failure of linear growth. The energy requirements of a newborn infant can exceed 120 kcal/kg/day (0.5 MJ) compared to the adult whose requirements are probably only about 40 kcal/kg/day (0.17 MJ). Similarly, the protein requirements of the newborn

Table 14.2.
Physical signs and clinical significance on abdominal examination

Physical sign	Clinical significance
Abdominal mass	Faecal mass in constipation
	Hydronephrosis
	Renal/Suprarenal tumour
	Crohn's disease
Splenomegaly	Haemolytic disease
	Neoplastic disease
	Portal hypertension
Hepatomegaly	Chronic liver disease
	Neoplastic disease
	Storage disease
Anal fissure	Chronic constipation
	Crohn's disease
Ascites	Hepatic cirrhosis
	Nephrotic syndrome
Abdominal distension	Air swallowing
	Malabsorption

infant are probably double those of the adult. Requirements are likely to be further increased in the ill child with cardiac or respiratory disease where there is an increased work of breathing, in children with malabsorption, where there is a loss of nutrients in the stool and in children with major trauma and sepsis where the increased requirements for tissue repair are compromised by the stress response which can lead to the inefficient use and breakdown of nutrients.

Case studies
Acute infective diarrhoea

Peter (aged 9 months) had been well until 36 hours previously. He was born at term and was bottle-fed exclusively until the age of about 3 months when weaning solids were gradually introduced by his parents. He is now on a full mixed diet and drinks formula milk from a beaker. His mother noted that he was slightly out of sorts, was a little irritable, and his previously formed stools had suddenly become loose and watery. His mother stopped his solids and gave him some juice to drink but over the subsequent 24 hours the diarrhoea continued, he became more lethargic and his eyes looked a little sunken.

Acute diarrhoea is a very common problem in young children and the history above is highly suggestive of an acute *gastrointestinal infection*. The most common pathogen is rotavirus (see Fig. 14.2) and over 60% of cases are associated with viral infections, although in older children bacterial pathogens become more common. Other commonly associated features of acute gastroenteritis are the development of vomiting and pyrexia and frequently there may be a family history of others with similar symptoms. In young children like Peter, however,

Summary
Causes of gastroenteritis

- Viruses
  - Rotavirus
  - Adenovirus
  - Astrovirus
  - Small round viruses
- Bacteria
  - *Campylobacter* sp.
  - *Salmonella* sp.
  - *Escherichia coli*
  - *Shigella* sp.

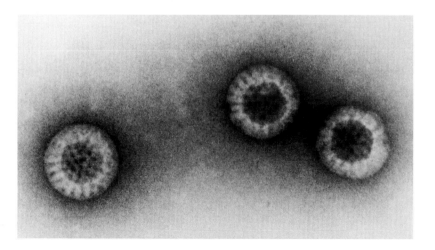

Fig. 14.2. Electronmicrograph of a rotavirus. (With thanks to Dr David A. Gregory.)

symptoms may be more severe as fluid loss from diarrhoea can quickly lead to the development of dehydration and the immature gut of the young child seems more susceptible to enteric infection, particularly in children who are malnourished or who have not been breast-fed.

Peter was seen and examined by his general practitioner Dr Ford and was found to have a slightly sunken anterior fontanelle and reduced skin turgor. Dr Ford advised that he be started on oral rehydration solution (ORS), to replace the fluid being lost to the body in diarrhoea. She explained to Peter's parents that treatment with antibiotics or antidiarrhoeal drugs was not indicated, and solid food could be re-started soon if Peter felt like eating. Arrangements were made for the health visitor to review Peter later that day.

The clinical findings suggest that Peter is between 3–5% dehydrated (see Table 14.3). The simplest way to rehydrate Peter is by the enteral route and he tolerated his ORS well. This is a specially prepared solution which contains sodium and glucose in a concentration of 60 mmol/l and 90 mmol/l, respectively, which through the action of a sodium/glucose co- transport protein on the brush border, greatly facilitates the absorption of water and electrolyte by the gut. As most

Table 14.3.
Clinical signs of dehydration

Dehydration	Sign
2–3%	Thirst
5%	Increased thirst Reduced skin turgor Sunken fontanelle Sunken eyes Oliguria
10%	All features above *plus* Hypotension Tachycardia Peripheral shutdown Apathy
15%	All features above *plus* Coma Anuria

Note. If the child is vomiting, > 5% dehydrated, or has poor social circumstances, admit to hospital.

enteric infections are self-limiting this allows the child to be rehydrated and supported without the need for intravenous therapy.

In children with more severe dehydration where vomiting precludes the use of oral rehydration solution, intravenous fluids are required and in the shocked child an infusion of plasma may be needed. Most dehydrated children have a normal serum sodium but if the child has received overconcentrated feeds there is a risk of hypernatraemia and great care is needed in the rehydration of these children as rapid electrolyte shifts may lead to convulsions. Traditionally, it has always been taught that children with diarrhoea should be fasted and gradually regraded back on to feeds. It has recently been shown that this in fact prolongs their illness and following rehydration the early introduction of liquid and solid feeds should be considered. Breast-fed infants should continue to feed through the rehydration process.

In the first 6 hours after starting ORS, Peter drank 500 ml and although his diarrhoea continued he became much more alert and his skin turgor improved. His bottle feeds were re-started, alternating with oral rehydration solution and shortly afterwards solids were re-introduced into his diet. Three days later his diarrhoea had resolved completely.

Most children make a complete uneventful recovery from acute gastroenteritis but in young children damage to the gut may lead to the development of a transient lactose intolerance or the child may become sensitized to cow's milk protein. Under these circumstances the child's diarrhoea is likely to persist beyond 3 or 4 days and the use of a diet free of milk (i.e. a soya or other hypoallergenic feed) may be indicated. In older children, dehydration develops less commonly but diarrhoea may be complicated by blood in the stool and with some strains of *E. coli* (e.g. *E. coli* 157) the release of a toxin may cause haemolytic uraemic syndrome which can progress to acute renal failure.

Chronic diarrhoea

The parents of John (age 3 years) become concerned because for 3 months his bowels were opening 3 or 4 times each day, passing stools which were occasionally soft and which sometimes contain undigested food matter, in particular peas and carrots. In other respects, John was a very well child who was growing entirely normally and did not appear ill to his parents. In particular, his loose stools have never contained any blood and were never precipitated by the ingestion of any particular foods.

Knowing exactly what bowel frequency is normal in young children is difficult as there is great variation in stool consistency, frequency, and sometimes also in

Summary
Management of diarrhoea

- Assess hydration
- Continue breast-feeding; stop milk feeds during rehydration (3–6 hours)
- Continue solids if wished
- Give oral rehydration solution
- Do not give antibiotics unless specific pathogen identified
- Do not give antidiarrhoeal drugs
- Inform parents of signs that would indicate dehydration

Summary
Causes of chronic diarrhoea

- 'Toddler diarrhoea'
- Post-gastroenteritis lactose intolerance
- Cow's milk protein intolerance
- Malabsorption
  - Cystic fibrosis (see Fig. 14.3)
  - Coeliac disease (see Fig. 14.4)

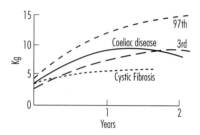

Fig. 14.3. Comparison of the pattern of growth in coeliac disease and cystic fibrosis.

colour. Prior to toilet training, when the parents are able to observe the stool readily, a lot of anxiety can be generated if the stool does not conform to their perception of what is normal. One of the commonest causes of chronic diarrhoea in this age group is a condition called 'toddler diarrhoea'. It has the features outlined above in a child like John who is healthy and thriving. Diagnosis will be made generally from the history but it is important that both past and present weights are accurately plotted on a growth chart so that normal growth can be confirmed. If the child is not thriving or atypical features are present, further investigation including full blood count, detailed dietary history, urine culture (to exclude urinary tract infection) and the measurement of antigliadin, antiendomysium, or antireticulin antibodies to screen for coeliac disease is required. If these antibodies are positive the diagnosis is then confirmed by jejunal biopsy.

John's parents were reassured after the nature of 'toddler diarrhoea' was explained to them. In particular, they were informed that he would almost certainly become toilet trained at the normal age and that this condition did not delay the attainment of continence. John was seen on one further occasion 3 months later where it was shown that his growth continued to be entirely normal and because his parents were reassured that there was no serious underlying cause of his loose stool, they were happy for follow-up to be discontinued.

Vomiting in infancy

Dawn, a 2-month-old infant, was breast-fed for 2 weeks then transferred to the bottle because her mother had sore nipples. Her mother has now brought her to the

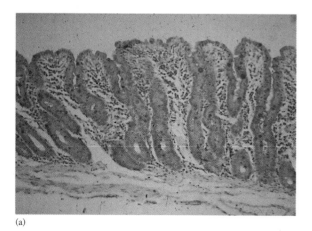

(a)

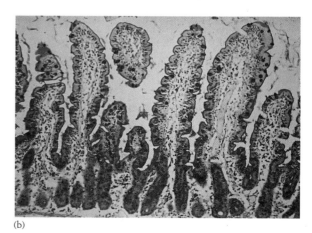

(b)

Fig. 14.4. Jejunal biopsy in a child with coeliac disease: (a) before and (b) following exclusion of gluten from the diet.

GP because of vomiting: she has a history of effortless regurgitation occurring almost immediately after feeds. Despite this she seems a happy baby and her regular weight checks have shown normal growth. The vomit has at all times been free of blood and she has not had any problems with her chest.

With a long history of regurgitation in a child who is otherwise well the most likely diagnosis is one of gastro-oesophageal reflux. This condition arises due to the immaturity of the lower oesophageal sphincter which allows the free regurgitation of milk in the young child. This is almost the physiological norm in young children and may be a source of anxiety for the parents where the regurgitation is more severe. The child may lose ingested calories and may therefore fail to thrive, or the regurgitation may cause oesophagitis or lead to aspiration of gastric contents. Severe reflux is often seen in children with cerebral palsy who may then aspirate because of the lack of a cough reflex. Diagnosis in most cases is based on the history. The presence of reflux may be confirmed by a barium meal or more accurately by a 24-hour oesophageal pH study, (see Fig. 14.5). The presence of oesophagitis can be assessed at oesophagoscopy. Investigations only tend to be required in more severe cases or where complications are suspected.

As Dawn was thriving well her mother was advised to thicken her milk feeds with a thickening agent and this resulted in some reduction in her regurgitation. The mother was also reassured that the natural history for gastro-oesophageal reflux was for their to be a natural resolution and as Dawn's diet became more solid and as she adopted a more upright posture it was likely that her reflux would settle spontaneously.

The treatment of children with more troublesome reflux is outlined in Table 14.4.

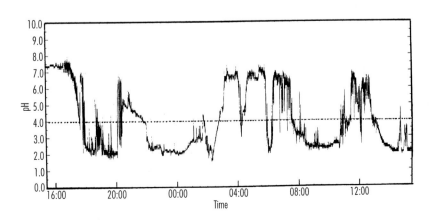

Summary
Coeliac disease

- Sub-total villous atrophy develops following the ingestion of gluten in the diet (see Fig. 14.4)
- Classically develops at about 9–12 months with weight loss and fatty stools
- This presentation is now uncommon due to the general reduction in the gluten load in children's diets
- Children present as toddlers with more vague symptoms such as anaemia, poor weight gain, or loose stools.
- Cured by treating with gluten-free diet

Summary
Differential diagnosis of vomiting in infancy

- *Overfeeding*
- *Oesophageal reflux*: effortless vomiting from birth
- *Pyloric stenosis*: projectile vomiting from 2–3 weeks after birth (male predominance)
- *Gastroenteritis* (associated with diarrhoea)
- *Bowel obstruction* (e.g. intestinal atresia leads to bile-stained vomitus and abdominal distension)
- *Metabolic disease*: family history, acidosis

Fig. 14.5. A segment of a 24-hour pH study showing prolonged episodes of reflux below a pH of 4.

Additional Information
Pyloric stenosis

- Projectile vomiting 2–3 weeks after birth
- Associated weight loss and constipation
- Diagnosed by clinical examination
- Cured by surgery (pyloromyotomy)

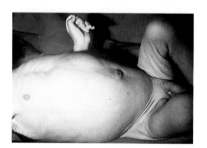

Fig. 14.6. In pyloric stenosis visible peristalsis is sometimes seen as 'golf balls' rolling across the upper abdomen from left to right. Note the visible peristalsis to the right of the midline and the obvious wasting in this infant.

Summary
Differential diagnosis of recurrent abdominal pain

- *Periodic syndrome*
- *Duodenal ulcer*: pain before meals and at night, eased by food, often family history
- *Chronic appendicitis*: (rare), signs suggestive of appendicitis
- *Inflammatory bowel disease*
- *Constipation*
- *Urinary tract infection*

Table 14.4.
Treatment of gastro-oesophageal reflux

Symptom	Treatment
Mild reflux	Reassure parents, *or*
	Feed-thickeners (Vitaquick or Carobel)
Moderate reflux	Feed-thickeners
	Alginates (Gaviscon)
	Prokinetic drugs (Cisapride)
Reflux with oesophagitis	H_2 blockers (ranitadine) added to above
Failure of medical treatment with the development of complications	Surgery (wrap or fundoplication)

In a child of this age with vomiting that develops acutely and leads to the development of dehydration and hypokalaemia one should consider the possibility of *pyloric stenosis* (Fig. 14.6). This occurs most commonly in males in the first month or two of life and can be diagnosed clinically by a test feed which reveals a pyloric tumour on abdominal palpation. If the vomit contains bile, assessment by a paediatric surgeon is urgently required. Vomiting may also develop secondary to infection from almost any source and will also occur in metabolic disease, cardiac failure, poisoning and following head injury.

Recurrent abdominal pain

Darren (aged 8 years) was brought to see his GP because of intermittent episodes of abdominal pain. This pain which occurred two or three times each week was central abdominal and generally colicky in nature. His bowel habit was normal and he seemed to be growing well. On closer questioning it was revealed that Darren was quite an anxious child and his mother reported that his pain did seem to be worse particularly first thing in the morning on weekdays but was absent over the weekends and on holidays.

Recurrent abdominal pain may affect up to 10% of children and in some, is brought on during stressful situations. In some children, the irritable bowel may be exacerbated by constipation, in others the ingestion of particular foods may lead to abdominal pain, while in some no clear cause can be found. The periodic syndrome is essentially a diagnosis of exclusion, although the features are quite characteristic in the presence of emotional turmoil. If the child's abdominal examination, full blood count, ESR, urea and electrolytes, and urine culture are all normal and there is a normal pattern of growth, this will rule out most major pathologies. If, however, the pain is atypical further investigation is required. Where the pain is largely epigastric, associated with wakening at night or a positive family history of ulcer disease, the diagnosis of duodenal ulceration or *Helicobacter pylori* gastritis should be considered and if the abdominal pain is association with diarrhoea or the passage of blood in the stool, inflammatory bowel disease should be excluded.

The simple investigations outlined above were all negative and it transpired that Darren was being bullied at school. Following a talk to his teacher by Darren's parents the problems were resolved and his pain slowly improved. He was encouraged to talk to his parents about problems encountered at school so as not to internalize them.

The basis for treatment of periodic syndrome is to recognize the factors that are causing and aggravating the abdominal pain and to try and eliminate them. In the above case it was alleviation of a stressful situation at school. In children with constipation this should be treated with laxatives or where it is suspected that certain foods are initiating the abdominal pain a trial of dietary exclusions should be considered. Very often the uncertainty of not knowing what is causing the pain creates alarm in both the parent and the child and reassurance may in itself help alleviate the child's symptoms.

Persistent jaundice (see also Chapter 8, p. 157)

At a 6-week check Mary (age 6 weeks) was noted by her GP to be jaundiced with yellow sclera. She was an otherwise healthy breast-fed baby who has grown normally and fed well from birth.

Jaundice is a concerning complaint at this age and should be taken seriously even if mild.

The commonest cause of persistent jaundice is related to breast-milk feeding. In this benign self-limiting condition the bilirubin is unconjugated and no specific treatment is required. Its importance lies in the fact that this should never be assumed to be the cause of the jaundice until other pathologies have been excluded. Where the bilirubin is conjugated a more serious pathology is likely to be present. In conjugated hyperbilirubinaemia the urine may be darker and on dip stick testing will contain bilirubin and where this is due to an obstruction to biliary flow the stools may be pale and lacking in pigment.

Summary
Differential diagnosis of persistent jaundice

First check if the jaundice is due to *conjugated* (bilirubin in urine and pale stools) or *unconjugated* hyperbilirubinaemia

- *Unconjugated*
  Breast-milk jaundice
  Hypothyroidism
  ABO or rhesus incompatibility

- *Conjugated*
  Biliary atresia
  Neonatal hepatitis
  Congenital infection
  Alpha-1 antitrypsin deficiency

Mary does indeed have dark urine and pale stools. She is referred urgently to hospital by the GP, where she is admitted for investigation. It is now important to make sure that she does not suffer from *biliary atresia* (a congenital absence of the bile duct) as this condition requires early recognition and treatment if long-term survival without liver transplantation is to occur. The diagnosis of biliary atresia is confirmed by a combination of liver biopsy, abdominal ultrasound, and an isotope excretion scan. Mary was referred to a centre with experience in hepatic surgery on children, for the formation of a porto-enterostomy (otherwise known as the Kasai procedure). Her jaundice cleared within a few weeks and she made a good recovery from surgery.

Although persistent jaundice due to serious pathology is fairly uncommon, it is essential that biliary atresia is recognized at an early stage (within 60 days of birth) if surgery is to have the best chance of success. When surgery is unsuccessful a progressive biliary cirrhosis leading to liver failure develops. Other pathologies which may present in a similar manner but which are not amenable to surgery include metabolic disturbance, (e.g. alpha-1 antitrypsin deficiency), chronic viral infections, or, as is often the case, an underlying cause cannot be found.

Failure to thrive (see also Ch. 5, p. 90)

Gemma, a 9-month-old infant, had been noted by her health visitor to be growing poorly and showed signs of having fallen from the 50th centile at 6 weeks of age to her present weight which lay just below the 3rd centile. Her mother, Jill, is a single parent living on her own in a council house with very little support; she has no relatives in the vicinity, does not now see the baby's father and is on income support. Jill had been brought up in foster care and the health visitor had been concerned about the feeding pattern. Gemma is not an easy feeder, she is fussy over solids, feeds at irregular times, and cries quite a lot. She wakes frequently at night and her mother takes her into her own bed. She is prone to upper respiratory tract infections and has had two episodes of diarrhoea when her mother was advised to stop solids. Jill, who is 19, finds her difficult to care for and frequently takes her to the GP. The health visitor requests an assessment by the community paediatrician.

Additional Information
Failure to thrive: definition

A fall off in weight across two centile lines on a 9 centile chart, rated to the baseline at 6–8 weeks and lasting over a month.

Failure to thrive is a common presentation but can be difficult to manage as it requires close attention to social, emotional, nutritional, and medical factors. Remember that infants' rate of weight gain varies and a low weight gain may be normal for some infants! Weight gain should be closely monitored in infants with feeding difficulty and the weight charted on the centile chart kept in the personal child health record. Large babies may *catch down* in weight to their genetic centile and the 6–8 week centile is a better predictor of weight at 1 year than the birthweight.

Most cases are not due to a medical condition as such and most often there is a combination of social, emotional and dietary factors which are responsible, through a mechanism of deficient calorie intake. Assessment and management require a team approach.

Gemma was assessed in the GP surgery by the community paediatrician and the health visitor together. The health visitor report was that Jill had a warm relationship with Gemma but was short-tempered and did not have patience with her at mealtimes. She ate irregularly herself and Gemma followed this pattern. Jill was tired from her disturbed nights and had no one who could take Gemma for short periods to give her a break. Currently, Gemma had no diarrhoea or vomiting but had a mild upper respiratory tract-infection. On examination, she was thin but well cared for. She did not appear pale and there were no specific abnormalities, in particular, no abdominal distension or evidence of congenital abnormalities. The paediatrician decided not to admit her to hospital but to perform limited outpatient *investigations*.

Urine microscopy and haemoglobin were both normal and it was decided not to investigate further as the picture strongly suggested a nutritional/social cause for the problem. The advice of a paediatric dietician with an interest in failure to thrive was sought, and she observed the feeding at home. She found that Gemma's calorie intake was seriously low as she had a limited intake of solid foods and could be difficult at mealtimes. Her milk intake was adequate but her solid food intake was irregular, as was her mother's, and as she had strong dislikes her mother mainly gave her low-calorie yoghurts and biscuits.

Jill agreed to alter her own eating pattern so as to have three regular meals and to introduce a wider variety of high calorie foods, using her own meals as a basis with the addition of snacks such as full-cream yoghurt and bananas. She was encouraged to take Gemma to the mother and baby group at the local family centre, where she met other mothers and learned from their experience. A baby-sitting service was also organized through the centre so that she was able to have some evenings out and enjoy a break from baby care.

A multidisciplinary approach is needed for the management of failure to thrive and it is essential that full attention is given to the support of the parents. In a

Summary
Differential diagnosis of failure to thrive

- *Inadequate intake*
  - Feeding mismanagement
  - Mechanical difficulty (e.g. cerebral palsy)
- *Excessive losses*
  - Vomiting
  - Diarrhoea
- *Malabsorption*
  - Cow's milk protein intolerance
  - Cystic fibrosis
  - Coeliac disease
- *Increased requirements*
  - Severe cardiac disease
  - Severe respiratory disease
- *Failure of utilization*
  - Metabolic or liver disease

Summary
Investigations in failure to thrive

- Urine microscopy for bacteriuria
- Haemoglobin level
- Stool culture *if chronic diarrhoea*
- Sweat test *if cystic fibrosis suspected*
- Antigliadin antibodies *if coeliac disease suspected*

small number of cases failure to thrive may be a presentation of child abuse and social services input should be sought, (see also Ch. 5, p. 90).

At follow-up 2 weeks later Gemma was gaining weight and her mother looked much happier. She was able to develop her own interests through contact with other mothers at the family centre and was also stimulating Gemma better. It was agreed that the health visitor would continue to watch the weight in the baby clinic and the dietician would review in 2 months time.

Summary
Differential diagnosis of soiling

- *Chronic constipation*
  History of constipation in infancy
  Often painful defecation before potty training
  Large amount of hard faeces in rectum
  No period of normal defecation

- *Primary behaviour disorder*
  Constipation not a prominent feature
  Other behavioural disturbances
  Faeces often hidden
  Often soiling is secondary— follows normal defecation

It is taken for granted that bringing up a baby is something that comes naturally to most mothers but particularly where the mothers themselves have had negative experiences from their own parents, where the parents are young, single or unsupported, or where there is evidence of mental illness or major tensions within the family this does not always occur. The most important aspects in the history of a child which fails to thrive, relate to the child's dietary intake and only when this has been confirmed to be normal should other causes be sought.

Faecal soiling (encopresis)

Mark, age 7 years (initial presentation given in Chapter 6, p. 108) is no better after advice on increasing his fruit and vegetable intake. His parents are becoming worried about the possibility of 'something wrong with his bowel'. His mother comments that Mark has never been fully toilet trained and that he is often very reluctant to go to the toilet. She also recalls that when he was younger he frequently screamed on defecation and on one or two occasions streaks of bright red blood were seen in his stool.

Soiling is stressful to the family and can be difficult to treat. Psychological factors are present in all cases although they are not always causal.

The history outlined above is classical of a child who has chronic constipation and subsequently developed rectal impaction with overflow incontinence. The initial constipation is compounded by the fact that defecation is painful and this sometimes leads to the development of anal fissures with bleeding. The discomfort on passing stools often leads to the development of a toilet phobia with further retention leading to continued constipation. With time, distension of the bowel leads to insensitivity of the rectum and liquid stool from above bypasses the impacted rectum leading to overflow and incontinence (see Fig. 14.7).

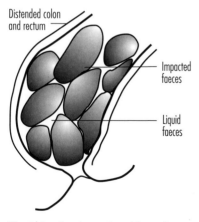

Distended colon and rectum

Impacted faeces

Liquid faeces

Fig. 14.7. Faecal retention with overflow.

Abdominal examination reveals palpable faeces, a loaded colon, and examination of the anus externally was normal although with a gentle rectal examination much faeces could be felt in the rectum. Paul was treated initially with quite high doses of Senokot and lactulose which helped empty his loaded bowel and the parents were given instructions on trying to encourage him to go to the toilet regularly. A 'star chart' and incentives were used to encourage regular toileting and when successful he was praised for his achievement. Over the subsequent 2 or 3 weeks Paul started again to go to the toilet and his soiling improved greatly. His parents were instructed to continue with the high roughage diet and to ensure that if further episodes of constipation developed they were treated quickly with courses of oral laxatives.

Most cases of simple constipation resolve quickly with treatment but if the problem is allowed to continue for a prolonged period or if the child is not motivated to get better (this often occurs in children with developmental problems), the soiling may become more chronic and relapse is quite likely. In addition, some children with severe constipation undoubtedly have a familial problem and in a small number, abnormalities of the enteric nerves in the bowel may be found. The most severe form of this type of problem is known as Hirschsprung's disease where there is aganglionosis which frequently leads to a presentation with constipation and intestinal obstruction in the first few weeks of life. However, Hirschsprung's disease usually presents in infancy and is rarely associated with faecal retention and overflow in mid childhood. Occasionally, a rectal biopsy may be required to exclude aganglionosis in intractable cases.

Further reading

Bisset, W.M. (1992). Disorders of the alimentary tract and liver. In *Textbook of Paediatrics*, (Ed. Campbell, A.G.M., McIntosh, N.), pp. 491–564. Churchill Livingstone, Edinburgh.

Milla, P.J. (1988). *Harries Paediatric Gastroenterology*. Churchill Livingstone, Edinburgh.

Mowatt, A.P. (1994). *Liver disorders in children*. Butterworth, Sevenoaks, Kent.

Walker-Smith, J.A. (1988). *Disease of the small intestine*. Butterworth, Sevenoaks, Kent.

15

Clinical genetics

Clinical genetics

Although a genetic disorder may affect an individual child, it is important to remember that the diagnosis of a genetic disorder has implications both for the immediate family and often for the wider family. It is this family orientated approach which characterizes clinical genetic practice. The human genome contains over 100 000 genes, mutations in which may cause specific genetic diseases. For some genes, different mutations cause different diseases. For example, different mutations in the dystrophin gene on the X chromosome band p21, can cause Duchenne or Becker muscular dystrophies, or X-linked dilated cardiomyopathy which has no skeletal muscle involvement. The range of possible genetic disorders is vast and although genetic disorders as a group are relatively common (about 1/20 individuals under the age of 25 have a genetic disorder), each individual disease may be rare. They also comprise a substantial paediatric workload: for example, congenital malformations, many of which are genetic, account for 2% of all livebirths but for 30% of bed-days occupied in a children's hospital.

The clinical geneticist fulfils three main roles in the care of children and families with genetic disorders. Firstly, he or she is involved with the clinical assessment and diagnosis of children with possible genetic disorders. Secondly, he or she is responsible for liaison with the cytogenetic and molecular genetic laboratories to arrange the appropriate investigations for the child and the family, and thirdly, and in parallel with this, he or she must counsel the family, explaining the diagnosis and its implications, both for the child and for other family members. Such counselling must include information about the options available to the family, to allow them to make appropriate health care and reproductive decisions, always remembering that for each family member, the decision is ultimately theirs alone to make on the basis of appropriate knowledge. This approach is known as non-directive counselling, and is fundamental to ethical clinical genetic practice.

Summary
Clinical genetics

- A family orientated approach
- 2% of all births have a congenital malformation (many are genetic)
- Human genome has > 100 000 genes

Summary
Role of clinical geneticist

- Clinical assessment and diagnosis
- Identify appropriate laboratory investigation
- Family counselling

Modes of inheritance
Single gene disorders

Genetic disorders may arise as a consequence of mutations in single genes (single gene disorders), from the interaction of single genes with other genes and environmental factors (multifactorial inheritance) or because of aberrations involving whole chromosomes or parts of chromosomes (chromosome aneuploidy). Single gene disorders are usually inherited according to the laws of Mendel. In autosomal dominant inheritance, the gene is located on an autosome, and only one of the two copies present is faulty. There is a 1 in 2 chance that the child of an affected person will inherit the faulty gene. Affected individuals in a family are often affected to different degrees (variable expression) and sometimes an individual may inherit a faulty gene but display no features of the disease (incomplete penetrance). Transmission of a genetic disorder from father to son implies autosomal dominant inheritance, transmission from father to daughter or from mother to son

Summary
Modes of inheritance

- Autosomal dominant
- Autosomal recessive
- X-linked
- Mitochondrial
- Multifactorial

KEY: symbols used in drawing
a family tree

☐ normal male

◯ normal female

■ affected male

● affected female

◪ carrier male (AR)

◖ carrier female (AR)

◉ carrier female (X-linked)

▣ normal transmitting male (X-linked)

◆ affected person (sex not specified)

◇ normal person (sex not specified)

or daughter may also occur but is also compatible with X-linked inheritance. In autosomal recessive inheritance, both copies of the gene must be faulty for the disorder to arise. The affected child therefore inherits one faulty copy from each parent, both of whom are therefore carriers of the condition. Autosomal recessive disorders are more common where there is parental consanguinity, and tend to affect members of one sibship only, with no previous family history. In X-linked inheritance, the faulty gene is on the X chromosome and therefore tends to affect boys severely (as they have only one copy of the X), and be transmitted by healthy females (who have a second, normal copy of the gene on their other X). Occasionally, girls may be affected by an X-linked recessive disorder because of Lyonisation. This is the process by which female embryos randomly inactivate one X-chromosome in each cell at around 12–16 days post-conception. If the embryo happens to inactivate mostly the X chromosome carrying the normal gene, then that female may manifest features of the disease. In addition to these forms of inheritance, some disorders arise because of mutations in genes coded on the mitochondrial genome. These disorders are often very variable and affect many different systems of the body (e.g. Kearns–Sayre syndrome) and are transmitted exclusively by the mother.

Multifactorial disorders

These arise from the interaction of many genes and environmental factors. Neural tube defects (spina bifida, anencephaly) and facial clefting fall into this category. Recurrence risks for siblings are usually between 2 and 10% depending on the disorder and its frequency in the population, and are much lower for more distant relatives.

Chromosome disorders

These may involve the loss or gain of a whole chromosome (e.g. the lack of one X chromosome in Turner's syndrome, or the presence of an extra chromosome 21 in Down's syndrome), or the loss of gain of part of a chromosome. Additional material present on a chromosome may arise from duplication of a segment, or because of translocation of material from another chromosome. If part of a chromosome is missing, it is said to be deleted. Part of a chromosome may also be turned around or inverted. Inversions often have no effect on the carrier, unless a gene is disrupted at the inversion breakpoints, but may cause reproductive problems because of pairing difficulties between a normal and inverted chromosome at meiosis. Translocations may be familial or sporadic (de novo). When two chromosomes have exchanged material, but none has been duplicated or deleted, the translocation is said to be a balanced reciprocal translocation. A balanced translocation usually has no clinical effects on its carrier, but that individual may suffer reproductive problems in terms of recurrent miscarriage or stillbirth, or the birth of a handicapped child if they transmit an unbalanced translocation (i.e. transmit one of the re-arranged chromosomes but not the other) to their potential offspring. A Robertsonian translocation occurs when two chromosomes become

Summary
Chromosome disorders

- Chromosome loss or gain (e.g. Turner's syndrome)
- Duplication of segment
- Deletion of segment
- Inversion of segment
- Translocation (usually no effect if balanced)
- Non disjunction (e.g. trisomy 21 Down's syndrome)

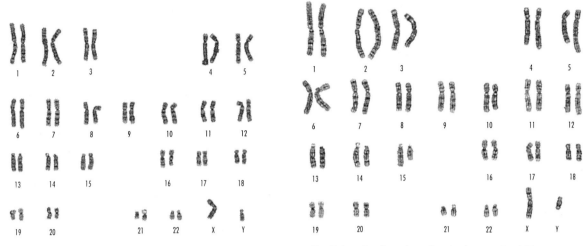

Fig. 15.1(a) Normal Karyotype. (b) Balanced reciprocal translocation between 5 and 15.

joined end to end. The clinically most important form of Robertsonian translocation involves chromosomes 14 and 21, as a parent carrying this translocation has a high risk of having a child with Down's syndrome. Translocation Down's syndrome accounts for only 5% of cases, however, the majority being due to failure of separation of the chromosome 21 pair at maternal meiosis (nondisjunctional trisomy 21).

Case studies
Marfan syndrome (autosomal dominant)

Jenny, aged 12 (III.1) presents with pectus excavatum, tall stature, and a 6 month history of low back pain, particularly on sitting for long periods at school, (see Fig. 15.2). Examination reveals that her height is just above the 97th centile, and she has a very mild lumbar scoliosis. She has mild joint laxity, although she has never suffered a dislocation, her thumb and 5th fingers overlap when wrapped around the opposite wrist (Walker–Murdoch wrist sign) and she can extend her thumb well beyond the ulnar border of the palm (Steinberg thumb sign) (see Figure 15.3). These last two signs are indicative of arachnodactyly. She also has a high arched palate. In the family history, her father (II.5) is tall and has myopia. His brother (II.2) died suddenly at work on his fishing boat at age 25. Another brother is tall, but well and two sisters are well. Their father died from a heart attack at age 58 and is known to have had poor eyesight.

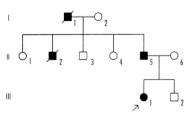

Fig. 15.2 Jenny's family tree.

Fig. 15.3 Steinberg thumb sign, (thumb extends beyond ulnar border of palm).

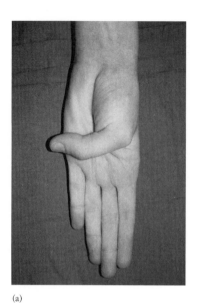

(a)

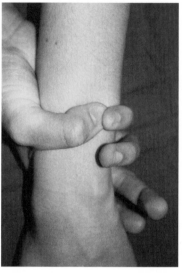

(b)

Jenny has many skeletal features suggestive of Marfan's syndrome. She has tall stature, arachnodactyly, scoliosis, and a high arched palate. This does not of itself allow a diagnosis of Marfan's syndrome to be made. Marfan's syndrome is an hereditary disorder of connective tissue with multisystem effects. There are many overlapping disorders with less serious consequences. Therefore, additional evidence should be sought at this stage from the family history, and from careful evaluation of other systems with expert examination and investigation where appropriate. The family history of tall stature, myopia, and sudden early death is highly suspicious. Investigation of the uncle's death certificate reveals that he died from a dissecting ascending aortic aneurysm. Clinical investigation of the father reveals that he too has skeletal features of Marfan's syndrome. This family history makes a diagnosis of Marfan's syndrome very likely, and the proband should therefore have an opthalmological assessment for evidence of lens subluxation or increased axial length of the globe, and a cardiac assessment including evaluation of the aortic root diameter by ultrasound, as aortic root dilatation is an early indicator of potential cardiovascular complications of Marfan's syndrome. When the diagnosis of Marfan's is confirmed on clinical grounds, consideration must be given to prophylactic therapy, as beta-blockers may slow the dilatation of the aortic root and therefore reduce the risk of aneurysm rupture. Regular surveillance by cardiac ultrasound will also be required.

Making a diagnosis of Marfan's also carries occupational and sporting implications, and is likely to cause life assurance difficulties which may affect her future ability to arrange a mortgage. It is therefore important both to avoid overdiagnosis and under diagnosis of the condition. Marfan's syndrome is an autosomal dominant disorder, although between 15 and 35% of cases are said to be due to new mutation. The fibrillin gene on chromosome 15 is implicated in most cases, but many different mutations have been found in different families. Mutation detec-

Summary
Marfan's syndrome

- Autosomal Dominant

- Up to 35% have no family history

- Fibrillin gene on chromosome 15

- Skeleton (tall, scoliosis, arachnodactyly)

- Ocular (lens dislocation, myopia)

- Cardiovascular (aortic dilatation, aortic aneurysm)

- Prophylactic beta-blockers, aortic surgery

- Echocardiographic follow-up

tion is therefore difficult, although linked DNA markers can be used to track the gene through a family when there is an extensive family history, and can sometimes be used to help confirm affected status in relatives. In Jenny's family, there appears to be a family history, and there is therefore a need to assess other family members, who may also have the disorder and in whom timely intervention might reduce morbidity and avoid premature death, so called tertiary prevention, (see Ch. 4, p. 68). On the other hand, the diagnosis may also cause financial difficulties for some family members, who may be angry at the bearer of bad news. There may also be conflicts of interest, if one family member demands that no details of his illness be disclosed to relatives, while his children nevertheless have the right to be informed of their risks and to seek appropriate medical advice and intervention. Following up the family history requires considerable tact and discretion.

Cystic fibrosis (autosomal recessive) (see also Ch. 22, p. 378)

Anita (III.3), aged 13 has cystic fibrosis (CF) and was diagnosed before the CF gene was identified. Her mother is an only child, but her father has a younger sister Jean (II.7) who is just married and would like a family. Jean is concerned about the risk of having a child with CF, (see Fig. 15.4).

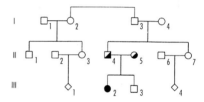

Fig. 15.4   Anita's family tree

When a child is diagnosed as having cystic fibrosis (CF), there are many ramifications in the wider family. Cystic fibrosis is an autosomal recessive disorder caused by mutations in a gene on chromosome 7 which encodes a membrane bound chloride ion channel. Over 400 mutations in the gene are known, but fortunately 4 account for 85% of all abnormal CF genes in the UK population (different populations have different mutation frequencies which has implications for carrier testing). 1/25 people with no family history in the UK are carriers of a CF mutation. In an autosomal recessive disorder, both parents are obligate carriers, there is a 1 in 4 chance that any child will be affected, and a 2 in 3 chance that any normal child will be a carrier. The family implications when cystic fibrosis is diagnosed start with the immediate family and then extend to the wider family. A major concern frequently expressed by the parents of the index case early on, is could my other child(ren) be affected? It may be that there are no clinical features of cystic fibrosis, but if the sibling is younger, then clinical features may not yet have developed—the average age at diagnosis was 3.5 years in one series but is coming down as awareness of the condition increases. Presentation may range from prenatal meconium ileus to recurrent chest infections in adult life. Congenital Bilateral Absence of the Vas Deferens, a cause of male infertility, is a related disorder often caused by mutation in the CF gene. It may be possible to reassure parents by demonstrating a negative sweat test. However, in most patients, analysis of the mutations in the cystic fibrosis gene on chromosome 7 will have formed part of the diagnostic process, and the result will be known to the parents.

Summary
Cystic fibrosis

- Autosomal recessive

- CF gene on chromosome 7 codes for a chloride channel

- 4 common mutations account for 85% of all mutations in North Europeans

- over 400 mutations known

- 1/25 North Europeans are carriers

Although Anita was diagnosed before the gene was identified (1989), she has been tested and found to be homozygous for the ΔF508 deletion, the most common mutation in the Northern European population. As expected, both her parents are heterozyous for this mutation (carriers). Anita's younger brother Andrew (III.4) frequently has a cough and her parents ask for him to be tested. He has previously had a normal sweat test.

It is unlikely that Andrew has CF if he has a normal sweat test, but his parents want additional reassurance. If we carry out DNA testing, and Andrew has at least one normal copy of the CF gene, then this will be very reassuring that he does not have CF. There is a 2/3 chance that we will find that he is heterozygous for the ΔF508 mutation. He will now have had a carrier test, whose result will be of no direct benefit to him until he reaches reproductive age, without formal informed consent. If this carrier result is disclosed, will there be stigmatization (unjustified as being a carrier has no adverse health implications) and how can the future transmission of the result to the child be assured with appropriate supporting information, at an appropriate age? Does this responsibility lie with the parents or the health professionals?

Anita's paternal aunt, Jean (II.7) has recently married and is now pregnant. When she learns that her niece has cystic fibrosis, she is keen to have carrier tests for herself and her husband.

Testing Jean is relatively straightforward—she has a 1 in 2 chance of carrying the mutation which has caused CF in her niece. If we know Anita's mutations, we can easily test Jean for them. Permission should be sought from Anita's parents to use information about their child's mutations in this way. Usually, parents will agree to this, but a conflict of interest issue could arise if they refuse. If Anita's mutations had not both been identified, it might still be possible to determine whether Jean is a carrier by linkage analysis. This is where we follow the inheritance of the chromosome 7 carrying the unknown CF mutation through the family using linked polymorphic DNA markers. A linked marker is a non pathogenic DNA variation which is physically close to or within a disease gene on a particular chromosome and can therefore be used to trace the passage of that particular copy of the gene from parent to child in a family. To carry out such an investigation, it is necessary to collect samples from Anita, her parents, her father's parents (I.3 and I.4) and from Jean herself. The older generation may be reluctant to become involved, as they may feel that blame will be apportioned if it is determined from which grandparent the mutation came. Once again, such investigations require careful counselling, tact and consideration. If Jean does prove to be a carrier, then her husband's carrier test becomes of great importance. If he is a carrier also, there is a 1 in 4 risk to the baby, but prenatal diagnosis by DNA analysis of a chorionic villous biopsy taken at 11–12 weeks gestation would be possible. If he is not a carrier, such testing is unnecessary. Unfortunately, for an individual such as Jean's husband, II.6, with no family history, the carrier risk can never be zero, because even once we have found no evidence for the 4 common mutations, he might still carry a rare mutation for which we cannot test. However, if he comes from the mainland UK white population, then his risk of being a carrier after a negative 4 mutation test will be only 1 in 150. The risk that the baby will be affected is then 1 in 600. This low risk is reassuring to most couples, but screening of the newborn infant using the IRT test (immuno-reactive trypsin) is advisable. The couple should be warned that the IRT test is sensitive but can occasionally produce false positives, such that follow-up with a sweat test at a later date might be required.

In the course of taking the family history, we learn that Anita's father's cousin, Anne (II.3) is also expecting a child. We ask the family if she should be contacted to be offered counselling and carrier testing, but the family refuse and warn us not to contact her.

Anne has a 1 in 8 risk of being a carrier. Why would the family not permit contact with Anne? It may be that there has been a family conflict, and that one branch does not want the other branch to be aware of their medical problems. Another possibility that should be borne in mind in all family tracing of this type is that of undisclosed non-paternity in any generation. The great-grandparents may be aware of such an event which has been a family secret for many years, and which may not be known to the younger generations. Such circumstances require very delicate handling. It is very rarely helpful to a family to open such secrets to public scrutiny after many years lying dormant.

X-linked disorders

Duchenne muscular dystrophy has a typical X-linked inheritance pattern, while Fragile-X syndrome illustrates the characteristics of an important class of genetic disorder—those due to unstable triplet repeat or 'dynamic' mutations. Different problems arise in genetic counselling as a result.

Duchenne muscular dystrophy (X-linked recessive) (see also Ch. 7, p. 129)

Duchenne muscular dystrophy is an X-linked muscle wasting disorder. Affected boys present with delay in learning to walk or with motor difficulties including difficulty keeping up with children of the same age in walking or running and difficulty going up and down stairs. It is a progressive disorder, associated with pseudohypertrophy of the calves (not always evident in the early stages). Respiratory muscle involvement leads to death by early adult life. In about 60% of cases, a deletion of the dystrophin gene at Xp21 is detectable by routine multiplex PCR testing of lymphocyte DNA. The remaining cases have other types of mutation in dystrophin which are more difficult to detect. A related disorder, Becker muscular dystrophy is clinically very similar, but is much milder and is compatible with continued mobility without a wheelchair for some years into adult life and with a longer lifespan. It is also due to mutation or deletion in the dystrophin gene. Muscle biopsy usually shows absence of dystrophin in Duchenne muscular dystrophy, but reduced quantity of abnormal dystrophin in Becker muscular dystrophy.

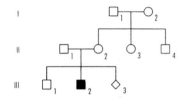

Fig. 15.5   Ben's family tree

Ben (III 2) was suspected of having DMD when his mother became concerned about his delay in walking. His clinical features (see Chapter 7), raised creatine kinase (CK) and muscle biopsy findings confirmed the diagnosis. His mother already has one other child but is pregnant again and wonders whether the pregnancy will also be affected, and whether it is possible to carry out prenatal diagnosis, (see Fig. 15.5).

There is no family history of DMD but Ben's mother may still be a carrier of DMD. Although Ben could be a new mutation, two thirds of mothers of DMD boys where there is no family history are carriers. This proportion arises from the mathematics of X-linked inheritance if affected boys do not have children of their own because of early death or other reasons. Some carriers of DMD have a mildly raised serum CK (because of Lyonisation) and therefore it may be useful to measure Ben's mother's CK. Unfortunately, many carriers of DMD have a normal CK, so a normal result is not reassuring. In addition, CK tends to fall in pregnancy, and is therefore even less useful in Ben's mother's case at the moment. However, if Ben has a dystrophin deletion detected on DNA analysis of his blood sample, then we can test his mother's blood by fluorescin in situ hybridization (FISH) or DNA doseage studies to see whether she has one or two copies of the part of the dystrophin gene deleted in Ben. If she is a carrier, there is a 1 in 4 chance that her pregnancy will be an affected boy, and prenatal diagnosis could be carried out by DNA analysis of a chorionic villous biopsy or cultured amniocytes. If no deletion can be detected in Ben, then carrier testing during pregnancy is not possible. Ben's mother could opt to terminate all male pregnancies, knowing that some or all of the male fetuses will actually be normal. We could also take a blood sample from Ben's healthy brother, and if the fetus is male, test using linked DNA markers to see if it has inherited the same X chromosome as Ben or as his brother. If it has the same chromosome as Ben, then it may be affected, but only if Ben's mother is a carrier. Because the dystrophin gene is very large, linked markers may give incorrect results because of crossover. There are some very difficult and complex issues to be dealt with, and Ben's mother may have a very

Summary
Duchenne muscular dystrophy

- X-linked recessive
- Allelic with Becker Muscular Dystrophy
- Dystrophin gene at Xp21
- 60% have detectable deletion in dystrophin
- 2/3 of mothers of an isolated case are carriers
- Test female relatives' carrier status by CK or DNA based tests

difficult decision to make. The importance of early DNA investigation and appropriate counselling and support can easily be seen.

As in all genetic disorders, the implications do not necessarily end with the nuclear family. Just as there is a 2/3 risk that Ben's mother is a carrier, there is a 1/3 chance that Ben's maternal grandmother (1.2) is a carrier, and also a possibility that Ben's maternal grandfather (I.1) has gonadal mosaicism for the dystrophin mutation/ deletion. In either event, Ben's aunt Sally (II.3) may be a carrier, and there is an obligation to discuss these issues with the wider family and to offer them carrier testing should they so wish. Because of the relentless progression of DMD from an apparently healthy baby boy, to a disabled child and then a dying teenager, families suffer immense emotional trauma which deserves sympathy and understanding from those involved in their management and counselling.

Fragile X Syndrome (X-linked dominant)

James (IV.5), aged 2 years, attends the clinic for assessment of developmental delay and 'hyperactive behaviour'. He is found to have marked speech delay, mild motor delay and a short attention span. He has macrocephaly with a prominent forehead and mild joint laxity. His mother Carol reports that her brother, George (III.6) had similar speech and behavioural problems as a child, and still has poor communication skills as an adult. He lives at home with his parents and attends a day centre. His disability has been attributed to birth injury. Carol's older daughter, Kelly aged 8 (IV.3) was also late in learning to speak, but following speech therapy, she attended a normal school. Because of learning difficulties, she has been kept back a year and has learning support, (see Fig. 15.6).

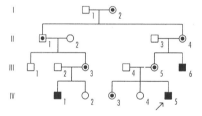

Fig. 15.6 James' family tree.

The family history of an affected boy, an affected maternal uncle and mildly affected girl suggests X-linked inheritance. The second commonest cause of learning difficulties after Down's syndrome is the X-linked disorder, Fragile X syndrome, which affects around 1 in 1000 liveborn males. Fragile X patients often have a characteristic behavioural and developmental phenotype (short attention span, speech and language delay, poor sensory integration, dysgraphia, dyscalculia). Physical features including macrocephaly, prominent forehead and chin, prominent ears, post-pubertal macroorchidism and sometimes joint laxity, mitral valve prolapse may be seen. There is overlap with other learning disorders and with the normal range, and the diagnosis is difficult to make solely on clinical grounds. Any child with a compatible learning disability, developmental or behavioural pattern should be tested for Fragile X, particularly if there is an X-linked family history. X-linked disorders usually affect boys and are passed on through healthy unaffected women. In Fragile X syndrome, affected females and normal transmitting males (male 'carriers') are often found, because of the unusual characteristics of the unstable mutation which underlies the disorder. The syndrome takes its name from the appearance, first noted in 1979, of the X chromosome in some cells of affected individuals when cultured under special conditions (usually folate deficient medium). This type of testing was unreliable, particularly for detecting female carriers and normal transmitting males. The discovery of the Fragile X A gene and the presence of a triplet repeat expansion mutation within it in 1987 revolutionized testing. Normal individuals have up to around 50 copies of a triplet code near the beginning of the gene, carriers (i.e. normal females or normal males who have affected descendents) have between 50 and 200 copies (known as a premutation) and affected individuals (male or female) have more than 200 copies (full mutation). A triplet repeat length of greater than 200 copies is associated with methylation of the control region of the gene, and no expression of its product. The variation in the severity of the disorder seen in families is associated with variations in the number of copies of the triplet repeat: when passed on by a female, triplet repeat lengths of more than 50 copies are unstable, and the child may have many more repeats than the mother. Full mutations also display somatic instability (different copy numbers in different tissues or in different cell lines in the same tissue), (see Fig. 15.7).

DNA testing of blood samples from James, Kelly, their mother Carol and maternal uncle George was carried out, (see Fig. 15.8). James and George each have a large expansion (full mutation) of the triplet repeat region of the Fragile X A gene, confirming that they both have Fragile X syndrome. Kelly has inherited a normal copy of the gene from her father and a moderately large expansion of the gene (full mutation) which is displaying somatic instability, from her mother. Her mother has a normal copy, and a mildly expanded copy (pre-mutation). Kelly's mother is therefore an unaffected carrier of a small Fragile X pre-mutation, but Kelly has a full mutation and therefore her learning difficulties are probably due to Fragile X syndrome, despite the fact that she is a girl. She has responded well to speech therapy and is therefore clinically less severely affected than her uncle George. Girls are usually more mildly affected than boys even when the mutation size is similar because of Lyonisation. Making the diagnosis was very helpful to this family—they felt relieved of guilt that James's, Kelly's and George's

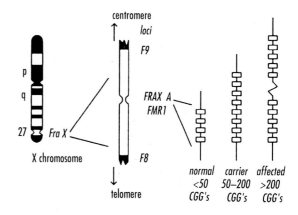

Fig. 15.7

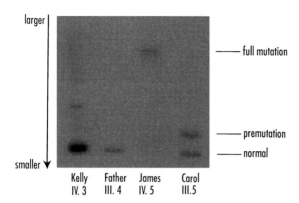

larger

— full mutation

— premutation
— normal

smaller

| Kelly | Father | James | Carol |
| IV. 3 | III. 4 | IV. 5 | III.5 |

Fig. 15.8 Southern blot of DNA from James' family. Lymphocyte DNA has been digested with an enzyme which cuts on either side of the triplet repeat region. After electrophoresis in which smaller fragments of DNA move further down the gel than larger ones, the DNA is blotted onto a nylon membrane and probed with a radioactive DNA probe which recognises and sticks to the triplet repeat region. The location of the probe, and therefore of DNA containing the Fragile X mutation is detected by exposure of the nylon membrane to an X-ray film (autoradiography). The nearer the top of the film the probe signal is, the larger the fragment, and therefore the greater the number of triplet repeats in that patient's DNA.

problems were somehow due to bad parenting or to things their mothers had done wrong in pregnancy.

Kelly has a younger sister, Katy aged 2 years (IV 5). She is developing normally, but the parents ask whether she could also have a Fragile X test. Should we agree to this request?

There is a difference between testing Kelly and testing Katy. Kelly has symptoms of speech delay and learning difficulties, and testing is therefore diagnostic: i.e aimed at finding the cause. Katy has no symptoms, and testing would therefore be to determine whether she is a carrier of Fragile X (rather than affected by Fragile X), with a risk of having affected children in the future. Because the issue is one of future reproduction, it can be argued that Katy should have the right to decide whether she should have a test when she is old enough to understand the issues. This puts a burden on both the family and the hospital to arrange appropriate follow-up for her in 10–15 years time.

James's second cousin, Craig (IV 1) is an 8 year old boy who, despite being kept back a year at school, is disruptive in class and has been very slow to develop reading and writing skills. The school feels that part of the problem is a lack of discipline at home, and the parents are very distressed by this.

DNA testing shows that Craig also has Fragile X full mutation, confirming that he has Fragile X syndrome. His disruptive behaviour is a consequence of his poor communication skills and short attention span. Appropriate schooling with specific attention to the characteristic learning difficulties of Fragile X syndrome will improve both his progress and his behaviour. Making the diagnosis comes as a great relief to his parents and eases conflicts with the school. Looking at the family tree, you can see that Craig is connected to Kelly's branch of the family through his maternal grandfather II 1. He is intellectually normal, and has a very small expansion of the gene, of premutation size. Premutations do not affect the individuals who carry them, and II 1 is therefore said to be a 'normal transmitting male'. The occurrence of such individuals can make the interpretation of family trees difficult at first glance, as they appear to contradict the rules of X-linked inheritance. DNA testing shows that this is not actually the case, but that it is the variability and instability of the Fragile X mutation when transmitted by a mother to her child that is causing the difficulty.

In investigating this family, you will have noticed that family history was the key to finding the diagnosis—the behavioural and learning difficulties are non-specific and can only be seen to be compatible with Fragile X syndrome in retrospect. When assessing a child with learning difficulties, always consider Fragile X syndrome and take a family history bearing in mind the difficulties illustrated by James's family.

There are a number of other genetic disorders which have been found to be caused by triplet repeat expansion mutations. These all have the characteristics of being variable within families, often demonstrating anticipation (earlier onset of more severe disease in succeeding generations), and being more likely to be more severe when transmitted by one parent in particular (e.g. mothers who have myotonic dystrophy are more likely to have severely affected children than fathers who have myotonic dystrophy: Huntington's disease is more likely to begin at an early age if it is inherited from the father). It is likely that other genetic diseases may yet be found to be caused by similar unstable mutations, which will mean that testing for those disorders will be easier than it is for disorders such as Marfan's syndrome, where each family has a different type of mutation, or cystic fibrosis where certain mutations are common but others are rare.

Additional Information
Disorders caused by unstable triplet repeat mutations

Disorder	Location	Inheritance
Fragile X A	Xq27	XLD
Fragile X E	Xq27	XLD
Spinobulbar muscular atrophy (Kennedy Disease)	Xq11–12	XLR
Huntington's Disease	4p16	AD
Myotonic Dystrophy	19q13	AD
Spinocerebellar ataxia type 1	6p	AD
Spinocerebellar ataxia type 3 (Majado-Joseph Disease)	14q24–32	AD
Dentatorubral pallidoluysian atrophy	12p13	AD
Friedreich Ataxia	9q13–21	AR

XLD = X-linked dominant
XLR = X-linked recessive
AD = Autosomal dominant
AR = Autosomal recessive

Additional Information
Genetic Counselling

- Family implications — Genetic disorders affect families, not individuals

- Whose information is it? — Confidentiality of genetic information is crucial to avoid stigmatisation

- Conflict of interest — Family disagreements may cause problems if one part of a family will not permit disclosure of information useful to another branch

- Social implications — Insurance and occupational problems can arise from prediction and diagnosis of genetic disease

- Responsibilities of health professionals with regard to family — Informed choice, non-directive counselling, appropriate follow- up, resolution of conflicts of interest

- Careful counselling, not indiscriminate testing

Further reading

Connor J.M., Ferguson-Smith, M.A. (1993). *Essential medical genetics*. Blackwell Scientific Publications.

Bradley, J., Johnson, D., Rubinstein, D. (1995). *Lecture notes on molecular medicine*. Blackwell Scientific Publications.

Harper, P. (1993) *Practical genetic counselling*. Butterworth-Heinemann Ltd.

Young, Ian, D. (1991). *Introduction to risk calculation in genetic counselling*. Oxford University Press.

Acknowledgements

Figures: Karyotypes: Courtesy of Dr David Couzin, Cytogenetics Laboratory, Department of Medical Genetics, Aberdeen Royal Hospitals NHS Trust

Fragile X DNA autoradiograph: Courtesy of Dr K. Kelly, Scottish Molecular Genetics Consortium, Aberdeen Laboratory, Aberdeen Royal Hospitals NHS Trust.

16

Growth and the endocrine system

Normal growth and puberty

The study of the process of normal growth and its variations is as crucial a part of the study of child health as intellectual development. It is important to remember that the most rapid phase of longitudinal growth is in the mid trimester fetus (Fig. 16.1), which may hold a clue to the little understood problem of children with chronic intra-uterine growth retardation who fail to achieve their adult height potential. Thereafter, rapid growth in infancy slows down through early

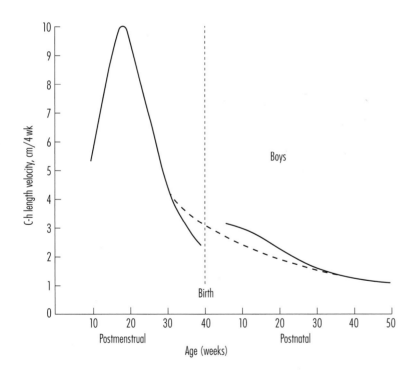

Fig. 16.1. Velocity curves for growth in body length in the prenatal and early postnatal period (boys). (——), actual length and length velocity; (– – –), theoretical curve if no uterine restriction took place. (Based on several sources of data and reproduced from Tanner, J.M., 1989. *Fetus into man*, (2nd edn). Castlemead Publications, Welwyn Garden City.)

childhood before a doubling of growth rate at puberty and virtual cessation of growth with long bone epiphyseal fusion in the mid teen years. Growth is so rapid in the first few months of life that weight increase is a handy marker of growth at that stage. After 6 months of age increase in length is much more relevant as a growth marker with height being measured after 2 years.

Growth measurements on an individual child need to be compared with standard norms for the child's peer group by plotting the results on a growth chart showing normal centiles for that group. British charts produced by Tanner and Whitehouse in 1965 have recently been superceded by new charts produced from recent data from all over Great Britain (Fig. 16.2). Similar charts are available for other countries. World-wide racial and genetic variation in growth shows that for example the Hausa of Nigeria and the Norwegians are among the tallest populations in the world. Populations from the Far East are shorter, but there has been a considerable secular trend to taller stature in Japan over recent decades, possibly due to dietetic factors.

The main influences on growth in fetal and early postnatal life are nutritional, rather than endocrine, with hormonal factors becoming progressively more important from infancy onwards. Severe psychological deprivation can adversely affect growth, both in infancy and childhood, (see Ch. 5, p. 90).

Sexual differentiation and development of the endocrine system occur in fetal life towards the end of the 1st trimester. Androgens from the fetal testis stimulated by placental chorionic gonadotrophin and luteinizing hormone (LH) and follicle stimulating hormone (FSH) from the developing pituitary are important in the developing fetus in imprinting a male pattern on the indifferent female-type precursor (Fig. 16.3).

After birth the gonadal hormone axis enters a quiescent phase until the onset of puberty. The first sign of puberty in girls is breast bud development occurring on average just before 11 years of age and in boys 3 ml testicular size found shortly after 12 years. *Pubertal development* is described conventionally according to stages with stage I being the prepubertal state. Testicular size is measured by comparing with a standard set of beads of graded volumes known as an orchidometer.

The peak height velocity of the pubertal growth spurt is just before 12 years in girls and 14 years in boys, making the average 12 year old girl briefly taller than her male peer. Menarche at average age 13 is a late event in female puberty after which the average girl grows only 6 cm. The hormonal contribution to the pubertal growth spurt is partly due to androgens from the adrenal cortex in both sexes and testis in boys, but more important is increased growth hormone production under the influence of sex steroids.

Variations in timing of puberty are common. Early pubertal development (precocious puberty) is defined as before age 8 in girls and 9 in boys. Early development in girls from age 6 and sometimes younger is relatively common and usually due to constitutional factors, whereas early development in boys is much more likely to be due to organic disease of the adrenal, testis or hypothalamus. Delayed

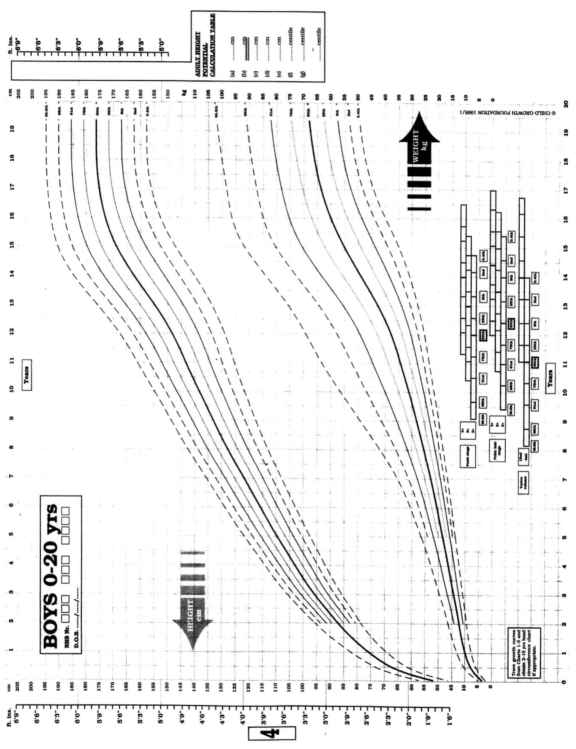

Fig. 16.2. UK G-CENTILE Growth chart. (From HArlow Printing Ltd, South Shields, UK)

Fig. 16.3. Diagrammatic sagittal sections of the internal genitalia of the 8-week fetus (a), at 13 weeks fetal age in the male (b), and female (c).

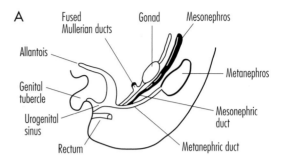

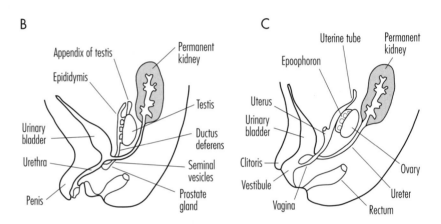

puberty is common in boys, a tendency often inherited from the father, but occurs less often in girls, where again mother may have passed menarche late. A boy with no sign of puberty at 14, or a girl at 13 warrants specialist assessment. A boy of 14 with little or no sign of puberty may suffer psychological stress on this account and may also be at lifelong risk of low bone density and other effects of late onset of androgen surge.

Case studies
Short stature

Rob (age 7 years) is the youngest child of a family of five. He is the shortest in his class at school entry measuring 103.0 cm at 5 1/2 years (3rd centile). A further measurement of 111.8 cm is taken at 7.1 years (3rd centile). His parents are unhappy to accept the explanation that he is just going to be short because his four big brothers, (ages 16–24 years) range in height from 172 to 177 cm. They ask their school doctor and GP for a second opinion.

The common *causes of short stature* include short parents, chronic intra-uterine growth retardation, delayed puberty, and psychological deprivation. Basic information to make a growth diagnosis thus includes parental heights, birthweight, timing of puberty in parents, and social background. In Rob's case he was of normal birthweight, there is no suggestion of social problem, parental puberty was at the normal time, his father is 172 cm tall and mother 150 cm. Calculation of mid parental height (Table 16.1) shows this to be 168 cm with a target centile range of 158–178 cm (0.4–50th centile). It is extremely important that height is measured accurately (see below).

Rob eventually proves to have an adult height of 163 cm and the conclusion is that he is a child of constitutional short stature taking after his mother whereas his older brothers took after their father in height (Fig. 16.4).

Accurate and careful *height measurement* (Figs 16.5, 16.6) is important if there is doubt about growth measurements being made at long intervals of 6 months to a year. Height measurement in the older child requires an accurate measuring device such as the portable Leicester height measure, or the Raven minimetre or the wall-mounted Holtain stadiometer or Raven magnimeter. Shoes, bulky socks, and hair slides need to be removed. Heels, buttocks, and shoulder blades should touch the vertical measure with feet together and upward pressure exerted on the mastoid processes to achieve a stretched height. Infants and very young children can be measured in the supine posture.

Tall stature is usually due to tall parents or early puberty, but there are a number of organic conditions to be considered at different ages.

Growth assessment must be accompanied by careful explanation, reassurance, and counselling of parents about their child's likely progress in height. Particularly in the case of constitutional delay in puberty in boys, a child at the time much shorter

Summary
Causes of short stature

- Constitutional (short parents)
- Psychosocial deprivation
- Delayed puberty
- Intra-uterine growth retardation (low birthweight for gestational age with lack of later catch-up)
- Organic disease (see below)

Summary
Technique of height measurement

- Accurate equipment
- Child erect
- Head in Frankfurt plane (horizontal line between middle of ear and outer canthus of eye)
- Heels together, knees straight
- Upward pressure under mastoids
- Height measurements at less than 6-month intervals not useful

Additional Information
Causes of tall stature

- Constitutional (tall parents)
- Early puberty
- Thyrotoxicosis
- Marfan's syndrome
- Homocystinuria
- Cerebral gigantism (presents in infancy)

Table 16.1.
Calculation of mid parental height (MPH)

(a) = father's height	172 cm
(b) = mother's height	150 cm
(c) = sum of (a) and (b)	322 cm
(d) = c ÷ 2	161 cm
(e) = d + 7 cm (MPH)	168 cm (*subtract* 7 cm for a girl)
(f) = mid parental centile [nearest centile to (e)]	9th centile
(g) = target centile range (MPH +/− 10 cm)	158–178 cm
	0.4th–50th centile

This calculation is not applicable if either natural parent is not of normal stature.

Fig. 16.4. Rob's growth chart. Rob's height at 5.5 and 7.1 years. Note that he is growing at a normal pace along the 3rd percentile.

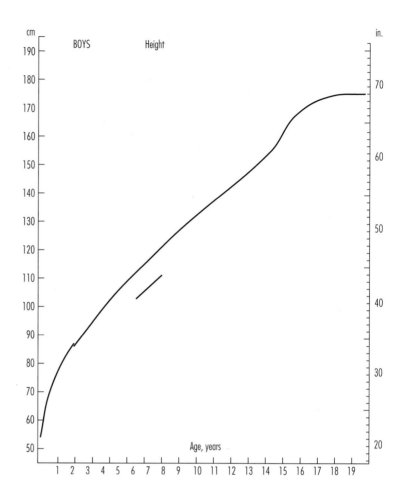

Fig. 16.6. Height measurement.

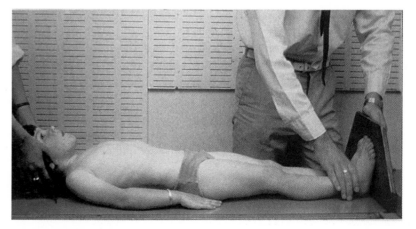

Fig. 16.5 Length measurement.

than his peers may turn out to be of normal adult height. It is important to remember that expectations of a child are very dependent on his height. Thus, although a short child may suffer on this account within his peer group less is expected of him than a child who is tall for his age and expected to have the intelligence and the maturity of a child several years older. This particularly affects the early developing girl who may appear physically like an 11-year-old at the age of 7 years.

Growth failure

Kirsty is 122.5 cm tall at 9.2 years (3rd centile). She complains that she has been gaining weight over the past year or two. Comparison with the measurement at 7.3 years (121.8 cm) shows that she has fallen away in height through two major centile lines (25th and 10th) on the growth chart. She is a little pale, has dry scurfy hair, and a small smooth goitre. Blood is taken for thyroid function and thyroid antibodies confirming the suspicion of auto-immune hypothyroidism.

Summary
Some causes of growth failure

- *Infancy*
  - growth hormone deficiency
  - dyschrondroplasia
  - Down syndrome
- *Mid childhood*
  - hypothyroidism
  - Turner's syndrome
  - chronic renal failure
  - Crohn's disease
- *Early teenage delayed puberty* (delay pattern also seen in asthma and severe eczema)
  - craniopharyngioma
- *any age*
  - psychosocial deprivation

Hypothyroidism is one of the disease conditions which can present with growth failure in mid childhood: the incidence probably about one-quarter that of childhood diabetes mellitus. In Kirsty, the thyroid underactivity may have been developing for 4 or 5 years. She has primary hypothyroidism in which total thyroxine or free thyroxine will be low in the blood with high pituitary thyroid stimulating hormone (TSH) because of lack of negative feedback. Treatment is daily thyroxine by mouth for life. Weight will reduce and height increase rapidly as will general level of activity. Hypothyroid children tend to overachieve at school (unlike thyrotoxic girls who commonly present with school failure). Treatment of moderate or severe hypothyroidism can lead to school problems for a year or two due to difficulties in concentration.

Neonatal hypothyroidism is diagnosed by high heel prick thyroid stimulating hormone (TSH) level, which is taken on all babies in the first week of life, (p. 155). Thyroxine is not needed in the fetus, but from birth to 2 years of age it is required for brain growth, so diagnosis and treatment of hypothyroidism in a neonate with absent thyroid gland or abnormal thyroxine metabolism prevents mental deficiency. On the other hand, children with hypothyroidism usually due to auto-immune thyroid disease diagnosed *after* age 2 have normal intellect.

Clues to the likelihood of an organic problem in growth feature can lie in the age at presentation. Psychosocial deprivation must be considered at any age. The commonest problem is apparent growth failure in a 12- to 14-year-old boy with delayed puberty who will regain his former height centile later in his teens. For failure to thrive in infancy, see Chapter 7, p. 256.

Classical growth hormone deficiency usually presents in the newborn with hypoglycaemia or with growth failure in infancy or early childhood. Children treated with cranial irradiation for cancer may lack growth hormone due to pituitary damage and other groups who may benefit from growth hormone treatment are children with Turner's syndrome or chronic renal failure. Growth hormone is given by daily injection and is expensive (£5000 annually for a 7-year-old at current costs). It may be of short-term benefit in improving growth rate in children with constitutional short stature and with delayed puberty, but it does not significantly improve adult height in these conditions.

Neonatal sexual ambiguity

Flora was born by normal delivery at full-term but at first it was not clear which sex she should be assigned. She had a 2 cm long phallus, a small apparent vaginal opening, and rugose genital folds with no palpable gonads.

Never assume the baby is a male in these circumstances. *Gender assignment* can be done later if there is real doubt. It is important to carry out investigations quickly and these include chromosomes, sex hormone levels, and pelvic ultrasound.

An urgent karyotype was reported as 46,XX and serum 17-hydroxyprogesterone was greatly elevated. Pelvic ultrasound and genitogram showed normal infantile uterus and vagina. She was started on hydrocortisone and fludrocortisone and arrangements were made for feminizing genitoplasty later in infancy and for genetic advice.

Summary
Gender assignment

- If you are in doubt as to the gender of a baby, say so. Taking a guess is completely unjustified and may be disastrous in the long term

- Most babies do not need to go to special care. The criterion for admission is the safety of the baby. Separation of mother and baby should be avoided if at all possible

- Inexperienced staff should involve more senior staff at an early stage

Summary
Congenital adrenal hyperplasia

- Lack of the enzyme 21-hydroxylase
- By negative feedback excess adrenal androgens
- Inheritance is autosomal recessive
- Treated with hydrocortisone (glucocorticoid) and fludrocortisone (mineralocorticoid)
- Risk of adrenal salt-losing crisis as in Addison's disease due to intercurrent stress or infection

Sexual ambiguity at birth is an unusual and complex, but important, problem. A virilised female due to 21-hydroxylase deficiency *congenital adrenal hyperplasia* is the commonest cause (incidence 1 per 10 000 live births). An undermasculinized male may need much more complicated endocrine, genetic, radiological, and surgical investigation, but a diagnosis must be vigorously sought and the sex of rearing agreed within a few weeks. The multidisciplinary team involved may include neonatologist, paediatric surgeon, paediatric endocrinologist, radiologist, clinical geneticist, and child psychiatrist.

Polyuria/diabetes mellitus

Andrew (aged 9 years) presented with a 3-week history of polyuria, polydipsia, and weight loss. He was a previously healthy boy with no family history of diabetes.

Polyuria and polydipsia in a child of this age should be assumed to be diabetes mellitus until proved otherwise. Urinary tract infection (Chapter 19, p. 323) should

always be excluded, and if the urine is negative for glucose then the child should be investigated for diabetes insipidus, in which there is a deficiency of production of antidiuretic hormone by the pituitary.

Random capillary blood glucose was 21.2 mmol/l and urine strongly positive for ketones. This is diagnostic of diabetes mellitus. Andrew was started on twice daily mixed short- and medium-acting insulin and was quickly brought under control. On his second day in hospital his breakfast was not given after his insulin so that a mild hypoglycaemic episode could be experienced.

It is important that a diabetic child learns how to recognize hypoglycaemia so that he will know what to do when it happens later. Andrew will be expected to carry with him a high-calorie snack to eat should a 'hypo' develop; his parents should have glucagon to inject in case of a severe episode.

Andrew was discharged after three days, he and his parents having been fully educated about diabetes and how to give give the insulin himself. He learnt this quickly. The diabetic liaison nurse visited the home and school to make sure that everyone knew about the condition, and passed on information to the school health service.

Diabetes mellitus usually presents in childhood in a well child with a short history of polyuria and weight loss. Secondary enuresis is also common as an indication for capillary blood glucose testing. A result of >12 mmol/l confirms the diagnosis and no other diagnostic tests are necessary. There must be no delay in starting insulin for severe ketoacidosis may develop rapidly (e.g. due to viral infection). Initial education involves the whole family (parents adapt to the diagnosis of a serious chronic disease less well than their child) and the diabetic hospital and community team (Table 16.2). Newly diagnosed diabetic children may be managed at home rather than in hospital if not dehydrated, drowsy or with ketoacidosis. It is important to remember the serious psychological impact of the diagnosis of diabetes on parents who often need a great deal of support and counselling over the first days and weeks. The affected child quickly realizes that diabetes does not stop him doing any of the activities of his peer group and will usually adapt to the diagnosis much more rapidly.

The diabetic child is managed with at least two insulin injections a day. Unlike the non-diabetic, where blood sugar rises and insulin appears in the blood after a meal, the diabetic takes insulin beforehand so the body thinks it has been fed. The main aspect of the diabetic lifestyle, apart from regular insulin injections and capillary blood glucose tests, is frequent food intake to catch-up with the exogenous

Summary
Diabetes mellitus

- The commonest endocrine condition of childhood
- Child diabetes is nearly always type I (requires insulin)
- Peak incidence at 12–14 years
- Involves a progressive destruction of pancreatic islet cells usually over several years
- 35% concordance in identical twins
- The incidence of type I diabetes in children in developed countries has increased strikingly over the past 40 years
- Population screening and prevention of diabetes may develop in the future.

Table 16.2.
Education at diagnosis of diabetes

INFORM	Consultant in charge of children's diabetic clinic	
	Dietitian	
	Diabetes Nurse Specialist	
TEACH	Insulin injection	
	Capillary blood glucose testing	
	Urinary ketone testing	
DEMONSTRATE	Induced hypoglycaemia by missing a meal and if necessary giving extra quick acting insulin	
PROVIDE	For hypoglycaemia	Glucose tablets
		Glucose gel
		Glucagon for injection
	For intercurrent ketosis	Urine ketostix
		quick-acting insulin

insulin (Fig. 16.7). The commonest insulin regimen remains two daily injections of mixed short-acting insulin (basically 6 hours) and medium-acting insulin (basically 12 hours).

The aim of diabetic management is to make the child and family self-sufficient in coping with most problems of diabetic control. Children become involved with their own diabetes from 5 years onwards and should be doing their own capillary blood tests and giving their own injections under supervision by 8 years, though not becoming completely independent in their diabetic management until aged 15. Children as young as 5 years old can understand the effect of food, insulin, and exercise on blood sugar level (Fig. 16.8) and how an event involving exercise, excitement, and perhaps forgetting food intake (e.g. a holiday or a football match) can be either managed with planned taking of extra food, or less insulin beforehand. The routine of a diabetic child involves sugar and a Talisman or a diabetic identity card to be carried at all times. Any absence from base for more than a few minutes requires food being available and for absences of more than a few hours insulin must be carried in case a snack or an injection time become due. Insulin resistance can be a problem during the pubertal growth spurt due to additional production of growth hormone. Increased doses of insulin are required during these years and control of diabetes becomes more difficult. Though diabetes can present at any age in childhood, the peak incidence is at 12–14 years when diabetes is more likely to become an adolescent rebellion (see Chapter 24).

Intercurrent illness requires careful management, and insulin should be continued in spite of loss of appetite.

In general, diabetic children do well at school and in settling into work. Very few careers (e.g. airline pilot or uniformed services) are not open to people with diabetes, but on the whole, they can cope with most professions and careers. After

Summary
Management of illness in diabetic children

1. Continue normal insulin at usual dose and times

2. Test more frequently

3. If capillary blood glucose > 20 mmol/l test urine ketones. If urine ketone level 16 or above, give extra Actrapid insulin 1/6th of the total daily dose. Check again at 3–4 hours and repeat this dose if indicated

4. Take carbohydrate-containing fluids if solids are not tolerated such as fruit juice or ordinary diluted juice

5. Persistent vomiting indicates hospital admission

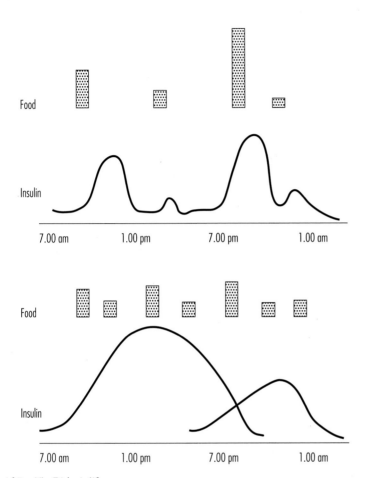

Food

Insulin

7.00 am   1.00 pm   7.00 pm   1.00 am

Food

Insulin

7.00 am   1.00 pm   7.00 pm   1.00 am

Fig. 16.7. The Diabetic life

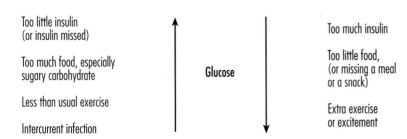

Too little insulin
(or insulin missed)

Too much food, especially
sugary carbohydrate

Less than usual exercise

Intercurrent infection

Glucose

Too much insulin

Too little food,
(or missing a meal
or a snack)

Extra exercise
or excitement

Fig. 16.8. Basic factors in control of glucose in a diabetic child.

20 years of the diabetic life, microvascular complications become an inevitable factor. Renal failure is less of a problem in the young than in the past, and retinopathy, if detected early, may be controlled with laser treatment. The major threat is of myocardial infarction/peripheral vascular disease during young and middle-aged adult life, the risks of which are greatly increased by smoking. A recent large trial in the USA demonstrated conclusively that better diabetic control over the years in young adults (as represented by lower levels of haemoglobin A1c, an index of the average blood sugar over the previous few months), led to fewer complications. This was balanced by an increased number of hypoglycaemic episodes. Screening for complications of diabetes should start in the children's diabetic clinic and include regular measurement of blood pressure, retinal examination, and testing for urinary microalbumin (as a screen for nephropathy), coupled with meticulous care of the feet (where microvascular disease often presents) and advice about smoking.

Further reading

Tanner, J.M. (1989). *Fetus into man*, (2nd edn). Castlemead, Welwyn Garden City, Herts.

Brooke, G.D. (1989). *Clinical paediatric endocrinology*. Blackwell, Oxford.

Court, D. and Lamb, W. (in prep.) *Childhood and adolescent diabetics*.

17

Haematology and oncology

Basic haematology

In the embryo, the initial site of haematopoiesis is the yolk sac. From there haematopoietic stem cells migrate to the liver and spleen which are the main sites of blood formation in the fetus, and finally to the medullary cavity of the bones which is the main site of haematopoietic tissue in extra-uterine life.

Bone marrow stem cells have the characteristics of self-replication, and division and differentiation (Fig. 17.1). The production of circulating blood cells is under the control of a variety of cytokines (growth factors). Erythropoietin is the main growth factor controlling red cell production—chronic hypoxia being the main stimulus to erythropoietin production by the kidney. The production of white cells and platelets is controlled by a variety of cytokines—'colony stimulating factors' and interleukins. The importance of cytokines lies not only in their physiological control of haematopoiesis but in their therapeutic potential. Many cytokines can now be produced by recombinant DNA technology and their therapeutic role is now being investigated. The role of erythropoietin in improving the anaemia of chronic renal failure is established, but the clinical role of granulocyte and megakaryocyte growth factors remains to be defined.

These controlling factors in haematopoiesis mean that the number of red cell, white cells, and platelets in the circulating blood remain within defined 'normal' limits in health. Peripheral blood is one of the most easily biopsied tissues of the body, as a result the normal ranges for the blood count have been established for different age groups. In adults, the normal range for the blood count is stable over years, whereas in children there is a considerable variation with age. This is seen particularly in the haemoglobin and red cell volume (MCV) (Fig. 17.2), but there

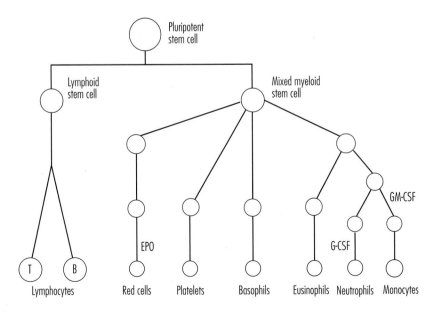

Fig. 17.1. Diagram of the bone marrow pluripotent stem cell and the cell lines arising from it. The growth factors of demonstrated clinical efficacy are shown; (Epo, erythropoietin; G-CSF, granulocyte colony stimulating factor; GM-CSF, granulocyte macrophage colony stimulating factor.)

Pluripotent stem cell

Lymphoid stem cell

Mixed myeloid stem cell

GM-CSF

EPO

G-CSF

T   B

Lymphocytes   Red cells   Platelets   Basophils   Eusinophils   Neutrophils   Monocytes

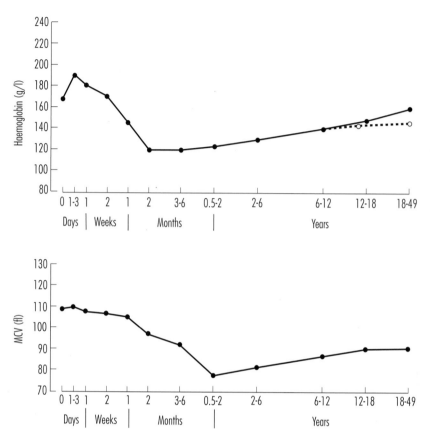

Fig. 17.2. Normal ranges for haemoglobin concentration and mean cell volume (MCV) of red cells, from birth to adult life. The broken line on the right shows the values for adult females.

is variation in the proportion and number of white cells, with lymphocyte predominance being normal at a young age and neutrophil predominance when older. Platelet count remains much the same. All these physiological, age-related variations must be taken into account when analysing a blood count result before deciding an abnormality is present and requires investigation. It is important to remember that a test result may be 'abnormal' (i.e. lie outwith the defined normal range) but does not necessarily indicate the presence of disease.

Many other investigations have age-related variations including coagulation tests, levels of coagulation factors, and levels of immunoglobulins.

The haemoglobin in red cells is composed of two alpha chains and two 'non-alpha' chains forming a tetrameric molecular structure with a varying affinity for binding oxygen. During later fetal life the main haemoglobin is fetal Hb (HbF, alpha-2, gamma-2) which has a relatively high oxygen affinity. Within a short period of birth HbF production 'switches' to adult Hb (HbA, alpha-2, beta-2) with a lower oxygen affinity more suited to extra-uterine life (Fig. 17.3). Alpha chains are present in all types of haemoglobin, therefore any abnormality of alpha chain structure or amount will be expressed during fetal life. Conversely, any abnormality of beta chain structure or amount will not be clinically apparent until 3–6 months of age as HbF production reduces.

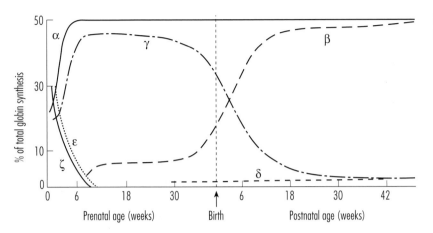

Fig. 17.3. Pattern of haemoglobin chain production in the fetus and first year of life.

As in other branches of medicine, the *clinical assessment* of a patient starts with the history as will be emphasized in the case studies. The family history can be of particular importance because of the significant numbers of hereditary forms of anaemia and *haemorrhagic disorders*.

Physical examination should assess features such as pallor and other signs of anaemia, character and distribution of bruising, site and nature of lymphadenopathy, presence of hepatomegaly or splenomegaly, and abdominal masses. There should be assessment for pre-existing disease, for example, the characteristic facial appearance of a child with Down syndrome where there is an increased incidence of leukaemia.

Case studies
Nutritional anaemia

Paul (age 2 years) is a first child. His parents were not concerned about his health but when his grandmother came to visit she was worried that he was very pale. She encouraged his parents to take him to his GP. Paul's mother indicated that he had been well with no change in his energy level and no increased frequency of infections. On reflection, she felt that he was probably paler than 3 or 4 months before. From the time he was a baby it had always been difficult to get him to settle at night and he usually had a bottle with cow's milk to get him to sleep. He was described as a 'picky' eater and took at least 500 cc (1 pint) of cow's milk a day. He took little in the way of solid food. Paul had not been breast-fed. There was no history of obvious blood loss or jaundice, and no family history of anaemia.

On examination Paul was strikingly pale but lively. There was no bruising or purpura, no enlargement of lymph nodes, liver or spleen, and no abnormality on

Summary
Clinical assessment

1. Remember the parents and siblings—they may give important clues as to the disorder in the child

2. Do not consider the haematological disorder in isolation—there may be important effects on the growing child

Summary
Clinical effects of blood disorders

- *Anaemia*: symptoms due to reduced oxygen delivery to tissues
- *Polycythaemia*: sludging of circulation
- *White cell abnormality*: risk of infection
- *Platelet abnormality*: risk of bleeding; possibly thrombosis
- *Haemostatic factor abnormality*: imbalance of haemostatic mechanism; risk of bleeding or thrombosis

Summary
Differential diagnosis of anaemia

- Iron deficiency
- Blood loss (e.g. epistaxis, rectal bleeding esp. Meckel's diverticulum)
- Blood dyscrasia (e.g. aplastic anaemia, haemolytic anaemia, leukaemia)
- Haemoglobinopathy (e.g. sickle-cell disease)
- Other congenital disorders (e.g. spherocytosis, thalassaemia)

Additional Information
Approach to anaemia

1. *Is the child anaemic?*
   Knowledge of 'normal' ranges
2. *What type of anaemia?*
   Morphological diagnosis
3. *What is the mechanism?*
   Pathophysiological diagnosis

Additional Information
Serum ferritin levels

- *Low*
  - Iron deficiency
- *Normal/Increased*
  - thalassaemia
  - haemoglobinopathy
  - sideroblastic
  - secondary

abdominal examination. The GP measured his Hb on a capillary haemoglobinometer and it was 7.0 g/dl.

This is a very low haemoglobin—the lower limit of normal at this age is 11 g/dl. Questions to ask should cover a careful dietary history, whether there has been any rectal or other bleeding (e.g. epistaxis), and any recent infections, as well as a family history.

In view of the low Hb, the GP requested a full blood count. This showed a haemoglobin of 7.2 g/dl, MCV 55 fl, and MCH 20 with the blood film confirming a severe hypochromic microcytic anaemia with rod forms, the white cells and platelets appearing normal.

In the assessment of a patient with anaemia the blood count and film appearances can give important information to guide subsequent investigation. In Paul's case the blood picture is suggestive of *iron deficiency* anaemia and this can be confirmed by measuring the *serum ferritin*. If Paul's family came from the eastern Mediterranean, India or the Far East it would be wise to look for thalassaemia or sickle-cell disease which can mimic iron deficiency.

When the diagnosis of iron deficiency anaemia has been established it is important to correct the underlying cause. Iron content at birth is affected by maternal anaemia, growth retardation, birthweight, and the time of clamping of the cord, among others (see Table 17.1). The type of milk used in feeding the infant is extremely important in determining iron stores after birth (see Chapter 3, p. 48). It is necessary to involve the dietician in assessing the iron content of the diet and advising the parents on appropriate dietary management to correct any deficiency and modify eating patterns (e.g. how to reduce milk intake and increase solids with a greater iron content). The specific correction of the iron deficiency is by administration of iron, usually in the form of sodium iron edetate (Sytron) which is a liquid preparation more likely to be tolerated by children. It is necessary to ensure the treatment has corrected the anaemia and the expected improvement in haemoglobin level would be 1 gm/dl/week. Once the haemoglobin has returned to normal, it is important to continue the iron treatment for a further 2–3 months to build up the iron stores in the bone marrow.

Paul was treated with iron and his Hb rapidly returned to normal. It seemed that the main reason for his anaemia was his early introduction to cow's milk and his poor dietary intake of iron-containing foods. Dietary advice was offered particularly

Table 17.1.
Factors affecting iron content at birth

	Increased	Decreased
Tissue iron	High birthweight	Low birthweight
	Haemolytic disease	
Blood volume	High birthweight	Low birthweight
	Late cord clamping	Early cord clamping
		Haemorrhage from cord
	Materno-fetal transfusion	Feto-maternal transfusion
	Feto-fetal transfusion	Feto-fetal transfusion
Cord haemoglobin	Growth retardation	Preterm infant
	Maternal anaemia	Haemolytic disease
	Maternal hypoxia	

Summary
Causes of iron deficiency

- Low birthweight
- Early introduction of doorstep cow's milk
- Late introduction of iron-containing solids
- Malabsorption
- Bleeding from intestines (e.g. Meckel's diverticulum)

emphasizing the use of vitamin C or orange juice to promote the absorption of iron from the diet (see Ch. 3, p. 48). A final check of his Hb in the surgery 6 months later showed that it remained in the normal range.

Sickle-cell disease (haemoglobinopathy)

John presents with anaemia at 2 years of age. His parents are from Nigeria and his mother was known to be a carrier of the sickle gene but father had been tested previously and was said to be normal. At birth, John appeared well but he now has a hypochromic microcytic red cell picture and characteristic sickle cells on the blood film. Haemoglobin electropheresis showed the bands of HbF and HbS with no HbA. He also had a period of jaundice which responded to phototherapy.

Summary
Sickle-cell disease

- Sickle-cell trait is symptomless
- Sickle-cell anaemia does not present until HbF replaced by HbS
- Presenting features are anaemia, abdominal pains, limb pains, swelling of fingers
- Pneumococcal infection common
- Skull X-ray characteristic
- Child is stunted in later years
- Splenomegaly marked in early years, disappears later

While the clinical history points to a sickle-cell disorder the presence of a hypochromic microcytic anaemia in a patient from a non-Caucasian ethnic background should raise the possibility of either a haemoglobinopathy or a thalassaemia. Both are inherited disorders that affect the production of globin chains. In thalassaemia, there are a number of mechanisms resulting in the reduction of production of a normal globin chain (e.g. alpha chains in alpha- thalassaemia and beta chains in beta-thalassaemia). A haemoglobinopathy results from the production of normal amounts of an abnormal globin chain (e.g. HbS in sickle-cell disorders). The ethnic origin of the patient can give a clue as to the likely underlying disorder (e.g. beta-thalassaemia in Mediterranean races, alpha-thalassaemia in Asians, and sickle-cell disorders in Afro-Caribbean races).

Fig. 17.4.   Blood film in sickle-cell disease

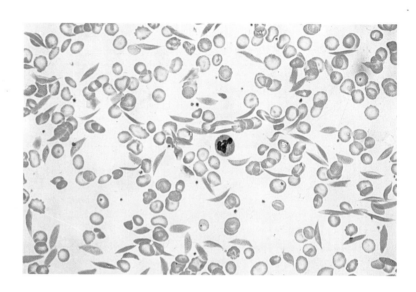

In the sickle-cell abnormality there is substitution of valine for glutamic acid at position 6 in the beta globin chain and this causes profound changes in oxygen binding by haemoglobin. When both genes for beta chain production are abnormal sickle-cell disease results and if only one gene is affected sickle-cell trait or carrier state results. The pathogenesis of sickle-cell disorders results from the insolubility of HbS at low oxygen tensions so that when HbS is deoxygenated it polymerizes to form strands called 'tactoids'. This results in the characteristic sickle cells seen in the blood film (Fig. 17.4), these cells losing the normal deformability of the biconcave red cells and causing microvascular obstruction. Depending which part of the circulation is involved there can be neurological symptoms, pain of tissue infarction such as in the bones of the hands, abdominal or chest pain, and pain from splenic involvement. The disease also causes renal tubular damage with difficulties in concentrating urine.

The clinical course of a patient with sickle cell disease is punctuated with different types of 'crises' the majority of which are precipitated by infection resulting in hypoxia and often aggravated by dehydration.

Summary
Management of sickle-cell disease

- Blood transfusion may be required for anaemia

- Pain relief and correction of dehydration in crisis

- Splenectomy if evidence of hypersplenism

- Folic acid to support bone marrow function

- Pneumococcal vaccine and penicillin prophylaxis

- Genetic counselling

John appears to have sickle-cell disease in which both genes controlling beta chain production are abnormal and there is no production of normal adult haemoglobin. His father was tested again and was shown to be a carrier of the HbS gene. There was thus a 1 in 4 chance of any child being affected by sickle-cell disease, and a 1 in 2 chance of a child being a carrier. This was important in advising the parents in planning future pregnancies and offering antenatal testing if this was acceptable, (see also Ch. 15).

The aim of *management* of a child with sickle-cell disease is to reduce the number and severity of crises, thus reducing long-term problems with growth and development. Bone marrow function is supported by supplemental folic acid. As infection is an important precipitating factor of crises it is important to try to prevent these if possible. Pneumococcus is a common infecting agent and John should be on prophylactic penicillin and immunized with pneumococcal vaccine.

Education of the parents, and subsequently the child, in avoiding situations which may lead to a crisis is a central aspect of management.

John was started on prophylactic penicillin and folic acid supplements. He was seen regularly at the clinic and immunized against pneumococcus as well as his routine immunizations. His parents were instructed in the signs of infection and other possible crises to look for and were given open access to the hospital.

Bruising and petechiae (thrombocytopenia)

Alison (age 3 years) presented to the accident and emergency department with a 3-day history of easy bruising. Her mother had noted bruising on various areas of Alison's body, many apparently without obvious trauma. On the morning of attendance there had been bleeding from the gums after Alison brushed her teeth.

One of the major considerations currently in a child with bruising is that it might be due to non-accidental injury (see Chapter 5, p. 86). In taking the history it is necessary to assess whether the bruising or other evidence of bleeding is consistent with that history.

Alison's mother reported that she had an infection about 2 weeks before with vomiting and diarrhoea. This had settled but the bruising started. There was nothing else of note in the history, in particular no drugs had been taken and there was no history of a bleeding disorder in the family.

On examination Alison appeared well and contented in her mother's presence. She was afebrile and the was a purpuric rash on her face neck and trunk. On various parts of her body there were bruises of different ages, these being most marked on her arms and legs (Fig. 17.5). In the mouth there were purpuric spots and some blood blisters on the buccal mucosa. There was no significant lymphadenopathy,

Summary
Differential diagnosis of purpura

- Capillary damage: meningococcaemia, bacterial endocarditis, scurvy, Henoch–Schönlein purpura
- Congenital capillary defect (e.g. Ehlers–Danlos syndrome)
- Thrombocytopenia: ITP[*], leukaemia, aplastic anaemia
- Abnormality of platelet function

[*] ITP, idiopathic thrombocytopenic purpura

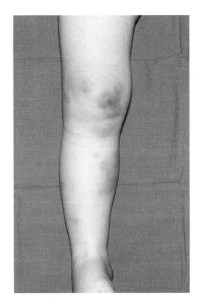

Fig. 17.5. The bruises in idiopathic thrombocytopenic purpura (ITP) are typically of different ages and can be extensive.

Additional Information
Components of normal haemostasis

1. Reaction of the vessel wall
2. Formation of the platelet plug
3. Activation of coagulation to form fibrin
4. Fibrinolysis

Additional Information
Comparison of bleeding in platelet and coagulation disorders

- *Platelet disorder*
  - Purpura
  - Easy bruising
  - Bleeding from mucosal surfaces (e.g. gums, nose bleeds)
- *Coagulation disorder*
  - Easy bruising
  - Bleeding into joints and muscles
  - Internal bleeding

Summary
Idiopathic thrombocytopenic purpura (ITP)

- Acute onset following upper respiratory tract infection
- Presents with bleeding or purpura
- Rarely, there may be intracranial haemorrhage
- Normally remits spontaneously but may become chronic
- Steroids used in severe cases

hepatomegaly, or splenomegaly, and examination of the cardiovascular and respiratory systems was normal. There were no haemorrhages on fundoscopy.

The assessment of a child with a bleeding disorder requires a basic understanding of haemostasis. The *major components of normal haemostasis* are intimately related in a dynamic process and any single part may become abnormal, the commonest being abnormalities of platelets or coagulation factors. There are different *clinical presentations* of these.

Alison's blood count showed a normal haemoglobin and white cell count but the platelets were markedly reduced at 10×10^9/l. The coagulation screen and urea and electrolytes were normal. Examination of the blood film confirmed the severe thrombocytopenia but no immature white cells were seen and there was no red cell fragmentation (which is seen in haemolytic-uraemic syndrome, see Ch. 14, p. 251).

Alison has severe thrombocytopenia which is most likely due to idiopathic thrombocytopenic purpura (ITP), given that the rest of the blood count is normal and the other initial investigations are normal. ITP usually follows an infection and tends to affect two age groups. In the younger age group, 2–5 years, the sex incidence is equal and in the majority of children the disease remits with no long-term problems. In the older age group, the majority of children affected are girls and the disease has a tendency to become chronic, tending to merge with adult auto-immune disease (Table 17.2).

There are still controversies about the management of ITP in children, the main ones being the need for bone marrow examination and the need for any form of treatment. In the presence of a normal haemoglobin and white cell count, it is

Table 17.2.
Idiopathic thrombocytopenic purpura (ITP)

	Acute	Chronic
Age	Younger	Older
Sex	Girls = Boys	Girls > Boys
History	Short (days)	Long (weeks)
Seasonal	?Winter/Spring	No
Preceeding infection	Frequent	Rare
Platelet auto-antibodies	Not implicated	Implicated
Proportion	85%	10%

highly unlikely that the thrombocytopenia could be due to bone marrow infiltration by diseases, such as leukaemia, but many clinicians prefer to perform a bone marrow test to exclude this possibility. ITP in children tends to recover spontaneously with no major risk of serious haemorrhage. If it is decided that treatment is indicated the initial choice is to give a short course of prednisolone, not lasting more than 2–3 weeks, leading to a prompt recovery of the platelet count. In children it is rarely necessary to consider other forms of treatment, such as intravenous immunoglobulin, splenectomy, or other more potent immunosuppressive agents, unless the thrombocytopenia becomes chronic.

In Alison, it was thought that the severity of the thrombocytopenia and bleeding warranted treatment. A bone marrow aspirate was performed and showed normal red cell and white cell production with an increased number of megakaryocytes. There was no abnormal cell population present. She was started on oral prednisolone and after two days there were no new bruises and her platelet count had risen to $27 \times 10^9/l$. Alison was allowed home to continue the prednisolone for 2 weeks, and to take part in normal activities. In an older child there would be restrictions on physical activities such as contact sports until the platelet count had recovered to near normal levels. Alison was seen for review at the end of 2 weeks of treatment when all the bruising and purpura had resolved and her platelet count was normal at $245 \times 10^9/l$. Alison was seen for review at regular intervals over the next 6 months and her platelet count remained normal.

Another important condition which can present with purpura is *Henoch–Schönlein* (or anaphylactoid) *purpura*, (Fig. 17.6).

Bruising (coagulation factor deficiency)

The parents of Norman (age 5 months) were concerned about bruises appearing on his body without obvious injury. He could roll around in his cot but it seemed doubtful that this would be sufficient to cause the bruising. Could his 3-year-old sister be hitting him because of jealousy?

Norman had been born at full-term and there had been no obvious bruising or bleeding related to the delivery or at separation of the umbilical cord. His routine immunizations had not resulted in bruising although there may have been some swelling after the injections. His development was normal. There was no family history of any bleeding disorders.

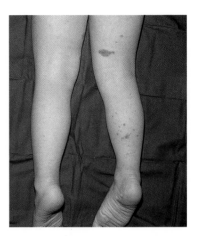

Fig. 17.6. Henoch–Schönlein purpura (HSP) are typically over the buttocks and back of the legs, and are more discrete.

Summary
Henoch–Schönlein purpura

- Often a preceding history of sore throat

- Rash is highly characteristic: purpura on buttocks, extensor surfaces of legs and arms, face spared (except in infants, when the disease is rare)

- May be painful swollen joints

- Haematuria is common, with glomerulonephritis in 20% (see Ch. 19, p. 330)

- Normally recovers completely, a small number develop long-term renal problems

On examination Norman appeared well and his length and weight were on the 50th centile. He had a number of bruises on his trunk and one on his right hand. The bruises were raised and had a firm central area. There were no purpura and no enlargement of lymph nodes, liver, or spleen. All his joints appeared normal. His head circumference was proportionate to his size and the fontanelle was not tense.

Bruising in a child of this age with no satisfactory history of how they might have occurred should raise the possibility of non-accidental injury. However in Norman's case the bruises were of abnormal appearance and suggestive of an underlying *bleeding tendency*. This always requires investigation. The earliest presentation is often at the time of the separation of the umbilical cord, although this did not happen in Norman.

Initial investigation showed a normal blood count, in particular the platelet count was normal. It was appropriate to assess the coagulation mechanism (Fig. 17.7). The coagulation screen gave a normal prothrombin time (PT) and thrombin clotting

Summary
Investigation of bleeding tendency

- Full blood count with platelets
- Prothrombin time (PT) (measures extrinsic pathway)
- Thrombin clotting time (TCT) (extrinsic pathway)
- Activated partial thromboplastin time (APTT) (intrinsic pathway)
- Specific factor assays

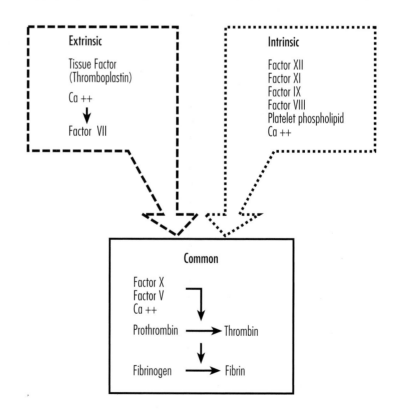

Fig. 17.7. The coagulation cascade and its laboratory assessment.

time (TCT) suggesting that the extrinsic clotting pathway and the ability to convert fibrinogen to fibrin was normal. The activated partial thromboplastin time (APTT) was prolonged at 115 seconds (normal range 35–45s), consistent with an abnormality in the intrinsic clotting pathway. Specific assays of factors VIII and IX showed the factor VIII coagulant level (VIII:C) to be 1%, with a normal factor IX level. A diagnosis of severe haemophilia A was made.

Summary
Haemophilia A

- Sex-linked recessive, deficiency of factor VIII
- Blood clots inadequately
- Tendency to bleed into joints
- Can be effectively treated with factor VIII

There are two main aspects to the care of a family with a child who has a bleeding disorder—the specific care of the child and the investigation of the family to identify other members who might be affected. The latter involves assessing the mother to decide if she is a carrier of the *haemophilia A* gene. This can be done with coagulation tests which can be predictive of the likelihood of her being a carrier and now the gene mutation can be looked for in both the child and the mother. If the mother is a carrier further family studies should be carried out to determine other family members who are affected, in particular identifying carriers who have a 50% chance of having an affected son (Fig. 17.8), (see also Ch. 15, p. 269).

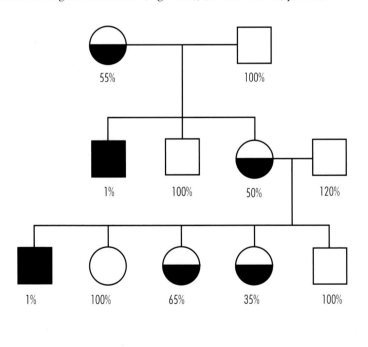

Key

○ Unaffected ♀   ◑ Carrier ♀

□ Unaffected ♂   ■ Affected ♂

Fig. 17.8. A typical family tree in haemophilia A, showing the X-linked mode of inheritance.

Summary
Management of haemophilia

- Education of parents and child
- Early treatment of bleeding episodes
- Treat any bleeding episode with factor VIII replacement
- Prophylaxis with factor VIII may be given at home

The specific *management* of the child with haemophilia starts with educating the parents to recognize the signs of internal bleeding, particularly as it affects joints and muscles, and to report for treatment. The *clinical manifestations* of the bleeding disorder depend on its severity (Table 17.3) and in Norman it is likely he will have spontaneous bleeding into muscle and joints, and have excessive bruising and bleeding from relatively minor trauma. Another important area is dental care to maintain the teeth in good condition and limit the need for extractions with their risk of bleeding. The child should be immunized against hepatitis B and monitored for the development of antibodies to other viral infections such as hepatitis C, human immunodeficiency virus (HIV), and parvovirus B19.

The treatment of an episode of bleeding involves giving factor VIII replacement to achieve a sufficient level in the blood to stop the bleeding. Until recently the main source of factor VIII concentrate was by processing plasma given by blood donors and this carried the risks of transmission of viral infections. The gene for factor VIII has now been cloned and recombinant factor VIII produced in vitro without the need for plasma is now becoming available. This greatly reduces the risk of transmission of infection. It is likely to become the treatment of choice within the next few years. Prophylactic therapy involves the administration of factor VIII concentrate on alternate days to maintain the factor VIII in the blood at a level to greatly reduce the frequency of severe bleeding episodes and thus reduce the long-term joint damage. This requires the parents to give the factor VIII concentrate at home and in young boys will necessitate the insertion of a semi-permanent intravenous device.

Table 17.3.
Clinical manifestations of haemophilia A or B

Coagulation factor activity (% of normal)	Clinical manifestations
< 2	Severe disease Frequent spontaneous bleeding episodes from early life Joint deformity and crippling if not adequately treated
2–5	Moderate disease Post-traumatic bleeding Occasional spontaneous episodes of bleeding
5–20	Mild disease Post-traumatic bleeding

Lymphadenopathy

Helen (age 4 years) had been less well for 3 months with frequent sore throats and increasing tiredness and pallor. Her parents had noticed swelling in her neck for about a month and over the previous 2 weeks she had developed a limp. Helen had previously been well with normal development and no problems with resistance to infection. There was no family history of note.

On examination Helen appeared pale and unwell. Her temperature was elevated at 38 °C and she had scattered small bruises of varying ages. Her throat was inflamed with enlarged tonsils and there was generalized superficial lymphadenopathy, the largest nodes being about 2 cm in diameter in the neck. Auscultation of the heart revealed a tachycardia with an ejection systolic murmur. In the abdomen the liver was enlarged at 3 cm below the costal margin and the spleen at 5 cm.

In assessing a child with lymphadenopathy one of the most difficult decisions is whether the lymphadenopathy is reactive or may indicate a more serious underlying disorder. As children start going to nursery school then primary school they are exposed to an increasing number of infections and a common feature of these is lymphadenopathy. In the history it is important to get an impression of how well the child has been in terms of general health, energy level, appetite, weight loss, fevers, and frequency of infections. It can be helpful to know how any siblings have been and if they have had similar problems. Children with reactive lymphadenopathy are usually well between episodes of swelling of the nodes, and there is no evidence of a progressive illness. Symptoms such as bruising or recurrent nose bleeds, and bone or joint pain with limp can indicate serious underlying illness such as leukaemia. It is easy to carry out initial investigations in the form of full blood count in primary care, but if there are suggestive features of a general disorder as here then the child should be referred to hospital for fuller investigations.

Helen's blood count showed a Hb of 5.6 g/dl, white blood cell count of $7.9 \times 10^9/l$, and platelets, $21 \times 10^9/l$. Serum biochemistry showed elevation of the uric acid and lactic dehydrogenase. Examination of the blood film confirmed the anaemia and thrombocytopenia, and demonstrated that the majority of the white cells were small, uniform blast cells suggestive of an acute leukaemia.

Acute leukaemia is the commonest form of cancer in children accounting for approximately one-third of all cases (Fig. 17.9), and *acute lymphoblastic leukaemia* (ALL) comprises 80–90% of the cases of leukaemia. Most of the others are acute myeloblastic leukaemia (AML) with a small number of very rare forms of leukaemia. The aetiology of acute leukaemia is a topic of extensive investigation with particular interest in environmental factors such as infection and radiation. However, it is known that certain genetic disorders carry an increased risk of developing leukaemia, including Down syndrome and certain immunodeficiency disorders.

Summary
Differential diagnosis of cervical lymphadenopathy

- Upper respiratory tract infection (notably, glandular fever, see Ch. 12, p. 228)
- Chronic infection (e.g. tuberculosis, see Ch. 22, p. 383)
- Malignant disease (e.g. leukaemia, Hodgkin's disease)

Summary
Clinical features of acute lymphoblastic leukaemia (ALL)

- Anaemia
- Increased tendency to infections
- Bleeding/bruising tendency
- Organ infiltration leading to splenomegaly, testicular enlargement
- Uric acid nephropathy may develop with treatment

Fig. 17.9. Pie chart demonstrating the relative incidence of various childhood cancers.

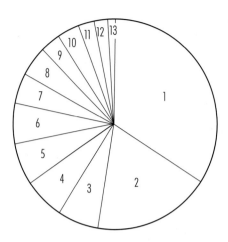

1. Leukaemia
2. Brain tumours
3. Wilms' tumour
4. Other tumours
5. Non-Hodgkin's lymphoma
6. Neuroblastoma
7. Rhabdomyosarcoma

8. Hodgkin's disease
9. Retinoblastoma
10. Osteogenic sarcoma
11. Germ cell tumours
12. Ewing's sarcoma
13. Histiocytosis X
   (the exact figure is not
   known and this is an
   estimate)

The clinical features of leukaemia are due to bone marrow failure with the predictable consequences of anaemia, risk of infection, and risk of bleeding (Fig. 17.10). Other features can include tissue and organ infiltration including lymphadenopathy, splenomegaly, and testicular enlargement in ALL, and skin rash and gum hyperplasia in monoblastic variants of AML. There can be biochemical abnormalities including elevation of the uric acid due to rapid cell turnover and this can be aggravated by treatment. Care has to be taken to avoid uric acid nephropathy in which deposition of urate crystals in the kidney can lead to renal failure.

Fig. 17.10. Pathogenesis of leukaemia—proliferation of malignant bone marrow cells with reduction in normal bone marrow function.

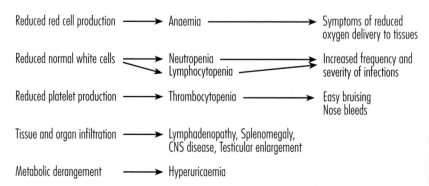

The diagnosis of leukaemia is established by examination of the bone marrow. At the same time a lumbar puncture is performed as the central nervous system can be involved by the leukaemic process in ALL. Assessment of the bone marrow includes examination of stained films under the microscope, the use of monoclonal antibodies to define cell surface antigens, and cytogenetic analysis of the chromosomes of the blast cells. By combining all the information a diagnosis is made and certain features can have prognostic significance (Table 17.4). Predictably, the majority of patients do not fit neatly into prognostic groups.

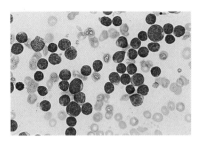

Fig. 17.11. Bone marrow. Appearance in acute lymphoblastic leukaemia. Note the large number of small uniform blast cells.

Helen was diagnosed as having acute lymphoblastic leukaemia (ALL) of the common type with an increased number of chromosomes in the blast cells; this combination of factors putting her in a favourable prognostic group. After full explanation to the parents, she was started on chemotherapy according to the current Medical Research Council protocol.

Discussion with the parents must be done sensitively and at leisure by someone who is skilled. Fortunately, the prognosis is much better now with a good prospect of a cure, so the disease does not have the grim associations it used to. Ultimately, the prognosis for an individual child depends on their response to treatment and the first 4 weeks of induction therapy are of great importance. By the end of the first 4 weeks it is hoped that the child will be in remission where the blood count is normal and the majority of the bone marrow consists of normal cells with very few or no obvious leukaemic cells.

The care of a child with leukaemia comprises two interrelated parts, general supportive care and specific therapy directed at the leukaemic cells. It is important that there is adequate supportive care available for the *management* of infection with appropriate antibiotic, antiviral, and antifungal therapy, transfusion support with red cells and platelets, and nutritional support if required. Also important is the psychosocial support given to the child, parents and siblings as they come to terms with the diagnosis.

Summary
Management of acute lymphoblastic leukaemia

- *General*
  - Transfusion of red cells
  - Transfusion of platelets
  - Antibiotics
  - Antifungal agents
  - Antiviral agents
  - Stimulation of neutrophil recovery

- *Psychosocial*

- *Specific*
  - Chemotherapy
  - Radiotherapy
  - Bone marrow transplantation

Table 17.4.
Prognostic factors in actue lymphoblastic leukaemia (ALL)

	Favourable	Unfavourable
Age	2–10 yrs	< 2 yrs and > 10 yrs
Sex	Female	Male
White count	$< 10 \times 10^9/l$	$> 50 \times 10^9/l$
Genetics	Hyperdiploid	Translocation

Helen started her chemotherapy with a combination of cytotoxic drugs given orally, intravenously, subcutaneously, and by lumbar puncture: the last being necessary as drugs given by other routes have relatively poor penetration through the blood–brain barrier. By the end of 4 weeks of treatment Helen was feeling much better and her bone marrow was in remission. She then went on to have blocks of intensification therapy directed at the central nervous system, and continuation therapy for almost 2 years.

Modern therapy for acute lymphoblastic leukaemia has improved the prognosis greatly and Helen has approximately a 70% chance of remaining free from disease long-term with the prospect of cure. With an increasing proportion of children with cancer surviving it is becoming apparent that there can be significant late-effects of treatment, for example, impaired growth and intellectual development after cranial irradiation, cardiac damage due to certain drugs and impaired fertility. Hence, all long-term survivors require long-term follow-up.

Abdominal mass

When Daniel was 2 years old, his parents took him to their GP when they thought they felt a lump in his tummy when they were bathing him. Daniel had previously been a healthy child achieving all his milestones at the appropriate time. He had been growing well and his parents had no concerns about his health until now. There was no past medical history or family history of note.

The GP confirmed there was a mass in the abdomen and referred Daniel immediately to hospital. On examination Daniel appeared well and not distressed. He was afebrile and there was no bruising or superficial lymphadenopathy. Examination of the cardiovascular and respiratory systems was normal. Daniel's abdomen was distended and there was a firm mass to the left of the midline extending down into the left iliac fossa. The mass was not tender, moved slightly with respiration, and there was no bruit audible over it.

The investigation of a mass in a child's abdomen has to answer *four questions*. The most likely organs of origin of a mass in the left side of the abdomen are kidney, spleen, and adrenal gland. The modalities of investigation are chosen to give the best chance of answering the questions. Blood tests may give a clue, for example, the blood count may be consistent with an abnormality related to splenomegaly

Summary
Differential diagnosis of abdominal mass

- *Kidney*: hydronephrosis, nephroblastoma, neuroblastoma
- *Liver*: metabolic disease (e.g. Gaucher's disease, secondary malignancy)
- *Spleen*: haemolytic disease (e.g. congenital spherocytosis), or malignancy (e.g. leukaemia)
- *Other*: (e.g. intestinal mass)

Additional Information
Questions in a child with a mass

1. What is the structure or organ of origin?
2. What is the likely tumour type?
3. Is there any other evidence of disease? (Staging)
4. Is the mass operable?

while serum biochemistry will give a crude assessment of renal and liver function. More sensitive tests can be used such as urinary catecholamine metabolite excretion and serum neurone specific enolase in neuroblastoma. However, radiological investigations are the most useful in assessment of abdominal masses. Plain abdominal films may show calcification of an adrenal mass suggesting neuroblastoma, while ultrasound may define the structure of origin of the mass. It usually requires the use of computerized tomograph (CT) or magnetic resonance imaging (MRI) scanning to define the mass more accurately, its origin, and its relationship to surrounding structures so that there can be some assessment of operability or the presence of metastatic disease.

Daniel had a CT scan which confirmed that the mass consisted of an enlarged kidney. There were no enlarged lymph nodes in the abdomen, the liver appeared normal, and there was no evidence of metastatic disease in the lungs. The most likely diagnosis was of nephroblastoma (Wilms' tumour) and Daniel was prepared for surgery.

The removal of the kidney containing the tumour serves two purposes, it removes the tumour and allows accurate staging of the disease which is important in deciding subsequent therapy and the prognosis.

At operation, the tumour was confined to the kidney and was removed in total without rupture of the capsule. There was no evidence of spread within the abdomen. Daniel therefore had stage 1 disease and was given 10 weeks of chemotherapy with vincristine. He has approximately a 90% chance of remaining free from disease long-term.

Current innovations in the treatment of Wilms' tumour include investigating the value of giving preoperative chemotherapy to assess whether this will reduce the size of the tumour and reduce the morbidity of surgery.

Further reading

Hann, I.M. and Gibson, B.E.S. (ed.) (1991). *Paediatric haematology. Bailliere's clinical haematology.* Balliere Tindall, London.

Hinchliffe, R.F. and Lilleyman, J.S. (ed.) (1987). *Practical paediatric haematology.* Wiley, Chichester.

Hoffbrand, A.V. and Pettit, J.E. (1993). *Essential haematology,* (3rd edn). Blackwell, Oxford

Nathan, D.G. and Oski, F. (ed.) (1992). *Haematology of infancy and childhood,* (4th edn). W.B. Saunders, London.

18

Infectious disease and immunology

- Background

- Case studies
  fever
  immune deficiency
  fetal infections
  meningitis

Throughout human history infectious diseases have been responsible for more death and morbidity in childhood than any other single cause. Causative pathogens interact with deprivation, malnutrition, overcrowding, and poor sanitation to give rise to human disease; consequently economic improvement, planned housing, clean water, and sewage disposal are as critical to controlling these diseases as immunization and antibiotics.

The eradication of smallpox in the 1970s remains the only complete victory of humans over a specific infectious agent. Against this solitary triumph have to be placed the rise in antibiotic-resistant organisms, the emergence of new infections (most notably the human immunodeficiency virus, HIV) as well as the reappearance of conditions such as polio and diphtheria when immunization rates have fallen. In addition, infections such as gastroenteritis, measles, pneumonia, and tuberculosis, which rarely kill in Europe or North America, still have high mortality rates in parts of the developing world. Measles has a mortality rate of 30% or more in some areas of Africa. The need for continuous vigilance means that most countries have systems for the *notification of infectious diseases* to public health authorities.

In the UK, infectious diseases are the commonest reason for a child consulting a primary care physician or being admitted as an emergency to hospital (Fig. 18.1a,b,c). Many such cases are relatively minor viral infections, but they may cause both misery to the child and great anxiety in some families. Serious infections such as bacterial meningitis, herpetic encephalitis, or staphylococcal pneumonia occur rarely but can kill or lead to significant disability. Recognition and early aggressive treatment of the child with a serious infection is essential to prevent the rapid deterioration which may occur in such cases.

Additional Information
Notifiable diseases (England and Wales)

• Typhoid	• Tetanus
• Paratyphoid	• Measles
• Dysentery	• Mumps
• Food poisoning	• Rubella
• Tuberculosis	• Viral hepatitis
• Whooping cough	• Malaria
• Scarlet fever	• Leptospirosis
• Meningitis	• Acute encephalitis
• Meningococcal septicaemia	• Ophthalmia neonatorum

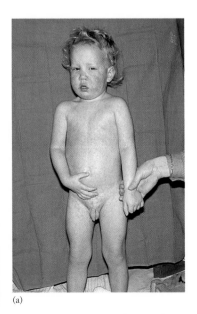
(a)

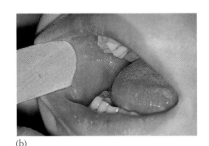
(b)

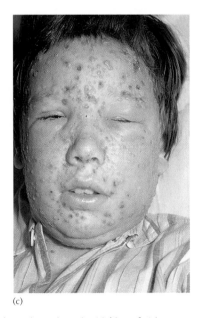
(c)

Fig. 18.1. Characteristic features of rashes in two infectious diseases. (a) *Measles*. Note the generalized maculopapular rash with bleary facial appearance and conjunctivitis. (b) Koplik's spots are typically seen inside the mouth—described as grains of salt. (c) *Chickenpox* (varicella). Note the vesicles and the different stages of evolution in the rash, with crusting in some lesions and vesicle formation in others—so-called cropping.

Summary
History for fever

- *Symptoms*
  - rashes
  - diarrhoea
  - vomiting
  - cough
  - nasal discharge
- *Background*
  - contacts
  - immunizations
  - pets
  - overseas travel
  - malaria prophylaxis

Case studies
Fever

Fiona, aged 18 months, is the only child of healthy parents in their early thirties. Her father is an engineer in the oil industry who has moved back to the UK after 12 months working in Abu Dhabi in the Middle East. Apart from uncomplicated varicella (chicken pox) at the age of 10 months Fiona has been previously well and is up to date with her immunizations including BCG. Three weeks after returning to the UK she wakes one morning with fever and is much quieter than normal. Her parents are anxious and contact their new GP.

Fever is the commonest manifestation of infection and infection the commonest cause of fever. The history must include presence of rashes, diarrhoea, vomiting, cough, or nasal discharge. It is helpful to know if she has been in contact with any infectious diseases. In this case, because she has recently returned from overseas the history should also include where the family were living and when they returned. Was she unwell whilst living abroad and did the family visit any other countries during the return home? What immunizations has she received and if relevant was she taking malaria prophylaxis?

The GP started the examination with inspection. She assessed how ill Fiona was: seriously ill children tend to be pale, floppy, and uninterested in the examination. She was not lying with her head retracted suggesting meningismus, nor with legs pulled over the abdomen suggesting abdominal pain. She was not distressed. The GP looked for cyanosis, pallor, and jaundice. There were no obvious respiratory symptoms, such as increased rate or work of breathing, nasal flaring, stridor, or wheeze. There were no rashes, and no obvious swellings to suggest a local infection, including lymphadenopathy, or joint swelling.

During the examination, use the least-threatening manoeuvres and watch her reaction.

On examination Fiona had a fever of 38 °C, a marked nasal discharge with injected tonsils but no pustules. There was no cervical lymphadenopathy.

Summary
Examination for fever

- Posture
- Colour
- Rash
- Respiration
- Ears, nose, and throat
- Lymphadenopathy
- Hepatomegaly
- Splenomegaly
- Meningism

Observe whether she is irritable when some limbs are moved but not others, suggesting a possible local infection either in soft tissue, bone, or joint; and whether the abdomen is soft or if there are areas of guarding. When listening to the chest remember that in young children, and sometimes older children, pneumonia may have no abnormal sounds. When examining the ears for injected tympanic membranes, remember that a crying child will often have injected ears without underlying infection. No examination is complete without seeking signs of meningitis, (neck stiffness and photophobia). If there are no focal signs then consider other causes of fever such as urinary tract infection.

> Since the examination suggests a viral nasopharyngitis, antibiotics are not prescribed.
>
> Fiona is treated with paracetamol and on review the following day is much recovered.

Because serious infections, such as meningitis, may not be easily diagnosed at an early stage, and children can deteriorate rapidly, be prepared to reassess after a short time. Make sure that the family know who to contact again if they are concerned, and that they are aware of the sort of symptoms which they should look out for, such as poor feeding or increasing drowsiness.

In this family there is also the question of travel abroad. Abu Dhabi is a very low malaria risk area for which prophylaxis is not routinely recommended. However, malaria should always be considered in anyone with a fever in the first few months after returning from a malarial area. It can present as a flu-like illness and missing the diagnosis can be lethal. Malaria may not show in the first blood sample, so be prepared to repeat it when the child is febrile if the diagnosis remains uncertain. Another disease which has to be remembered in any child coming from a developing country is tuberculosis, although the onset tends to be rather more insidious than in Fiona, and is also unlikely because she has had BCG immunization.

Distinctive features of the main viral infections are in the box.

Immune deficiency

> Fiona's mother returns to work in the same company as father and Fiona attends the firm's creche. Over the next six months Fiona presents to the GP on four occasions with febrile illnesses, three with coryzal symptoms and injected pharynx, as well as one episode of otitis media which responds to antibiotics. The parents express concern over the number of infections and feel that 'something must be wrong'.

It is always worth exploring what particular 'something' may be worrying the parents. It is not uncommon for parents' true fear to be that their child may have

Additional Information
Features of viral infections

Virus	Features
measles	incubation 10–14 days Koplik's spots in mouth generalized rash and fever conjunctivitis cough
mumps	incubation 14–21 days fever parotid swelling may affect pancreas and testes
rubella	incubation 14–21 days low grade fever generalized rash occipital lymph nodes
herpes simplex	gingivostomatitis conjunctivitis encephalitis
varicella (chicken pox)	incubation 10–21 days generalized itchy rash may become infected

All these infections are very severe in a person with depressed immunity.

Additional Information
AIDS

- caused by HIV
- In child usually vertical transmission (intrauterine or breastfeeding)
- In UK, usually children of IV drug users
- 15% of infants of infected mothers become HIV positive
- HIV positivity does not indicate infection <18/12 owing to circulatory maternal antibody
- clinical presentation is with lymphadenopathy, recurrent infection, Pneumocystis pneumonia

Additional Information
Some immunodeficiency states

Primary
- Selective IgA deficiency
- IgG subclass deficiencies
- Cyclic neutropenia
- Agamma-globulinaemia (X-linked)
- Severe combined immune deficiency syndrome (SCIDS)
- Deficiency of complement

Secondary
- Acquired immune deficiency (AIDS) (see box)
- Chemotherapy for cancer
- Anti-rejection treatment for transplants
- Steroid treatment

leukaemia, or a condition such as diabetes mellitus, having heard that these conditions may present with repeated infections or fever.

The infections have been mild and have probably been viral in origin. There is no history of problems with her immunizations and she had uncomplicated varicella at the age of 10 months, an infection which can be life-threatening in the presence of cellular immunodeficiency states. Hence, it is possible to reassure the parents that the infections are occurring with a normal frequency.

Recurrent upper respiratory tract infections are common in preschool children especially those who attend playgroups or equivalents as in this case. Studies have suggested that normal children under the age of 5 years may have at least 6 infections per year and some as many as 12. Thus, it is common to see children who would appear to have a permanent 'cold', so short are the gaps between infections. Some children described as having frequent coryzal infections in fact have an allergic rhinitis but typically, unlike the child here, they do not have fever.

Given the number of infections in normal children, who should be investigated? A reasonable approach is to reserve investigations for:

- dramatic and very rare situations, such as vaccine induced polio;
- children with 2 episodes of invasive meningococcal infection or more than one episode of pneumonia within a 12 months period;
- any child with recurrent staphylococcal infections such as discharging ears, lymph node abscesses, skin or other local infections;
- infants with intractable diarrhoea, failure to thrive, and erythroderma. They may have a severe deficiency of both B and T lymphocytes (severe combined immune deficiency syndrome, SCIDS) which can be fatal, but which can be cured by bone marrow transplantation.

Immunodeficiency states may be primary or secondary. Primary immunodeficiency syndromes vary in incidence from the relatively common, such as selective IgA deficiency with a prevalence of about 1 per 600 of the population, to the extremely rare. In clinical terms, there are four main components of the immune system which can be considered when planning investigations:

1. Phagocytosis (both granulocytes and macrophages)
2. Immunoglobulins
3. Cellular immunity
4. The complement system

The specific tests depend on the type of infections being investigated. One of the commonest deficiencies relates to IgG subclasses. For example, in a child with recurrent pneumonia due to *Streptococcus pneumoniae* or *Haemophilus influenzae*, the possibility of IgG2 deficiency should be considered because the bacterial wall of these organisms generates antibodies in the IgG2 subclass. Measurement of total immunoglobulin levels alone will often not show up a sub-class deficiency.

Fetal infections

Following the series of infections Fiona seems to settle and is not see again until 6 months later when she develops a febrile illness with a rash. Her mother is now 8 weeks pregnant. She has heard about the devastating effects of rubella in pregnancy and is very worried.

Fevers with rash are common in childhood and the main concern is whether the organism might be capable of affecting the fetus if mother contracts the infection. The acronym 'TORCH' covers most pathogens (but not all) which may affect the fetus: toxoplasma, syphilis, rubella, cytomegalovirus, and herpes. The range of effects varies widely between infections and the timing of infection during pregnancy is also important in determining the outcome.

It is the possibility of rubella that causes most concern and diagnostic problems. The clinical diagnosis of rubella is unreliable and must be confirmed by serology.

Fiona has been immunized against rubella so this diagnosis is unlikely.

A serological diagnosis in Fiona can be made by showing a threefold rise in antibody titres from paired sera taken at an interval of 10–14 days. An alternative for rubella is to show the appearance of specific IgM antibodies. These may be detected for about 4–6 weeks, followed by a specific IgG response. If rubella is clinically suspected in Fiona then measuring IgM antibodies to rubella is helpful as it involves only a single blood test and can give a rapid answer. If her mother was susceptible to rubella (i.e. she had no detectable rubella IgG), and Fiona had rubella IgM, then her mother (and therefore the fetus) could potentially catch rubella from Fiona. Serology on the mother would then have to be negative 2 and 6 weeks later to make the chance of infection unlikely.

On examination the rash is a non-specific maculopapular rash with mild nasopharyngitis and no lymphadenopathy. Serology shows Fiona to be have no rubella IgM antibodies but rubella IgG antibodies, consistent with prior immunization. The pregnancy proceeds normally to the delivery of a healthy full-term infant boy.

Meningitis

Fiona's baby brother is duly christened James at 3 months of age. A few hours after the service he is noted to be febrile and vomits his evening feed. He is unsettled and

Summary
Organisms that can affect the fetus (TORCH)

Toxoplasma

Other (syphilis)

Rubella

Cytomegalovirus

Herpes

also
Parvovirus B19

Summary
Meningococcal infection

- Rapid onset
- High case fatality rate
- Purpuric rash
- Decreased level of consciousness
- Septicaemia without meningitis has worst mortality
- Thrombotic complications compromise digits and limbs
- Pre-hospital penicillin may be life-saving
- Meningitic cases need audiological follow-up

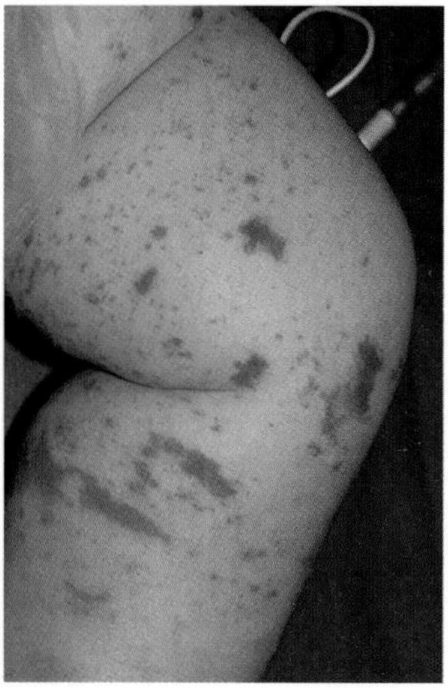

Fig. 18.2. Typical purpuric rash of meningococcal septicaemia

remains feverish overnight. The following morning he vomits twice further and refuses all feeds. When the GP comes she notes a rectal temperature of 39 °C but is unable to find any focus of infection. While she is in the home James has a generalized clonic-tonic seizure lasting approximately 2 minutes. James is given rectal diazepam and referred as an emergency to the local paediatric department.

James has had a fit in association with a fever. It is not a simple febrile convulsion because he is too young (see Chapter 20, p. 343), and other diagnoses must be sought. On arrival at the accident and emergency department James is febrile, pale, hypotonic, and irritable on handling. James is now an ill, infected infant needing urgent examination, investigation, and treatment. The history is focused on the presenting condition; details of family history and social background can wait.

Examination is directed towards confirming or excluding meningitis. Are there signs of meningismus or the presence of purpura (Fig. 18.2)? Is the fontanelle open and if so is it full or tense? One has to be cautious with this as it may bulge if the infant is crying. A high-pitched cry and irritability on handling may be the only signs suggesting meningitis. Neck stiffness is the classical sign in meningitis but may be absent in early infancy or masked by anticonvulsant medication. Inability to straight leg raise is rarely helpful under the age of 18 months. Although meningismus without meningitis may be found in some children with other infections (such as tonsillitis) a lumbar puncture must be carried out. The arguments for delaying or even omitting lumbar puncture are covered in Chapter 20, p. 350. The different types of meningitis occurring in children are shown in the box.

As well as examination of the cerebrospinal fluid (CSF), James must have cultures of blood and urine. If there is a purpuric rash it is possible to prick the lesion with a lancet or needle to produce a small blob of serum or blood which can be both cultured and examined with a gram stain for meningococci.

Urinary tract infections should always be considered and at this age are more common in boys than girls, sometimes presenting as an severely ill infant with septicaemia. Urine culture is important and to avoid delays in starting treatment or missing the diagnosis a suprapubic bladder tap is sometimes indicated in children under a year.

Since the only sign of pneumonia may be fever (and many feverish children are tachypnoeic without having pneumonia), a chest X-ray is also indicated. Additional investigations might include a full blood count with white cell differential, electrolytes (for biochemical evidence of dehydration), blood glucose (to check for hypoglycaemia and to compare with the value in the cerebrospinal fluid), and blood gases (if there is a possibility of respiratory failure). C-reactive protein (CRP) or erythrocyte sedimentation rate (ESR) may be measured as a marker of an inflammatory process, but of the two the ESR is slower to rise. Neither should be elevated in a viral infection.

The cerebrospinal fluid (CSF) shows white cells $500 \times 10^6/l$, all polymorphs with no red cells. The CSF glucose is 0.8 mmol/l (blood glucose 4.1 mmol/l), and protein 9.1 g/l. Gram stain shows Gram-negative diplococci but the antigen test is negative for meningococcus. The circulating neutrophil count is $23.4 \times 10^9/l$ with platelets $560 \times 10^9/l$. A few purpuric lesions appear on the foot prior to lumbar puncture.

The diagnosis was meningococcal meningitis on the basis of the culture of CSF and blood; the antigen test is often negative. The differentiation of bacterial and viral meningitis is usually possible on the basis of the cell count together with protein and glucose levels (Table 18.1). If there is doubt then it is safer to treat as a bacterial meningitis.

Jamie is treated for 7 days with an intravenous third-generation cephalosporin, and dexamethasone for 2 days.

The length of the antibiotic course in meningitis depends on the organism but in the case of meningococcal infection is usually about 7 days. It should be followed by rifampicin to clear carriage because:

- the infection does not confer immunity;
- the treatment itself may not clear carriage; and
- re-infection can occur even in immunologically normal children.

Dexamethasone is often given to reduce the rate of complications such as deafness.

Meningococcal infection has a significant mortality rate. Meningococcal septicaemia in the absence of meningitis carries a worse prognosis than when meningitis coexists: in this situation, the virulence of the infection is such that the patient presents during the initial septicaemic phase before the organism has penetrated the CSF. Additional poor prognostic features are low white cell count and thrombocytopenia.

Additional Information
Types of bacterial meningitis

Age	Types
under 3/12	Group B streptococcus E. coli Listeria monocytogenes
1/12–6 years	Neisseria meningitis Streptococcus pneumoniae Haemophilus influenza
over 6 years	Strep. pneumoniae Neisseria meningitidis

Viral meningitis occurs at any age and is usually less severe than bacterial. Common causes are mumps (now infrequent owing to immunization), enterovirus and Epstein–Barr).

Table 18.1.
Cerebrospinal fluid findings in meningitis

	Cells	Protein	Glucose
Normal	$< 2 \times 10^6/l$	< 0.05 g/l	
Bacterial	Neutrophils	Normal	40% blood glucose
Viral	Lymphocytes		Normal

In a serious life-threatening situation, such as bacterial meningitis, communication with parents is of central importance. Parents cope better when they have some understanding of what meningitis means, the nature of the treatment, and the most common short- and long-term complications. Individual parents often have different ways of coping and different requirements from their respective partner so that it is important to speak to both parents and answer their very different questions and concerns. It is also helpful to have written information.

Close contacts need prophylactic treatment to try to clear carriage of the organism; this is normally the responsibility of public health authorities. Rifampicin in four doses is currently the most widely used prophylaxis but a single dose of ceftriaxone is an effective alternative. Penicillin is not effective in clearing carriage.

Three months later Jamie was reviewed for neurodevelopmental progress, which was normal. His parents wanted to know whether he could get meningitis again, and wanted the result of the hearing test performed after discharge.

There can be no guarantee that Jamie will not get meningitis again, even with the same organism although probably not *H. influenzae*, against which he has had routine immunization, (see Chapter 4). However, meningitis due to any cause is rare. Hearing impairment is an important complication: transient deafness may occur during the initial few weeks, or there may be permanent sensorineural loss. Children should therefore undergo formal audiological assessment.

Further reading

Davies, E.G., Elliman, D.A.C., Hart, C.A., Nicoll, A., Rudd, P.T. (1995). *Manual of childhood infections*. W.B. Saunders, London.

Watson, J.G. and Bird, A. G. (1990). *Handbook of immunological investigations in children*. Wright, London.

19

Renal disease

Disease of the urinary tract in children is common, important, and may have far-reaching effects into later life. Reflux uropathy remains a depressingly frequent reason for end stage renal failure in adults. Congenital abnormalities of the renal tract are relatively common, and some may be detected on routine antenatal ultrasound. Unfortunately, the ability to detect congenital disease outstrips current knowledge of how best to manage it postnatally.

Renal disease may present with an obvious symptom such as haematuria, but is frequently silent, and may only be identified during investigation of a presenting symptom, such as failure to thrive. Renal disease may occur alone or it may be a more variable part of a multisystem syndrome. In childhood, the two commonest ways in which renal disease may come to light is through the detection of haematuria and/or proteinuria.

Surveys of school-age children have suggested that between 3–5% of girls and 1–3% of boys will have at least one urinary tract infection (UTI) in childhood. In spite of being so common, UTI generates more uncertainty and controversy than almost any other area in general paediatrics.

UTI is also non-specific in its presentation, so young children with symptoms such as fever or abdominal pain commonly have their urine investigated for infection. Such infections need treatment in their own right, but may have an underlying cause such as a congenital abnormality of the urinary tract, and infection prompts investigations designed to elucidate this.

Summary
Renal disease: presenting patterns

- Haematuria
- Proteinuria
- Systemic hypertension
- Acute renal failure
- Effects of chronic renal failure
  - failure to thrive, short stature
  - malaise and tiredness
  - anaemia
  - renal osteodystrophy
- Effects of tubular dysfunction
  - acidosis
  - amino-aciduria
  - hypokalaemia
  - hypophosphataemia
  - normoglycaemic glycosuria

Case studies
Urinary tract infection (reflux uropathy)

Laura is a 3-year-old girl whose parents contact their GP after she has been generally unwell for a few days with intermittent abdominal pain and lethargy. Having been dry at night for about 8 months she has also started to wet the bed again. They have become concerned because during the night she has been feverish and having shivering episodes. She is seen later that morning by a doctor with no focus of infection found. Her parents are asked to collect a sample of urine for analysis and culture.

When there is no obvious focus of infection, it is particularly important to investigate for UTI. Secondary nocturnal enuresis has many causes (see Chapter 6, p. 108), including psychosocial problems, diabetes mellitus, and UTI. In this child, the possibility of a UTI requires to be specifically considered because of the abdominal pain, rigors, and the lack of any obvious focus of infection.

It is easy to over interpret inflammation of the tympanic membranes in a crying child as 'otitis media' and fail to consider the possibility of UTI. It is difficult to elicit tenderness in the renal angle in a child who is resisting examination. There may be a history of renal angle pain but this will depend on Laura's ability to describe her discomfort. Various other possible presentations of UTI at this age are outlined in Table 19.1.

The sample of urine is collected by her mother and taken down to the GP's surgery. One part is sent to the local laboratory for culture and microscopy the remainder is checked by a reagent strip and shows moderate blood and protein but no glucose. She remains fevered and is started on an antibiotic pending the culture results.

Urine cultures are potentially problematic. Where there is delay between urine collection and its arrival with the GP any growth is impossible to interpret. In an ideal world, the urine would be collected and transferred at < 5 °C directly to the laboratory, but even then the loss of pus cells can be rapid. A number of strategies for urine collection and transfer have therefore evolved. Urine culture can be obtained by dipslides which are rectangular culture plates which are wetted in the urine flow, then sent for incubation at the local laboratory. An alternative is to use sterile containers with boric acid, a reversible bacteriostatic, which is sufficiently effective to allow postal transfer, or direct immediate microscopy of the urine. Where the child is unwell as in this case, it is reasonable to start an antibiotic pending the result of urine culture.

Laura's urine culture has arrived in the laboratory a few hours after being taken. Microscopy shows numerous pus cells, red blood cells, and organisms.

Table 19.1.
Features of urinary tract infection in childhood

Age < 12 months	Age > 12 months
Asymptomatic	Abdominal pain
Failure to thrive	Back pain
Febrile convulsion (> 6 months)	Change in urinary frequency
Fever	Febrile convulsion
Septicaemia	Unexplained fever
Irritability	Visible haematuria
Persistent neonatal jaundice	Pain on micturition
Poor feeding	Rigors
Vomiting	

The findings are consistent with a urinary tract infection but still require culture for confirmation. It should be remembered that pus cells in the urine may occur with fever, particularly in girls. Additionally red cells may contaminate the urine if there is an associated vulvo-vaginitis. The combination of pus cells and organisms is statistically more likely to be associated with a UTI than either alone. Furthermore, although pus cells are usually present in the urine at the time of a first UTI this is not the case in subsequent infections where they may only be seen in about 50% of cases.

> The urine culture is reported as *E. coli* > 100 000 colonies per ml. The organism is sensitive to the antibiotic that has been prescribed and within 48 hours she is afebrile and much improved. Her parents are asked to continue giving the antibiotic for 10 days.

The length of antibiotic courses for UTI has been the subject of some controversy but both single high-dose treatments and short (5 day) courses are associated with a small proportion of relapses.

> Laura is reviewed after completing her course of antibiotics and a repeat urine culture is sterile. The GP arranges a renal ultrasound scan at the local X-ray department which is reported as normal. After 3 months she has a second UTI very similar to the first, and then a third a few months later. Her parents are very worried that these may be affecting her kidneys and she is referred to her local paediatrician.

Laura has recurrent UTI. This problem is most common in girls and can occur despite an anatomically normal renal tract and no vesico-ureteric reflux. The mechanism is not fully understood but the most likely explanation is a dysfunctional voiding pattern.

The incidence of UTI in the preschool child remains unknown, but it is in this group that infections complicated by vesico-ureteric reflux may cause progressive renal scarring. Where such damage is sufficiently severe it can in turn lead to progressive glomerulosclerosis of the remaining kidney substance with gradual loss of function despite precautions to prevent further scarring due to infection.

After many years, chronic pyelonephritis may lead to hypertension or chronic renal failure. The proportion of children who will eventually go on to these serious complications is unknown. However, the experience of adult renal dialysis/transplant programmes suggests that childhood reflux nephropathy is still the commonest cause of end stage renal failure in adults under the age of 50. Paediatricians aim to prevent these late sequelae by early diagnosis and treatment of childhood UTI.

Summary
Possible lesions in children with UTI

- Renal scarring
- Vesico-ureteric reflux
- Obstructive uropathy
- Renal calculi
- Renal polycystic disease
- Any congenital abnormality of the urinary tract

The optimal investigation of UTIs in childhood is a heatedly debated area, which reflects the limitations of currently available methods. The one unequivocal conclusion from a substantial number of studies is that renal ultrasound alone is not an adequate investigation. The main aims of investigation are to:

(1) determine the presence or absence of renal scarring; depending on age and the presence or absence of scarring, exclude vesico-ureteric reflux;
(2) exclude an obstructive lesion;
(3) screen for the rarer underlying problems such as renal calculi or evidence of autosomal dominant polycystic kidney disease.

It is currently believed that initial renal scarring in association with vesico-ureteric reflux occurs in the preschool child (under 5 years). However, when scarring has already occurred, and where vesico-ureteric reflux continues, further scarring may occur up to the age of 10 years. Hence, in children under 5 years of age it is necessary to know both whether there is scarring and whether there is reflux, while in older children the critical feature is the presence or absence of scarring.

Scarring is best demonstrated by the dimercapto-succinic acid (DMSA) isotope scan. Its ability to show up abnormal renal scan patterns during or immediately after infections has led to the suggestion that it may be possible to avoid imaging for reflux in children over 12 months who have a normal DMSA scan. However in children under 12 months reflux must be sought. This is the basis of the algorithm for renal investigation of UTI shown in Fig. 19.1. Traditionally, the micturating cysto-urethrogram (MCUG) has been used to demonstrate vesico-ureteric reflux, but children with bladder control can be imaged successfully using an indirect micturating isotope scan. A 5-point scale is used to describe reflux. This ranges from 1 for reflux into the ureter only to 5 for gross dilatation with reflux up to the calyces and loss of the normal papillary impression.

The DMSA scan shows that Laura has scarring in the upper pole of the left kidney which contributes 25% of the total functioning renal mass. MCUG shows grade III vesico-ureteric reflux on the left side.

The *management of vesico-ureteric reflux* is either medical with prophylactic antibiotics, or an attempt to correct the reflux surgically. Studies to date have not shown a clear difference between medical and surgical strategies. Children with evidence of renal scarring require long-term follow-up, maesurement of blood pressure, urinary surveillance for new episodes of infection, and repeat imaging of the kidneys and urinary tract.

Summary
Managing vesico-ureteric reflux

- Antibiotic prophylaxis
- Regular urine cultures
- Repeat renal imaging
- Regularly check blood pressure
- Consider surgery
- Long-term follow-up

Why should one investigate a child having his or her first proven UTI after the age of 5 years when they are at a low risk for developing scars even if reflux is present?

All children with UTI require investigation. The problem lies in the concept of first proven UTI. It is all too likely that many children with UTI may not have the correct diagnosis recognized before they were 5 years old. Furthermore, it is now

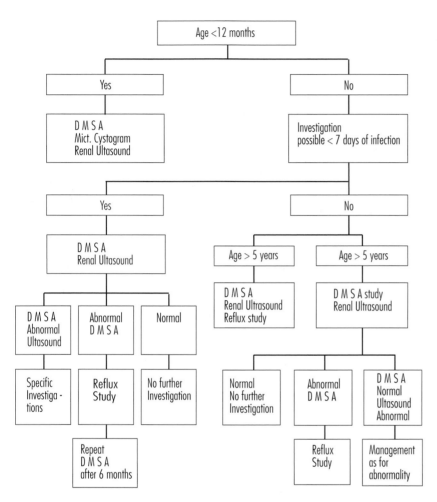

Fig. 19.1. Algorithm for the investigation of urinary tract infection.

clear that vesico-ureteric reflux can be an inherited disorder with an autosomal dominant pattern and that investigation of other siblings should be considered.

All children with proven UTI require investigation.

Urinary tract infection (obstructive uropathy)

Liam had been well until aged 10 days when he had started to scream between feeds and was started on an anticolic mixture which failed to settle him. By 2 weeks of age, feeding was poor with a weight which had fallen below his birthweight. He had started to vomit and to become increasingly lethargic. His GP referred him to hospital as an emergency at 3 weeks of age with possible pyloric stenosis.

Liam presents with poor feeding, failure to thrive and screaming episodes followed by vomiting. All of these complaints may be singly or in combination found in infants with UTI, while not in themselves being particularly specific.

On admission he looks poorly nourished but has no other abnormality. Urinalysis shows moderate blood and protein on reagent strip testing and urine microscopy shows large numbers of pus cells, and organisms. His plasma sodium is 124 mmol/l (normal 135–144) with a potassium of 4.9 mmol/l (normal for age 3.5–5.1). A spot urine sample shows a sodium concentration of 31 mmol/l.

There is marked hyponatraemia. Although plasma sodium levels may fall due to salt loss from vomiting, in such cases renal conservation of sodium causes the urinary sodium to become very low or almost undetectable. With significant urinary sodium concentrations, there must be renal loss. Although at this age the possibility of adrenal enzyme defects must be considered, they are rare, and hyponatraemia may complicate UTI because of a transient pseudo-hypoaldosteronism. This is often found in infants with an obstructive uropathy. Hence the management is to check a urine culture whilst replacing sodium and excluding a true adrenal (aldosterone) defect.

A urine culture collected from a suprapubic tap grew *E. coli* > 100 000/ml which was treated with intravenous antibiotics. A blood culture subsequently also grew *E. coli* with the same antibiotic sensitivity. Within 72 hours he was feeding well and putting on weight. Renal ultrasound showed a left-sided hydronephrosis with hydroureter subsequently shown to be due to a pelvi-ureteric obstruction. MCUG showed no evidence of vesico-ureteric reflux nor of posterior urethral valves. The obstruction was corrected surgically and 3 years later he has good bilateral renal function.

Summary
Haematuria

- Haematuria is the presence on microscopy of more than 5 red blood cells per high power field

- Normal urine contains a small number of erythrocytes

- Urine positive for blood on reagent strips may contain haemoglobin but no red cells

Table 19.2.
Characteristics of urinary tract infection

Age (months)	Female:Male ratio	% admitted to hospital	Total
0–6	0:5	69%	45
7–12	0:9	68%	31
13–18	1:7	57%	28
19–24	5:3	56%	25
> 24	3:1	7%	574

Unpublished data from the Royal Aberdeen Children's Hospital.

At this age, UTI is predominantly a male disease. Female preponderance takes over after the first year (Table 19.2). Younger children are more commonly admitted as an emergency, and neonates may have an accompanying septicaemia (see Fig. 19.1).

Haematuria (post-streptococcal nephritis) and glomerulonephritis

Allan is a previously healthy 7 year-old-boy whose mother brings him to his GP with a sore throat. The doctor finds inflamed tonsils with some tender enlarged anterior cervical lymph nodes. She says that antibiotics are not required and that he should be given paracetamol for his fever and discomfort. A few days later he feels better and returns to school. Twelve days after first noticing the sore throat his parents find his face looking rather puffy around the eyes and he is taken back to his GP. The presumptive diagnosis is hay fever and he is given an antihistamine. The following day the facial swelling is worse: he is taken back again to the doctor, who requests a sample of urine. The urine looks 'smoky' and on testing shows blood + + + and protein + + +. He is referred to the local hospital for further assessment.

Examination in hospital shows Allan to have some ankle oedema as well as facial swelling. His blood pressure is 125/85 mmHg. There is no rash and no cardiac murmur. Examination is otherwise normal. Urine microscopy shows red cell casts (Fig. 19.2).

Measurement of blood pressure is of central importance in any child presenting with a possibility of renal disease (including UTI).

The clinical picture together with the urine findings are consistent with an acute *glomerulonephritis*. The preceding history of a sore throat raises the possibility of a post- streptococcal glomerulonephritis (PSGN). The two main possibilities when macroscopic haematuria complicates an upper respiratory infection are PSGN and IgA nephropathy. IgA nephropathy often presents at the time of the infection while PSGN follows it after a lag period usually between 7–14 days although sometimes it may be delayed for up to 42 days.

Investigations show Allan's urea to be 11.4 mmol/l (normal < 6.5) and creatinine 85 mmol/l (normal for age up to ~70 mmol/l). Plasma albumin is 35 g/l (normal 36–48). A throat swab grows normal upper respiratory flora only. His antistreptolysin

Additional Information
Some causes of haematuria

- Exercise
- Benign familial haematuria
- Fever-associated
- Trauma
- Coagulation disorders
- Infection: any UTI, TB
- Glomerulonephritis
  - post-streptococcal
  - crescentic
  - membrano-proliferative
  - IgA nephropathy
- Henoch–Schönlein purpura
- Haemolytic uraemic syndrome
- Hypercalciuria (some with calculi)
- Autosomal dominant polycystic kidney disease
- Tumours: Wilm's, rhabdomyosarcoma of bladder
- Factitious (Munchausen by proxy)

Summary
The history in haematuria

- Preceding episodes of haematuria
  - dysuria, loin pain
  - skin rashes
  - joint pains
- Family history of haematuria, renal failure, kidney stones, or urological surgery

Fig. 19.2. Initial assessment of red urine for possible haematuria.

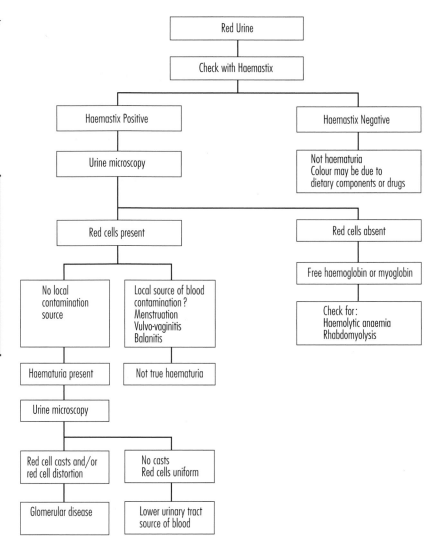

Summary
Examination in
haematuria/renal disease

1. Oedema: peripheral and periorbital

2. Blood pressure

3. Skin: rashes

4. Heart: enlargement, mitral murmur

5. Abdomen: palpable kidneys or bladder

6. Vulvo-vaginitis or a local injury

Summary
Glomerulonephritis

- Usually presents with haematuria

- Urine contains casts, protein, and red cells of glomerular origin

- Common causes are post-streptococcal and IgA nephropathy

- May be complicated by hypertension

- Post-infective causes have a good prognosis

O (ASO) level is elevated at 320 i.u./ml (normal < 160). Complement studies show a C3 of 23 mg/dl (normal 88–200) and a C4 of 32 mg/dl (normal 12–40) (Fig. 19.3).

PSGN follows infection with group A alpha-haemolytic streptococci which is most commonly pharyngeal but can be associated with impetigo. Not all strains of group A alpha-haemolytic streptococci will cause nephritis, but because subtyping is costly and time-consuming, empirical treatment with penicillin is justifiable. It is common for the streptococcal infection to have cleared by the time of the nephritis

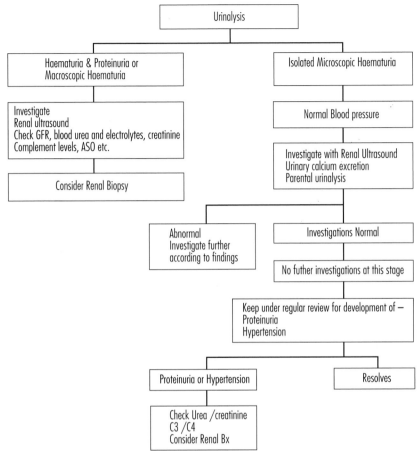

Fig. 19.3. Further assessment of microscopic haematuria.

so that Allan's normal throat swab does not exclude the diagnosis of PSGN. The ASO concentration is consistent with the suspected diagnosis but it is more useful to show a rising titre over time. The final and important finding is in the complement pattern which shows a low C3 and normal C4, suggesting alternate complement pathway activation which is typical (although not specific) in PSGN.

Allan's blood pressure rises to 130/105 mmHg.

The normal upper limit of blood pressure at this age is about 100/70 mmHg so he seems to have significant hypertension. Before starting treatment it is essential to ensure that a big enough cuff (covering at least two-thirds of the upper arm) has been used as smaller sizes give falsely high readings. It is also wise to ensure that several readings are taken before diagnosing hypertension as an elevated blood pressure may be related to the stress of being in hospital rather than to renal disease itself.

Summary Investigations for glomerulonephritis

- Urine culture
- Plasma urea, creatinine, electrolytes, albumin, immunoglobulins
- Urine culture
- Urinary protein: creatinine ratio (early morning)
- Timed protein excretion
- C3 and C4 components of complement
- Renal ultrasound
- Coagulation screen
- Antistreptolysin O level (ASO)
- Auto-antibodies
- Renal biopsy

Allan's hypertension is confirmed. He is treated with fluid restriction, a low salt diet, and nifedipine. Five days after admission his urine output suddenly increases and the blood pressure returns to normal. Blood biochemistry also returns to normal. His urinalysis still shows blood to be present at +++ but with protein now +.

The pattern so far is still consistent with PSGN. The next stage is to see whether the complement pattern has returned to normal. This usually occurs within 6 weeks of the initial disease. Persistence beyond that period of a low C3 raises the question of the alternative diagnoses including crescentic or membrano-proliferative glomerulonephritis or systemic lupus erythematosis. A similar picture can also be seen in bacterial endocarditis.

Six weeks after his admission the urine still shows ++++ blood but only a trace of protein. His blood pressure and C3 are now both normal.

The return of his complement concentration to normal is consistent with PSGN. Proteinuria tends to clear quickly, while microscopic haematuria persists beyond 6 months in around 50% of cases and may take up to 2 years to settle completely.

Six months after his admission the urine shows 2+ of blood but is negative for protein. At 10 months the urinalysis is negative for both blood and protein and remains so 2 years later.

The long-term prognosis should be good with end stage renal failure occurring in less than 1% of cases. Rarely, a child may present with a nephritic/nephrotic picture: this is a combination of nephritis as above combined with massive pro-teinuria resulting in the nephrotic syndrome (see below). This combination may occur in a number of other acute glomerulonephritic conditions such as Henoch–Schönlein nephritis and is associated with more glomerular damage and a higher risk of developing end stage renal failure.

Proteinuria and nephrotic syndrome

Anne is a previously healthy 3 years old girl who is taken to her GP with a 2-day history of increasing facial swelling. On examination her doctor notes that she has ankle oedema and ascites. A sample of urine shows protein ++++ but with no blood.

Summary
Benign causes of proteinuria

- Orthostatic (postural)
- Fever
- Exercise

Intermittent proteinuria is a fairly common and benign finding. Persistent proteinuria is less common, and is more likely to be associated with a significant underlying condition. Massive proteinuria with symptoms is always pathological.

Anne is sent to hospital for further assessment. Her blood pressure is found to be 80/55 mmHg. Her urea is 6.1 mmol/l, creatinine 58 μmol/l, and albumin 17 g/l (normal 36–50). Complement C3 and C4 are normal, as is a renal ultrasound scan. Her overnight urine protein excretion is 48 mg/m^2/hr (normal < 4) with a protein: creatinine ratio of 1500 mg/mmol (normal < 20).

The diagnosis is *nephrotic syndrome*, and there are no atypical features such as hypertension, haematuria, or renal failure. The definition of nephrotic syndrome is massive proteinuria causing hyopoalbuminaemia with consequent oedema. Hyperlipidaemia (low density lipoprotein fraction) is the fourth and less well-understood component of the syndrome. In over three-quarters of children under the age of 5 years this is due to light 'negative' or 'minimal change' glomerulonephritis (MCG) where the glomerular appearances on both light and electron microscopy are normal apart from the non-specific features found in any condition with massive proteinuria.

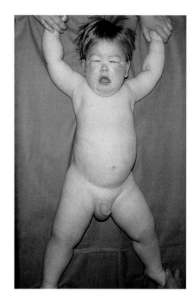

Fig. 19.4. Nephrotic syndrome. Note the marked generalised oedema. Facial oedema is usually more marked in the mornings. Ascites and hydrocoeles can also develop.

When the initial investigations have been completed, Anne is started on prednisolone (2 mg/kg ideal weight/day) and penicillin. She becomes progressively more oedematous but her blood pressure remains normal and urine continues to show protein only. After 13 days the proteinuria falls to + and then the following day is a trace only. The oedema clears and her albumin is found to be back to normal.

Children with nephrotic syndrome are intrinsically vulnerable to infection, particularly from *Strep. pneumoniae*, during the active phase when their immune system is compromised. It is therefore usual to prescribe prophylactic penicillin. It is important to monitor blood pressure as hypotension may be a consequence of the fluid leak from the vascular compartment.

Screening for proteinuria is most easily carried out using a urine reagent strip. This is accurate for albumin but will miss lower molecular weight proteins found in some renal diseases such as tubular disorders. An overnight timed urine collection is easier to perform but correlates well with a 24-hour collection. For regular monitoring or screening purposes an early morning sample collected as soon as possible after the child has got up is adequate.

A simple test for orthostatic proteinuria is to collect two urine samples, one in the early morning and a second later in the day and check the protein:creatinine

Summary
Nephrotic syndrome

- Massive proteinuria
- Hypoalbuminaemia
- Oedema
- Hyperlipidaemia

Additional Information Glomerular and tubular proteinuria

Glomerular
- Glomerulonephritis (damage to basement membrane)
- 'Minimal change' glomerulonephritis (MCG)
- End stage renal failure

Tubular
- Fanconi syndrome
- Pyelonephritis
- Lowe's syndrome
- Heavy metal poisoning

ratio in both. In addition the family can be asked to check two urine samples in a similar way with dip sticks and record the results.

Where investigations show *persistent proteinuria* there are many possible causes. It is always important to remember that glomerular and other diseases may have an orthostatic component so that it is important that the early morning protein level is normal and not just lower than the later sample. It should also be noted that in children with renal impairment urinary protein excretion tends to increase before other monitoring parameters such as plasma creatinine making it a useful monitoring tool. The stages in the investigation of proteinuria are shown in Table 19.3.

Anne's response to steroids was good, but if no improvement had been seen by 28 days of treatment she would have been considered steroid-resistant. Diagnoses other than MCG would then be considered, particularly focal segmental glomerulosclerosis and mesangial proliferative glomerulonephritis. The possibility of these diagnoses is increased if features such as haematuria, hypertension, or renal failure are present. These diagnoses can only be made on a renal biopsy.

Anne has responded well and is now ready for home. Her parents need to be aware that there is a high risk of relapse (~80%) which needs to be identified early and treated. Parents can be asked to check the urine on alternate days with a protein dip stick: if it shows ++ or more it should then be checked daily and if the proteinuria remains at or above that level, to start steroids again at the recommended dose.

Table 19.3.
Investigation of proteinuria

Primary
- Urine culture
- Urinalysis for blood and glucose
- Urine microscopy
- Early morning and late in the day urine for protein:creatinine ratio

Secondary
- For history of renal disease, abnormal physical findings, abnormal primary investigations
- Plasma urea, creatinine, electrolytes, phosphate, calcium, alkaline phosphatase, and plasma proteins
- Urine biochemistry
- Glomerular filtration rate (GFR)
- Timed urinary protein excretion
- ASO level
- Complement C3 and C4
- Renal imaging

Tertiary
- For heavy proteinuria, reduced GFR, persistently reduced C3, or possible familial nephritis
- Renal biopsy

Early treatment in steroid sensitive nephrotic syndrome often helps to limit the severity of relapses.

Some children relapse frequently, and in time may become either steroid-dependent (relapsing when steroids are reduced below a certain dose), or resistant to steroids. In these children, a single course of cyclophosphamide can restore steroid sensitivity or allow a reduction in steroid dose. The cyclophosphamide cannot be repeated because of its cumulative toxicity. Cyclosporin can also be used to maintain remission, but is also nephrotoxic.

Over the next 5 years Anne has three relapses which are treated at home without problem. She has subsequently gone into a longer-term remission of three years.

The risk of end stage renal failure in MCG is very low (<1%), so the long-term prognosis is very good. Even those who are frequent relapsers or steroid resistant eventually go into remission although this may take many years and can persist into early adult life.

Further reading

Holliday, M.A., Avner, E.D., Barrat, T.M. (1994). *Paediatric nephrology*. Williams and Wilkins, Baltimore.

Postlethwaite, R.J. (ed.) (1994). *Clinical paediatric nephrology*. Butterworth Heinemann, Sevenoaks, Kent.

20

Neurology

At least 20% of children admitted to hospital have a neurological problem either as the sole or associated complaint. In recent years, great strides have been made in our treatment of neurological conditions in children, and in our awareness of the aetiology of some of the cerebral palsies and the congenital myopathies. The development of more sophisticated neuroimaging techniques such as magnetic resonance imaging (MRI), the judicious use of a host of neurophysiological techniques (evoked potentials, videotelemetry, ambulatory electro-encephalographic monitoring), the advent of needle muscle biopsy, and rapid advances in neurogenetics have added new dimensions to the discipline. Despite this explosion in diagnostic techniques, paediatric neurology is first and foremost a clinical subject and therein lies much of its attraction and most of its challenge.

Many of the childhood neurological disorders have a genetic basis and others, such as the neuronal migration disorders, have their beginnings in the earliest days of fetal development. Other congenital malformations such as myelomeningocele (spina bifida, see Ch. 7, p. 138) evolve in early pregnancy. Fetal cerebrovascular accidents, intra-uterine infections (such as cytomegalovirus, rubella, toxoplasmosis, see Ch. 18, p. 317), and inborn errors of metabolism may all produce a legacy of neurological impairment in extrauterine life.

Cerebral palsy in 90% of cases has prenatal origins and only 10% have postnatal causes; less than a fifth of cases have their origin in intrapartum hypoxia. Infectious diseases (meningitis, encephalitis), trauma, and tumours all take their toll in childhood and most of the neuromuscular and neurodegenerative disorders present in early life. Certain neurological conditions, such as febrile seizures, occur exclusively in childhood whereas more adult neurological diseases, like multiple sclerosis, can sometimes take us by surprise by appearing in our young patients.

Paediatric neurology has a large network of interconnections: neurosurgeon, neuroradiologist, neurophysiologist, child psychiatrist and psychologist, ophthalmologist, geneticist, orthopaedic, and ear, nose, and throat surgeon are all close professional colleagues of the discipline. Treating the child with a neurological disorder is often best seen as a multidisciplinary exercise that transcends the artificial barriers of hospital and community services, and of course we must not forget that it is the parental role which is usually pivotal to success.

(See Chapter 7, pp. 125, for cerebral palsy and Chapter 18, p. 317 for meningitis.)

Case studies
Convulsions

Dougal, who was 6 years old, was sent up to the outpatients with a history of a 'funny turn' at night. Dougal shared a bedroom with his 10-year-old brother, Angus, who had been awakened at night by a noise. Angus found Dougal threshing around the bed and he hurriedly awoke his parents. At this point, all the parents

Additional Information
Conditions which can be
confused with seizures

- Breath holding attacks
- Simple faints
- Acute confusional migraine
- Cardiac arrhythmia
- Narcolepsy
- Night terrors
- Reflex anoxic seizures
- Benign paroxysmal vertigo
- Neonatal jitters
- Sleep myoclonus
- Benign infantile myoclonus
- Tics
- Benign paroxysmal ataxia
- 'Pseudo-seizures'

could describe was that Dougal could not be wakened, he appeared rather stiff, and both his arms and legs were jerking. By the time the GP arrived it was all over and Dougal just wanted to go back to sleep.

The GP spent most of his time allaying the parents' anxieties. While awaiting a hospital appointment, it was decided that Jim, his father, who was a very light sleeper, should share Dougal's room in case there was a recurrence. Three weeks later there were further attacks and Jim, a garage mechanic, was able to provide a good description of these which were identical. He was woken by Dougal making a grunt-ing noise (laryngeal spasm), there was an odd movement of Dougal's face with his mouth pulled to the left, and he was drooling. He was trying to say something but couldn't get his words out. Shortly after this Dougal seemed to lose consciousness, went stiff, and then there was jerking of all four limbs. The whole episode lasted 3–4 minutes. When Dougal recovered he could talk and said his face had felt funny at the beginning, but now he was very sleepy. He went back to sleep and the next morning he was entirely normal.

The story would fit well for a seizure, but it is important not to jump to this con-clusion too quickly: there are various other diagnoses whose *manifestations can be confused with seizures*. Most children with suspected seizures get referred to hospital for further evaluation, so Dougal was seen in the outpatient clinic.

Mother described Dougal as a previously well child whose perinatal history was normal, who had had no head injuries or serious illnesses and who had only once been in hospital (for a hernia repair in the first year of life). He had been perfectly well and normal on the nights before these attacks had occurred. There was no rel-evant family history but mother volunteered that her brother's child had had febrile seizures. Dougal was entirely normal on both general and neurological examination.

We are not always fortunate in getting a clear witnessed account of the 'funny turn' in question. Often, the parents have been too frightened by the event to observe the details, or sometimes the event has happened at school. The history is of utmost importance and here we were given the description of an initial simple partial seizure (consciousness retained, left-sided facial twitching) which became secondarily generalized (consciousness lost and tonic clonic seizures of the whole body following). This is classical of the condition known as benign Rolandic epilepsy and is found in up to 20% of children with epilepsy. The electro-encephalogram (EEG) showed the typical features of this condition. The purpose of performing the EEG was not to make the diagnosis of seizures, because the diagnosis was based on the clinical history. The EEG was used to define the type of epilepsy, and the location of abnormal discharges.

Epilepsy is the tendency to have recurrent seizures.

We must always be certain about the diagnosis before labelling a child with this disorder as it has profound implications for both the child and his or her family. About 5 schoolchildren in every 1000 have epilepsy. Most of these will have primary or idiopathic epilepsy (i.e. no underlying cause will be evident) but some will have *secondary epilepsy* due to a cause such as head injury, meningitis, birth asphyxia, tuberose sclerosis, etc.

There are many ways of considering the diagnosis of epilepsy. Not only do we need to think about an aetiological diagnosis, we also need to consider a seizure-type diagnosis. The most recent classification of the epilepsies is shown in Table 20.1. It divides the epilepsies into those that are generalized (where the whole of the brain is involved) and the partial (or focal) epilepsies (where the aberrant activity involves only a part of the brain). Another useful concept is to consider the epilepsies as age-related phenomena and Table 20.2 shows an example of this approach.

Summary
Secondary epilepsy

- Head injury
- Meningitis
- Encephalitis
- Birth asphyxia
- Tuberose sclerosis
- Inborn errors of metabolism

Table 20.1.
International classification of epileptic seizures

I.   Partial seizures (seizures beginning locally)
A.   Simple partial seizures (consciousness not impaired)
(1)   with motor symptoms
(2)   with somatosensory or special sensory symptoms
(3)   with autonomic symptoms
(4)   with psychic symptoms
B.   Complex partial seizures (with impairment of consciousness)
(1)   beginning as simple partial seizure and progressing to impairment of consciousness
(2)   with impairment of consciousness at onset
C.   Partial seizures becoming secondarily generalized
II.   Generalized absence seizures (bilaterally symmetrical and without localized onset)
A.   (1)   absence seizures
(2)   atypical absence seizures
B.   Myoclonic seizures
C.   Clonic seizures
D.   Tonic seizures
E.   Tonic-clonic seizures
F.   Atonic seizures
III.   Unclassified epileptic seizures

Adapted from *Commission on Classification and Terminology* (1981).

Table 20.2.
Epilepsy as an age-related phenomenon

- *A neonatal period* which extends to 3 months of age, during which seizures are related to a structural pathology and the prognosis is poor
- *3 months–4 years* in which the seizure threshold of the central nervous system is low and reactive seizures, especially with fever, are common. Serious epileptic syndromes such as infantile spasms (West's syndrome) and the Lennox–Gastaut syndrome occur in this epoch
- *4–9 years.* Here there is a predominance of primary or idiopathic generalized and partial epilepsies (e.g. childhood absence epilepsy and benign Rolandic epilepsy). Complex partial seizures secondary to structural brain abnormality also become better defined in this epoch
- *9 years onwards.* Primary generalized epilepsies, such as well-defined juvenile myoclonic epilepsy, occur while complex partial seizures are more frequently encountered.

From Aicardi (1992) and O'Donohoe (1995).

Summary
Prolonged seizures ('status epilepticus')

- Medical emergency
- First aid: place in recovery position
- May be treated with rectal diazepam (by parents or GP)
- May need intravenous barbiturate in hospital

Prognosis in the childhood epilepsies is very variable. There are those clearly benign conditions such as *petit mal* (generalized absences), Rolandic epilepsy, and juvenile myoclonic epilepsy, which, if treated properly, have an excellent prognosis. Most of the refractory epilepsies are found in children who have multiple disabilities or children who have acquired one of the malignant epilepsies (e.g. West's syndrome or the Lennox–Gastaut syndrome), which are associated with an inexorable decline in intellectual ability. *Prolonged seizures* ('status epilepticus') are seen most often in this group. The less common and less well-recognized 'non-convulsive status', where the child seems perpetually 'far away' and may have lost communication skills, also requires urgent attention.

Dougal's parents are anxious to know what further investigations might be needed, whether to restrict his activities, and whether he should be given treatment. They want to know what to do if he has another fit, and whether he needs to be followed up at the hospital.

Treating epilepsy is about far more than controlling fits. At this stage, it is not clear whether, if left untreated, Dougal's epilepsy might evolve to include daytime seizures as well. A full explanation must be given, indicating that in this case, apart from an EEG, no further investigation is required. The need for *further investigations* is determined by the seizure type (e.g. focal seizure), and the presence of positive findings on examination. Most parents wish to start medication, although benign Rolandic epilepsy spontaneously resolves in the second decade whether treated or not. It is important to acknowledge the trauma the parents suffered on witnessing Dougal's first fit (most parents in this situation think their child might die). Giving them basic first aid instructions about how to deal with possible future episodes improves their confidence. It is also important to discourage them from developing overprotectiveness, which is always counterproductive. Side-effects of the chosen anticonvulsant need to be discussed. Providing the parents with literature and showing the videos supplied by the Epilepsy Association and pharmaceutical firms is extremely helpful.

Dougal needs an explanation of what is happening to him and why he needs to take medicine. The school doctor and schoolteacher (with his parents' permission) should be informed, just in case he has a seizure in class. Children with epilepsy occasionally become the butt of teasing or are bullied by their peers and it is important to be mindful of this so that the problem can be tackled. Dougal must be encouraged to have a full, normal childhood, and with a *few restrictions in activities*, he and his parents must be left feeling the world is his oyster (just as it was for Julius Caesar and Napoleon, both of whom had epilepsy).

Both the complexity and fascination of childhood epilepsy lie in its heterogeneity. At one end of the spectrum we will find the child who attends normal school and is achieving, and at the other is the child with multiple disabilities, the seizure disorder being only one of them. Somewhere along the spectrum is the child who shows unblemished neurodevelopment until 18 months and then develops infantile spasms, almost guaranteed to blight his future potential, or the child who develops absences (*petit mal*) at 6 years, only to outgrow it at 10 years. It is a disease which was well known to Hippocrates and across the centuries has been mystified and stigmatized and even modern day parents consider the diagnosis with dread. The explosion of new anti-epileptic drugs, new developments in surgical approaches to treatment, and new discoveries in genetics have brought a welcome optimism to the management of the child with epilepsy.

Febrile convulsions

Andrew, aged 2 years, developed a running nose and cough on a Saturday night, and became progressively more feverish on the Sunday. His parents called the GP and a member of the deputizing service saw him, at which point his temperature was 37.9 °C and he was quite well. On the maternal grandmother's advice he was given a drink, and put to bed well-wrapped in blankets, close to a radiator. An hour later they heard a strange noise, and going to him they found him stiff and blue. They attempted mouth-to-mouth resuscitation and called an ambulance, which brought him to the accident and emergency department. On arrival there he was pink, not responding to his parents, and having jerking movements of all his limbs.

Summary
Further investigations of seizures

- Neuroimaging: CT, MR, ultrasound in infants
- Biochemical investigations
- 24-hour EEG
- Videotelemetry

Summary
Restrictions in activities

- swimming only with supervision
- not riding a bicycle in traffic
- not climbing heights

Andrew is clearly having a seizure. From the ambulance crew and the parents it appears that the fit has lasted nearly 20 minutes.

The duty senior house officer requests some diazepam to give rectally. Just as it is produced the fit stops spontaneously.

The senior house officer (SHO) finds out more information.

Andrew has been a totally normal child neurologically, with no previous history of epilepsy. The seizure clearly accompanied a high temperature, and was always seen to be generalized (although the onset of the seizure was not witnessed). The grand-parents volunteer that Andrew's mother took two short seizures with a fever when she was about the same age. Examining Andrew reveals no other physical sign apart from his temperature (now 38.9 °C) and his evident coryza.

Summary
Simple febrile convulsions

- Neurologically normal child
- No previous history of fits (except simple febrile convulsion)
- Fever
- Age 6 months–6 years
- Duration less than 30 minutes
- Generalized fit

The SHO therefore decides that he has probably had a *simple febrile convulsion*, and spends some time explaining about how to manage a hot child with a tendency to febrile convulsions. This advice includes:

(1) the use of anti-pyretics (paracetamol, not aspirin because of its rare association with Reye's syndrome);
(2) the avoidance of over-heating;
(3) resisting the temptation to strip all clothing off and expose the child in a cold room or give a cold bath—because cutaneous vasoconstriction results, and core temperature can actually rise further;
(4) explaining that Andrew may well have inherited the tendency to have febrile convulsions, and advising basic first aid (recovery position, and calling an ambulance after 10 minutes if the fit does not stop by itself).

As with any other first fit, the parents have been terrified; consequently they need a great deal of understanding and support. Written information, which they can absorb at leisure, and discuss later with their own GP, is invaluable.

An hour later, Andrew is playing happily in the waiting area, dressed only in a nappy and a T-shirt. His parents can see that there is nothing to be gained by admit-ting him to hospital, and take him home.

Children usually make a rapid recovery, generally because it has just been an acute viral infection, but need to be observed for a while to make sure the fit is not associated with a serious illness such as meningitis or septicaemia. The tendency to febrile convulsions often runs in families, sometimes in an autosomal dominant pattern, and this may emerge when grandparents can contribute a family history. A first febrile convulsion may be the precursor to others, but prophylactic treatment with anticonvulsants is ineffective in the short term, and does not modify the outcome in the long term. The natural history of simple febrile convulsions is benign: children grow out of them by school age, and there is no long-term risk of epilepsy.

Headache and migraine

Jamie is 9 years old. His headaches had started 3 or 4 months ago, had become increasingly frequent and latterly they had been accompanied by vomiting. Several medicines had been prescribed 'which made no difference' and now the GP was worried, the parents were worried, and, of course, Jamie was worried. An urgent outpatient appointment was requested.

In the clinic, Jamie described his headaches as 'being all over, or sometimes over one eye', occurring perhaps 2 or 3 times a week, starting at any time of day and usually persisting for several hours. He found it impossible to describe the nature of the pain but volunteered that it sometimes made him cry. Sometimes he felt sick when he had the pain and on 2 occasions he had vomited. He had never had any unusual visual experiences such as spots before the eyes or blurred vision, never had abdominal pain and had no other odd sensations. Neither James nor his parents recognised any precipitating factors but noise made his headaches worse and lying down in a quiet, darkened room often helped. Sleep made the headaches disappear.

The history is strongly suggestive of migraine but we need to elicit a number of other key points.

Jamie had been a previously well child, but had had mild eczema in infancy. There had been no previous head injury (ruling out post-concussion syndrome) and his parents described him as an outgoing child. He was well settled in school, of above average ability, and had the usual, normal, ambivalent relationship with his two siblings. There appeared to be no marital discord or financial worries in the family (making psychogenic headaches unlikely) and his father, an accountant, was in good health. His mother, a housewife, suffered with occasional bad headaches (but had never thought of them as migraine) but her father had severe migraine. Both parents reported that Jamie appeared entirely normal between his headaches, they had

Summary
Examination for headache

- Examine the fundi and the cranial nerves
- Measure the head circumference
- Measure the blood pressure
- Percuss and auscultate the skull (for an arteriovenous malformation, which often has a bruit)
- Ataxia suggests a space-occupying lesion, or hydrocephalus
- Inspect the skin for evidence of neurocutaneous syndromes

Summary
Defining migraine

The headache is paroxysmal and has 3 of the following features:
- throbbing nature
- unilateral location (less common in children)
- relief after sleep
- presence of an aura
- associated abdominal pain, nausea, or vomiting
- family history of a similar condition

Additional Information
Rarer forms of migraine

- Acute confusional migraine
- Basilar migraine
- Benign paroxysmal vertigo
- Hemiplegic migraine
- Ophthalmoplegic migraine

noticed no clumsiness, no wobbliness, no slurring of speech, no change in his personality, and there was no loss of skills.

We are not just amassing evidence in favour of this being migraine and reducing the likelihood of other possible diagnoses such as psychogenic headache, a space-occupying lesion, etc. We must recognize that we are dealing with a family with two problems: the first is Jamie's headaches and the second is the tremendous unvoiced collective anxiety that the headaches may have a sinister underlying cause (e.g. cerebral tumour).

Thorough examination will include measuring the head circumference, percussing, and auscultating his skull (for an arteriovenous malformation, which often has a bruit). The cranial nerves must be examined and of course the fundi. The presence of ataxia would suggest a space-occupying lesion, or hydrocephalus. When the neurological *examination* has been completed, the skin must be inspected for evidence of neurocutaneous syndromes and the blood pressure measured.

Where there are no abnormal physical findings on full neurological examination, as in Jamie's case, we can be very reassuring to the parents, indicating that much the most likely diagnosis here is *migraine*. When the parents have witnessed a thorough examination of their child, they almost always accept this (with great relief) and it is rare for even the most anxious to demand some form of neuroimaging.

We then need to offer Jamie an appropriate explanation for his symptoms, sympathizing with his headaches but acknowledging that while they may be very severe, they are not serious and are not associated in any way with a problem like a brain tumour. (Nine-year-olds are more sophisticated in their superficial medical knowledge since the advent of television soap operas!)

The severe throbbing headache of migraine is associated with dilatation of the cerebral vasculature. In cases of 'classical' migraine, this is preceded by an aura, thought to be associated with cerebral vasoconstriction. Jamie's migraine is the more common type, not preceded by an aura. Attacks may be provoked by stress, exercise, head trauma, dietary factors, menstruation, etc. In childhood, and especially in young children, migraine commonly has features such as cyclical vomiting, recurrent abdominal pain, and other periodic phenomena such as recurrent fever and paroxysmal limb pain that have been described as the 'periodic syndrome' (see Ch. 14, p. 254). Often, it is only when such episodes are replaced by migraine that the earlier episodes are appreciated as migraine equivalents. Some *rarer forms of migraine* may also present in childhood and these can present a considerable diagnostic challenge.

Having explained the various precipitants or triggers to migraine headaches, it is useful to ask Jamie and his parents to keep a diary of his headaches over the next 8 weeks and record his dietary intake for 4–8 hours prior to each headache. When Jamie returns to the outpatient clinic in 2 months, he may have identified some food such as cheese, cola, or chocolate, which has consistently precipitated his headaches. Removing the culprit from the diet is the best form of treatment, but obsessional elimination dieting must be avoided. It is more likely that no particular food can be incriminated. If some stressful situation at school has been identified as a trigger, it can be helpful to enlist the help of the school doctor in order to try and relieve this. Regular meals and sufficient sleep should be recommended and a lifestyle of 'moderation in all things'. In the long term childhood migraine has a good prognosis. Long remissions occur and many children outgrow their attacks. Commonly, child and parents report that the headaches are still happening but that 'they just don't bother him any more'.

If the headaches are still troublesome, as they will be in about 10% of children, and further examination reveals no abnormal signs or any other problems, then *medication* will need to be discussed. Many children do not respond to antimigraine treatment and it should be further explained that Jamie's lack of improvement when treated with various medications was entirely compatible with the diagnosis.

Table 20.3.
Some causes of headache other than migraine

Brain tumour:
Intermittent, often nocturnal. Vomiting, neurological signs

Vascular malformation
Fixed location headache:
Neurological signs, seizures

Malignant hypertension:
Throbbing pain. Seizures. Transient visual disturbance

Hydrocephalus (and congenital malformations):
No characteristic quality. Large head

Paranasal sinusitis:
Fullness, pressure, tenderness over sinus

Intracranial abscess/Chronic meningitis:
Neurological or meningeal signs

Benign intracranial hypertension ('pseudo-tumour cerebri'):
Non-distinctive. Papilloedema, diplopia

Cluster headaches:
Periodic. Facial pain, lacrimation. Horner's syndrome

Psychogenic headache:
Pressure, aching, tightness. Anxiety and/or depression

Post-concussion syndrome:
History of head injury. Dull aching headache. Anxiety/depression

Ophthalmological problems can cause headache
(rare)

From Aicardi (1992).

Summary
Managing migraine

- Identify triggers
  - food
  - stress (e.g. school anxiety)
- Promote regular lifestyle
- Discuss drug treatment for attacks
- Consider prophylaxis if attacks are frequent, or losing significant time off school

Additional Information
Medication for migraine

- *Treating the attack*
  - paracetamol
  - ergotamine tartrate (oral or rectal)
  - chlorpromazine (rectally if vomiting)
  - buclizine (e.g. Migraleve)
  - sumatriptan
- *Regular prophylaxis*
  - pizotifen
  - clonidine
  - propanolol
  - calcium channel blockers.

Summary
Symptoms of brain tumour

- Headache:
  - all ages except infancy
  - sometimes paroxysmal, mimicking migraine
  - worse with coughing, straining, exercise, and sleep
- Change in personality
- Clumsiness or loss of motor skills
- Nausea, vomiting
- Slowing of growth

Summary
Some important causes of childhood encephalopathy

- Trauma
- Poisoning
- Hypoxia-ischaemia (near-miss SIDS, near drowning)
- Meningitis, encephalitis
- Mass lesions (haematoma, abscess, tumour)
- Fluid & electrolyte disorders (severe dehydration)
- Blocked CSF shunt
- Status epilepticus
- Hypertensive encephalopathy
- Hemiplegic or basilar migraine
- Endocrine dysfunction (hypoglycaemia, diabetic ketoacidosis)
- Renal failure
- Hepatic failure (Reye's syndrome)
- Inborn errors of metabolism
- Haemorrhagic shock/ encephalopathy
- Iatrogenic (overrapid correction of dehydration; drug overdosage)

Other headache

Headache is an important symptom in childhood because it is extremely common. Some 4–5% of children suffer with migraine. The challenge of the clinical problem lies in separating the rare headache of severe organic cause (*brain tumour*, cerebral abscess) from the mass of benign ones (Table 20.3)

'Psychogenic' or 'tension' headache is often described as a feeling of pressure or tightness which increases as the day wears on. This continuing low-intensity headache has no associated symptoms or signs. It is usually caused by stress (which the child is unlikely to recognize) and the appropriate treatment is to clarify the underlying psychological cause and attempt to relieve it.

At all ages (except infancy) headache is an important symptom of brain tumour, especially those in the posterior fossa where the majority of childhood brain tumours arise. Sometimes, the headache is paroxysmal with intervals of remission, mimicking migraine, but it is exacerbated by factors which influence intracranial pressure such as coughing, exercise, and sleep. Sometimes, the headache of brain tumour may precede any other signs or symptoms but certainly by 4 months after its onset almost all children will have neurological signs, a change in personality, a slowing of growth, or other features.

Coma (encephalopathy), encephalitis

Kirsty, 13 years old, was found very drowsy in bed that morning by her mother, a single parent, who called the GP. Kirsty had gone to school the day before, but on her return complained of headache and refused her tea. She insisted on going to her friend's house, as previously arranged, but came home early because of worsening headache. She was given 2 paracetamol tablets and went to bed early. On further questioning it appeared that Kirsty had been rather withdrawn and not herself for a day or two and had vomited her school lunch the day before.

Any state which involves a change in the level of consciousness is termed an 'encephalopathy'. A large number of disorders may lead to encephalopathy and the primary pathology may arise within or outside the central nervous system. The commonest cause of acute non-traumatic child encephalopathy is intracranial infection but this is closely followed by hypoxic ischaemic insult.

The GP found that Kirsty's level of consciousness was diminishing at an alarming rate and arranged an emergency admission via the accident and emergency department.

The hospital management of a child with a depressed level of consciousness is a team effort. It is important to make a rapid general assessment of the patient immediately to ensure that no cardiorespiratory support is necessary at this stage. Provided there is no evidence of shock and respiration appears adequate, one member of the team can set about inserting an intravenous line and draw blood for baseline biochemical investigations while a nurse inserts a nasogastric tube and empties the stomach.

A careful history often provides clues to the diagnosis (Table 20.4).

Table 20.4.
Encephalopathy: clues from the history

Immediate: recent viral infection (encephalitis); found lifeless in cot (near-miss sudden infant death); empty drug bottle nearby (drug/substance abuse)

History of pre-existing disorder: diabetes, epilepsy, migraine, sickle-cell disease, CSF shunt

Family history: tuberculosis, epilepsy, previous unexplained deaths in infancy

Social history may alert to possibility of non-accidental injury

Kirsty had been a previously well child, there was nothing in the family history of relevance and mother felt that the possibility that she might have indulged in substance abuse was remote.

In Kirsty's case, the history was not particularly informative. Acute deterioration suggests a metabolic disturbance, poisoning, and cerebrovascular accidents. Deterioration over days or weeks is in keeping with infections, chronic ingestions, or raised intracranial pressure.

Kirsty was a well grown, mildly febrile girl but there were no clues to the cause of her encephalopathy on examination (Table 20.5). On neurological assessment,

Table 20.5.
Encephalopathy: clues from the examination

Breath odours: diabetic ketoacidosis, glue sniffing, aminoacidopathies

Bruises, abrasions, burns, torn frenula, bruised genitalia: non-accidental injury

Skin abnormalities: neurocutaneous disorder; minor abrasion (toxic shock); cold sore (herpes); petechial rash (meningococcaemia); bleeding into skin/orifices (haemorrhagic shock/encephalopathy); parotid swelling (mumps meningo-encephalitis)

Head: large circumference, splayed sutures, crackpot sound on percussion (hydrocephalus); bulging fontanelle (raised intracranial pressure); bruits (arteriovenous malformation); shunt present (blockage)

Meningism: bacterial meningitis, intracranial bleed

Midline pit over spine: CNS infection due to dysraphism

Eyes: proptosis, unilateral exophthalmia

Papilloedema: raised intracranial pressure from any cause

> Kirsty opened her eyes to pain, made grunts but not words when spoken to and was able to localize pain, giving her a score of 9 on the Adelaide Paediatric Coma Scale (Table 20.6).

The examination is sometimes very revealing. We were able to conclude that although Kirsty was encephalopathic, she was not deeply unconscious, there was no meningism, brainstem function was intact (corneal, pupillary, and oculo-cephalic reflexes were normal, and respiratory pattern and heart rate were normal), there were no focal signs and the presence of severely raised intracranial pressure was unlikely (Table 20.7).

Should an urgent lumbar puncture (LP) be performed? It is well recognized that some deaths in bacterial meningitis are associated with 'coning' (herniation of the brainstem through the foramen magnum) following LP. Most children with acute meningeal signs come to no harm from LP. In the seriously ill child whose consciousness is impaired, where there are fundoscopic signs of raised intracranial pressure, focal signs or other signs of incipient coning, it is wise to treat with antibiotics first and defer the LP for 24 hours.

Table 20.6.
Comparison of Adelaide Paediatric Scale (APS) and Glasgow Coma Scale (GCS: is appropriate for older children)

Paediatric scale (APS)		Adult scale (GCS)
Eyes open		
Spontaneously	4	
To speech	3	As in paediatric scale (APS)
To pain	2	
None	1	
Best verbal response		
Orientated	5	Orientated
Words	4	Confused
Vocal sounds	3	Inappropriate words
Cries	2	Incomprehensible sounds
None	1	None
Best motor response		
Obeys commands	5	
Localizes pain	4	
Flexion to pain	3	As in paediatric scale (APS)
Extension to pain	2	
None	1	

Table 20.7.
Symptoms and signs of acutely raised intracranial pressure

All age groups:
- Decreased conscious level
- Lethargy
- Vomiting
- Convulsions
- Decorticate/decerebrate posturing
- Respiratory arrest

Infants:
- Reluctance to feed
- Irritability
- Full fontanelle
- Spreading of sutures
- 'Sunsetting' eyes
- Crackpot sign
- Rapid increase in head size

Toddlers and older children:
- Anorexia
- Headache
- Nausea
- Diplopia
- Papilloedema
- False localizing signs

Although suspicion of bacterial meningitis was not high (no high grade fever, no meningism) Kirsty was treated with antibiotics and acyclovir (an antiviral agent) immediately. An urgent CT scan revealed a 2 cm dense lesion in the left temporal lobe which was surrounded by some oedema (Fig. 20.1). There was no midline shift and no signs of raised intracranial pressure. This was suggestive of *Herpes encephalitis* and a lumbar puncture was carried out.

The fluid showed a mild lymphocytosis with normal glucose. The protein was moderately elevated. *Herpes simplex* titres in the blood rose from 32 initially to 64 some 5 days later, suggesting a recent infection. Cerebrospinal fluid *H. simplex* titres were 1 in 8. Kirsty's EEG demonstrated an abnormality with high voltage slow wave activity seen over the posterior left hemisphere consistent with a localized encephalitis.

Kirsty was fully supported and very carefully monitored. She was treated with acyclovir intravenously for 7 days. She developed no seizures and made steady, slow progress. Throughout the admission Kirsty's mother required a great deal of support.

It is important to be aware that when dealing with a sick child close contact must be maintained with the parents to keep them informed of events. It is important to spend plenty of time with parents, encouraging them to talk and express their

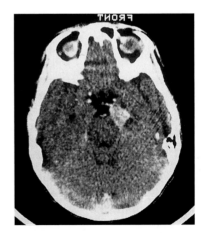

Fig. 20.1. Transverse computerized tomography (CT) scan showing an enhancing dense lesion approximately 2 cm in diameter in the left uncus with 1 cm of surrounding oedema. This is characterristic of *Herpes encephalitis*.

Summary
Herpes encephalitis

- Caused by *Herpes simplex*
- Presents with encephalopathy
- Has characteristic, but not consistent, EEG changes
- May be diagnosed by blood and CSF antibody titres
- Is treated with acyclovir
- Children often require intensive care and respiratory support
- With early aggressive treatment, usually has a good outcome

thoughts and concerns. Kirsty's mother felt bewildered and guilty and felt she should have realized the severity of Kirsty's illness sooner.

At discharge Kirsty was well and had no neurological impairment. Herpes simplex encephalitis can have devastating effects on cerebral function and commonly did before the advent of acyclovir, but Kirsty was diagnosed early and treated promptly and adequately. Psychometric assessment carried out some months later revealed an average IQ and long-term follow-up did not reveal any delayed sequelae such as epilepsy.

Further reading

Aicardi, J. (1992). *Diseases of the nervous system.* MacKeith Press, London.

Brett, E. M. (ed.) (1991). *Paediatric neurology.* Churchill Livingstone, Edinburgh.

O'Donohoe, N. V. (1995). *Epilepsies in childhood.* Butterworth Heinemann, Sevenoaks, Kent.

21

Orthopaedic problems

Orthopaedic disorders comprise 10–15% of children's health problems. Eliciting an accurate history may be difficult; examining a screaming child is impossible. The approach must always be slow and gentle. Use of the unaffected limb for comparison of movement is useful—joint movement varies with the age of the child. In this chapter, some of the common problems will be discussed, their presentation and treatment. With many conditions early diagnosis and treatment are essential to prevent permanent bone damage leading to long-term disability.

Case studies
Limping in a toddler

Emma (age 16 months), has limped since she walked on her own at 14 months. Her parents initially thought she was unsteady, and only worried when the limp persisted, when they decided to take her to see their GP. The limp did not distress Emma. The parents were uncertain as to whether she had had a hip test at birth as she was a premature baby. The mother had noticed that the left leg did not flop out like the right leg when changing her nappy.

A limp is an abnormal gait pattern which varies with the causative agent. There are several common *causes of limping*.

Relevant factors include the age and sex of the child, the general health, the onset of the limp, the presence of pain. The most common cause of a painful limp is trauma, but the history may not be forthcoming—the younger child forgets, the older child may prefer to forget!

Emma walked in quite happily but with an obvious dipping gait—the Trendelenburg gait. When the weight is transferred on to the unstable hip, the pelvis dips onto the opposite side. Examination of the hips confirmed shortening of the right leg with tightness of the right adductor muscles. The telescoping test for instability of the right hip was positive. There was no history of trauma, nor was there any pain, or limitation of movement to suggest infection. The GP arranged to have the hip X-rayed, and then a referral to the orthopaedic surgeon. The X-ray showed that the left femoral head was outside and above the acetabulum.

This is a classical late presentation of *congenital dislocation of the hip* (CDH).

The Ortolani test for CDH is routine in the newborn. (See chapter 8, p. 154) The hip ligaments are lax, and a positive test, producing a clunk, indicates a

Summary
Causes of limp

- *Pain*: trauma, infection
- *Instability*: congenital dislocation of the hip
- *Shortening* of one limb
- *Spasticity* of the muscles
- *Weakness* of the muscles
- *Ataxia*

Summary
Differential diagnosis of the toddler with a limp

- Congenital dislocation of the hip
- Trauma
- Infantile coxa vara
- Infection
- Talipes
- Cerebral palsy

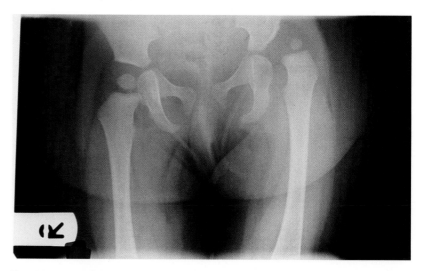

Fig. 21.1. Congenital dislocation of left hip. Note upward and outward displacement of the femoral head. (L) femoral head is smaller than (R). (L) acetabulum is shallower and more open or sloping.

Summary
Congenital dislocation of the hip (CDH)

- Occurs 1–2/1000 live births
- 8 × more common in girls
- 50% bilateral
- Most common in breech presentations

dislocatable hip. If positive, this would be followed by the application of an abduction splint to stabilise the hip until the capsule tightens. Unfortunately, a significant number of cases are missed; so untreated CDH remains a problem. Any limp in a toddler demands urgent referral and x-ray.

Emma was admitted for treatment with her mother. She was nursed in a cot, head down at a 45 degree tilt, with fixed traction on the left leg and weighted traction on the right. This gradually stretched out the tight soft tissues and the hip came down opposite the acetabulum. At this stage, the hip could be gently reduced under general anaesthetic, dividing the tight adductor muscles subcutaneously first. Plaster was then applied with both hips flexed, abducted, and externally rotated.

Emma will spend the next 15–18 months in plaster—having it changed every 2–3 months to allow for growth–until the X-rays confirm the femoral head and acetabulum are developing normally. The longer the hip is out of joint the longer the treatment.

The prognosis for congenital dislocation of the hip at this age, if treated by gentle closed reduction, is that 80% will have a successful outcome. If treated at birth, the hip should become completely normal hence early diagnosis is extremely important.

Limping in a young child

Peter (age 6 years), is a fit child who walked at 1 year. Two months ago, Peter started limping. Occasionally he said his left knee was sore, but he continued to play happily. His parents could remember no injury. Eventually they took him to the GP, who referred him to hospital. Initially, Peter's limp was hardly detectable, but after walking up and down it became more obvious.

Lying on the couch, he held the left hip semi-flexed and externally rotated. The knees were normal. He had marked restriction of abduction and internal rotation in the left hip—the classical signs of *Perthes' disease* which were confirmed on X-ray.

Perthes' disease (pseudo-coxalgia) is an avascular necrosis of the upper femoral epiphysis—aetiology unknown; but possibly a combination of genetic factors and trauma to the hip causing an intra-articular effusion. This increases the intra-articular pressure and compromises the circulation to the epiphysis. This leads to necrosis of the epiphysis. The dead bone is then gradually removed, and the new bone laid down producing a fragmented appearance on X-ray (Fig. 21.2). During this stage, the femoral head is rather plastic and can be moulded by outside pressure. The final phase is that of bone healing—any residual deformity will be permanent.

Summary
Differential diagnosis of a limp in a young child

- *Hip*
  Perthes' disease (pseudo-coxalgia)
  Trauma
  Infection
  Irritable hip syndrome
  Juvenile arthritis
  Missed coxa vara
- *Knee*
  Discoid meniscus
  Baker's cyst
- *Feet*
  Talipes
- *General*
  Shortening of the limb
  Cerebral palsy
  Early manifestation of a dystrophy

Additional Information
Perthes' disease (pseudo-coxalgia)

- Perthes' disease occurs between the ages of 3 to 11
- 4 × more common in boys
- 15% of cases are bilateral but not necessarily simultaneously

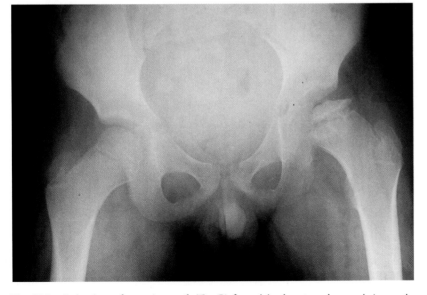

Fig. 21.2. Early phase of necrosis, age 6. The (L) femoral head appears denser relative to the decreased density of the metaphysis. The ossific nucleus is smaller, but the cartilage space is thicker.

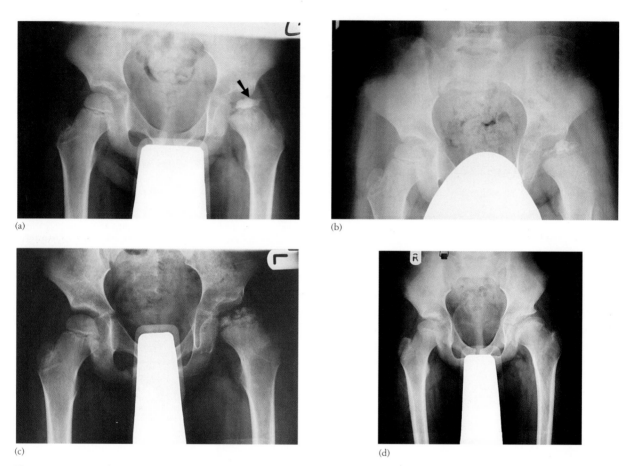

(a) (b) (c) (d)

Fig. 21.3. (a) phase of revascularization, age 6.75. The central dense area (arrowed) is where new bone is being laid down on the dead trabeculae on the nucleus. (b) The head is now looking 'fragmented' as new bone is laid down and dead bone is resorbed, age 6.75. (c) The healing phase. The head is 'filling in'. Note how the metaphysis has broadened a little, the femoral head has broadened and become partially subluxed, age 7.75. (d) Healing complete (age 12). Note the residual deformity—broadening and flattening of the femoral head. The articular cartilage is normal.

Peter was admitted the following day. Routine blood tests were normal. He was put on to traction, gradually increasing the abduction and internal rotation of the hip. After 3 weeks, this was almost normal, so Peter was allowed up gradually. No further limp developed so after 1 week he went home. He has to sleep with a pillow between the legs to abduct the hips, he cannot jump or play games but he can go to school. This regimen will continue until his X-rays show the left hip has healed— this may take up to 1 year.

In a young child like Peter revascularization is relatively rapid (see Fig. 21.3), so less moulding of the femoral head will occur; and minimum treatment is necessary.

In the 7 to 8-year-old group, it takes much longer and some protection for the femoral head may be necessary. This is done by taking the weight off the affected hip or by allowing weight bearing with the hip abducted—this equalizes the stresses on the femoral head. The methods vary with the individual surgeon.

Over the age of 8 years, the prognosis is not as good, because of the length of time the head will take to heal. Surgery to improve the 'containment' of the femoral head may be appropriate here.

Limping in an adolescent

John (age 13), has been 'walking badly' for several months. A keen fan of cowboy films, he was thought to be emulating his heros—walking with his feet turned out, rolling from side to side. His parents had not been worried, and John had never complained. It was a friend, a nurse, who suggested John might have a problem. John was somewhat overweight, with a tendency to female distribution of fat. General examination revealed underdevelopment of his genitalia.

Examination of the hips revealed loss of internal rotation, limitation of abduction, and as the hips were flexed, the legs rotated outwards suggesting *slipping of both upper femoral epiphyses* (SUFE). This was confirmed by X-ray, particularly the 'frog view'—with the hips abducted and externally rotated (Fig. 21.4). Both upper femoral epiphyses had slipped downwards and backwards in relation to the femoral neck. This was a chronic bilateral SUFE, and required surgery.

Summary
Slipped upper femoral epiphysis (SUFE)

- SUFE is most common in boys
- Age group: 10–15 years
- 30% will be bilateral
- Girls: often tall and thin
- Boys: often overweight

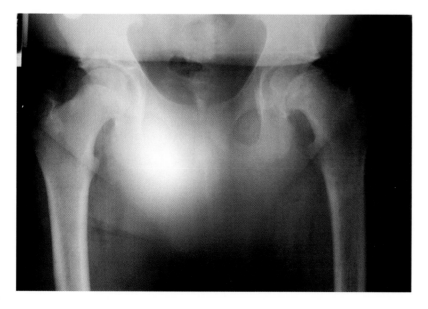

Fig. 21.4. Slipped upper left femoral epiphysis. Note the more prominent appearance of the femoral head, a consequence of its downward 'slip'. Bone 'build-up' around the epiphysis indicates that this is a chronic slip.

Fig. 21.5. (a) SUFE—lateral or 'frog' view with normal hip for comparison. (b) SUFE showing placement of nail to prevent further slip.

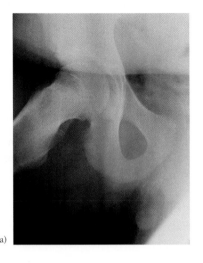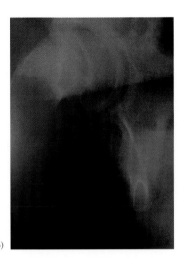

(a) (b)

All cases of SUFE will require surgery. This may be simple pinning of the epiphysis where the slip is minimal; or correcting the alignment of the femoral neck by osteotomy then pinning. This was necessary in John's case. Any attempt to manipulate the head back onto the femoral neck would have dire consequences on the blood supply and would lead to avascular necrosis.

After operations on both hips and 3 months in bed, John is now mobilizing well. The end result is still in the balance.

SUFE can occur as an acute condition and can be treated by manipulating the hip and pinning the epiphysis. There is minimal risk to the blood supply and results are good.

It can also present as an acute episode on a chronic slip—the most difficult to diagnose, but the X-rays will enable a trained observer to differentiate. The danger would be if it were treated as an acute slip and manipulated. This would almost certainly damage the remaining blood supply which has already been affected by the chronic slip.

Because of the risk of the other epiphysis slipping, many surgeons pin it prophylactically. Its aetiology is thought to be related to an imbalance between the growth hormone (too much) and the sex hormone (too little), which weakens the epiphyseal plates predisposing to the slipping.

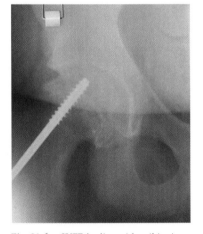

Fig. 21.6. SUFE-healing with nail in situ.

Painful limp

Philip is 5 years old. Three days ago he fell and grazed his right knee. Two days later he became unwell, developed a temperature and refused to walk, so his parents brought him to the surgery. Philip was obviously ill and pyrexic. Philip pointed to the pain on the inner side of his knee—there was some local swelling just below the knee and the overlying skin was warm. He would not allow anyone to touch his knee.

Although a limp may not be painful—a painful limb will always produce a limp. Any child with a painful limp who is unwell, needs to have an infective cause investigated as soon as possible.

Suspicious of *osteitis*, the GP sent Philip up to hospital immediately where he was admitted and put to bed. His left leg was splinted and elevated; blood was taken for cell counts and culture and the leg X-rayed—to exclude a fracture in a febrile child. Intravenous flucloxacillin was started with intravenous fluids.

Blood tests confirmed a raised white cell count and a raised erythrocyte sedimentation rate. X-rays were normal. After 12 hours, there was very little improvement; suspecting that Philip was developing a subperiosteal abscess, he was taken to theatre and the abcess drained. The drains were left in until drainage ceased in 48 hours. By the third postoperative day his temperature had settled, so his antibiotics were changed to oral on the next day. His blood cultures were negative, but fortunately he had responded to the flucloxacillin; culture of the pus from the small abscess grew *Staphylococcus aureus*. The antibiotic was continued for three weeks.

Philip eventually made a full recovery.

The white blood cell count and sedimentation rate are raised—blood culture may be negative. In 90% of cases the causative organism is *Staph. aureus*. The portal of entry may be mild skin trauma or sepsis of the upper respiratory tract, particularly infection in the ear. The antibiotic should be continued for 3 weeks and then stopped but only if the erythrocyte sedimentation rate has returned to normal.

Septic arthritis

This presents with a similar clinical picture to osteitis and can complicate osteitis. In the older child the disease is acute and rapidly progressive. Pyrexia associated with severe joint pain, muscle spasm, and virtually no movement at the joint because of pain are the presenting signs. There may be a visible joint effusion. Immediate antibiotic treatment must be given and diagnostic aspiration performed. If pus is obtained, the joint is opened and cleaned out then immobilized.

Summary
Differential diagnosis of painful limp

- Osteitis
- Trauma
- Septic arthritis
- Cellulitis
- *Rheumatoid arthritis*
- Rheumatic fever

Summary
Osteitis is a disease of growing bone

Symptoms:
- Pain
- Pyrexia
- Local skin tenderness
- Local skin warming
- Bone extremely tender to percussion
- Associated restriction of joint movement
- Minimal local swelling in the early stages
- X-rays initially normal, become abnormal after 7–10 days

Additional Information
Rheumatoid arthritis (Juvenile Chronic Arthritis)

- 9% systemic (Still's disease) with acute illness and high fever, anorexia, rash, splenomegaly, anaemia
- 16% polyarticular seronegative
- 3% polyarticular seropositive adult type
- 49% pauciarticular, usually knees
- multidisciplinary team required in management

Failure to remove the pus from an infected joint leads to destruction of the articular cartilage and irreversible joint damage.

In the *infant*, the presentation is less acute—possibly an irritable infant with restriction of movement in the joint. Fever may be mild and the blood counts may be only slightly raised. An X-ray may show distension of the joint and some displacement of the femoral head in the case of the hip. As with the older child, treatment involves aspiration to confirm the diagnosis, followed by arthrotomy and cleansing of the joint, antibiotics, and immobilization.

Knock-knees in a young child

Peter (age 5), attends the clinic as his parents were worried about his knock-knees and the way he wears his shoes. The knock-knees were noticed about a year ago and seemed to be getting worse. He is a fit child, although rather overweight, as is his mother.

Peter walks with an obvious knock-knee and is wearing his shoes down on the inner side of the heels. Standing, the gap between the malleoli is 3 cm, sitting it improves to 2 cm. Also when standing, it is noticeable that his heels roll in-over into *valgus*. (Valgus: the line of joint angles away from the midline; Varus: the line of the joint angles towards the midline).

Peter has a physiological knock-knee and at his age will probably grow out of it. However, he has the problem of the valgoid heels—often associated with knock-knee. It will therefore be worth putting valgus wedges under the heels on the inner side to control this. His mother will also be given appropriate dietary advice.

This regimen should correct the problem in 2–3 years, but if there is no improvement in 2 years, or he deteriorates, simple night splints to correct the deformity may be considered.

Knock-knees in an adolescent

Mary (age 13), came with her mother. She has been knock-kneed for years and has 'not grown out of it' and wanted her legs straightened. Mary is not overweight, she is an attractive girl with a very obvious knock-knee. The inter-malleolar gap is 5 cm on standing, 3 cm when sitting. Her feet are normal. Mary's growth years are limited; she is very co-operative and is an ideal case for night splints which should correct the deformity, as there is no pathological cause for her knock-knee.

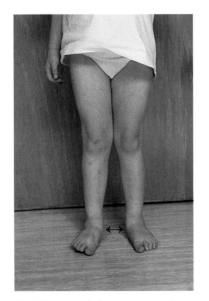

Fig. 21.7. Knock knees. Arrow indicates intermalleolar separation.

Knock-knees are part of normal child development—most improve spontaneously. Most children are born with some varus of the knees, which usually corrects by 2 years, some then overcompensate to produce the valgus or knock-knee. Knock-knees are a manifestation of general ligamentous laxity, in this case the medial collateral knee ligaments, hence tend to look worse when standing.

Most cases are physiological and resolve spontaneously by the age of 7. They are more common in overweight children and improve if the weight is reduced.

Assessment

The standard *measurement* is the intermalleolar separation measured standing and sitting, with the knees just touching. The intermalleolar separation is the gap between the medial and lateral malleolus.

Treatment of knock-knees

In *physiological cases*, the aim is to prevent deterioration and overstretching of the medial ligaments. Mild cases require no treatment. Moderate cases probably require no treatment under the age of 7, unless they are deteriorating or have associated valgus heels (about 50% of cases). In these children, there is a place for valgus wedges in the shoes. Night splints are usually reserved for the severe cases, especially in older children.

In *pathological cases*, any metabolic causes would need to be treated. This group may require surgical correction when growth ceases. Apart from knock-knee being an unattractive condition, in adult life it does increase the risk of degenerative joint disease in the knee.

There is still a great divergence of orthopaedic opinion as to whether knock-knee should be treated.

Intoeing in a toddler

Sarah (age 18 months), walked at 13 months and has always turned her toes in—occasionally tripping over them. Studying Sarah's gait, her ankles and knees point forward, but her forefeet turn in. Non-weight-bearing, the adduction (varus) occurs at mid foot level and corrects passively. This is the *postural metatarsus varus*, which occurs in 1 per 1000 births. Most correct once shoes are worn. As Sarah's had not, the physiotherapist showed her mother some spine stretching exercises and Sarah was given a night splint to stretch the inner borders of the feet. As Sarah slept prone with her feet tucked under her, Mum was advised to alter her sleeping position. In 6 months, her feet were normal.

Additional Information
Rarer causes of knock-knees

Metabolic: e.g. rickets (vitamin D deficiency or renal)

Congenital: arachnodactyly

Traumatic: previous epiphyseal or high tibial fractures

Summary
Intermalleolar separation

Standing measurement:
1–2 cm: mild
2–4 cm: moderate
over 4 cm: severe

Summary
Differential diagnosis of intoeing

True congenital metatarsus varus: much rarer with a more rigid deformity. The forefoot is adducted and supinated, with a high arch, the 'sickle-shaped' foot. Not usually noticed until the child starts walking, by which time the deformity is well established. Treatment will involve serial stretching, strapping, possible plasters, and splintage. In resistant cases, metatarsal osteotomies may be necessary.

Bow legs (tibia varum): very common in babies. It is a combination of varus and internal torsion of the tibia, and if it persists the child will have an intoeing gait. With these children, the knees point forward when walking but the ankles and feet turn in. Occasionally a night splint to externally rotate the feet may be necessary together with advice about the sleeping posture.

Congenital talipes equinovarus (see below)

Intoeing in a child

Michael is 6 years old and has always intoed. He is very fond of watching television and always sits on the floor with the hips and knees internally rotated and his feet out to the side—the TV or W position (Fig. 21.8). When walking, Michael's knees, ankles, and feet are turned in (i.e. the problem is with the hips: *persistent femoral neck anteversion*).

Due to ligamentous laxity, there is increased internal rotation at the hip with decreased external rotation. The child sits in the TV position as it is more comfortable, but this aggravates the condition.

Michael's mother was reassured that the condition would improve spontaneously, but she must stop him sitting with his legs turned in—if possible to get him to sit with his legs crossed 'tailor' fashion.

At birth, the femoral neck is anteverted 50 degrees on the femoral shaft. This gradually decreases towards the adult range of 15–20 degrees. Very rarely is corrective osteotomy necessary. In some children a compensatory external rotation of the tibia develops, the knees remain turned in, but the feet are straight, the so called 'squint knees'. No treatment is necessary.

Fig. 21.8. 'TV' position.

Foot deformity (talipes)

Sean Brown was born on the maternity ward: normal birth, birthweight 3.25 kg. It was obvious on delivery that the feet were not normal. The feet were pointing down, the heels turned in, and forefeet were rotated in (supinated) so that the little toes were almost touching.

As soon as possible the baby was seen by a paediatric orthopaedic surgeon. He found that the right foot was very rigid and resisted passive correction. The left foot, although it looked the same was more mobile, and corrected moderately well when passively manipulated.

Baby Brown had a severe right talipes and a moderate left talipes. Treatment was begun immediately.

Summary
Features of club foot (talipes equinovarus)

1. Equinus at the ankle: foot points down

2. Varus at the subtalar joint: heel turns in

3. Adduction and supination of the forefoot

4. Internal torsion of the tibia

Of the congenital foot abnormalities, *talipes equinovarus* (the club foot), is the most common, occurring in 2 per 1000 live births (Fig. 21.9). It is more common in boys and 50% are bilateral. The severity of the condition is judged by the rigidity of the deformity, not the appearance. The deformity cannot be corrected manually—this differentiates the condition from the postural equinovarus.

The deformity begins in early embryonic development. The muscles of the posterior and medial compartments of the calf are short and the affected joint capsules become fibrosed and contracted on the concave side. This produces secondary changes in the developing bones and joints.

Treatment begins at birth and requires a regimen of stretching, strapping, and splinting, carried out by specially trained physiotherapists. If the foot has not corrected by 3 months, surgical intervention becomes necessary to release contracted soft tissues and joint capsules and lengthen shortened tendons. In the older child, corrective bone surgery may be necessary. Even when the foot is well corrected the child may be left with a shorter foot and also some shortening of the tibia— usually 1–2 cm.

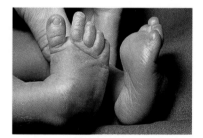

Fig. 21.9. Talipes equinovarus

Talipes calcaneo-valgus

This deformity is mainly postural and is due to intra-uterine pressure; with the foot folded up to the tibia. Most respond to passive stretching and possible bandaging. The condition is often ignored and in the more severe cases may cause delay in walking due to the over stretching of the Achilles tendon. When the child stands, the feet roll markedly into valgus (to be differentiated from infantile pes planus). The use of supportive boots with heel lifts and valgus wedges will encourage walking as the lax tissues tighten.

Congenital vertical talus

This is uncommon and difficult to treat. The hind foot is held rigidly in equinus and the forefoot is dorsiflexed and everted—producing the 'rocker-bottom' foot. X-rays show that the talus lies vertically and the navicular dislocates on to its dorsal surface. Surgery is necessary to correct this serious condition. It is often associated with spina bifida or arthrogryphosis.

Flat feet (infantile pes planus)

The majority of children will look 'flat-footed' when they begin walking. This is due to the presence of a fat pad under the arch of the foot. This normally disappears by the age of 3–4 years.

Flat foot (dropping of the longitudinal arch) must be differentiated from the valgus foot (*pes valgus*: where the heel rolls inwards), in which there is an appearance of a flattened arch. Persistent flattening of the arch is helped by exercises and arch supports. Pes valgus requires the insertion of valgus (medial) wedges in the heels of the shoes to correct the heel deformity.

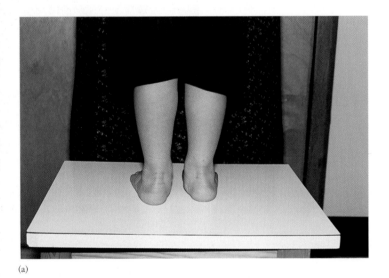

(a)

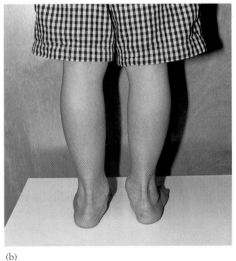

(b)

Fig. 21.10. (a) Valgus ankles, showing 'roll-in' effect of ankles. (b) Infantile pes planus. Heels are neutral.

Further reading

Renshaw, T.S. (1986). *Paediatric orthopaedics.* W.B. Saunders, London.

Salter, R.B. (1983). *Textbook of disorders and injuries of the musculoskeletal system.* William & Williams, Baltimore, MD.

Williams, P.F. and Cole, W.G. (ed.) (1991). *Orthopaedic management in children* (2nd edn) Chapman & Hall, London.

Respiratory problems

- A global perspective

- Respiratory system growth and development

- Clinical assessment

- Case studies
  croup (laryngotracheobronchitis)
  bronchiolitis
  asthma and wheezing; cystic fibrosis
  pneumonia

- Tuberculosis
- Sudden infant death

A global perspective

Respiratory disease is the single commonest cause for admission to medical paediatric beds in industrialized countries and is alongside acute gastroenteritis (World Health Organization estimates) the commonest cause of death in poorer countries of the world. In the under 5-year-old ago group alone it has been estimated that more than 700 children die each day from *acute respiratory infections* (ARI) (Fig. 22.1).

Although the agents responsible are somewhat different with more bacterial infections causing problems in poorer countries, viral lower respiratory infections are common particularly in infants. Every year, a predictable and remarkably constant winter epidemic of viral-induced bronchiolitis, often caused by the respiratory syncitial virus, affects each new cohort of young infants (Fig. 22.2).

Risk factors of serious morbidity and death are well known and some are potentially remediable.

A remarkable feature in developed countries has been the rise in acute wheezing illness and asthma alongside an increase in atopic disease (Fig. 22.3). This has yet to be explained but possible culprits include indoor (house dust mite and gas cookers) and outdoor air pollution (car and industrial), passive smoking, and changes in the diet (low vitamin C and other antioxidants) (see Ch. 1, p. 10–11). To a certain extent it is also attributable to changes in diagnostic fashion, as what was once referred to as 'wheezy bronchitis' or 'pneumonia' is now more likely to be labelled as asthma.

Respiratory system growth and development

The lungs develop as a ventral part of primitive foregut at approximately 8 weeks of gestational age and airway development is complete by 16 weeks. Alveolar

Summary
Acute respiratory infections (ARI)

- Mortality increases 10-fold with malnutrition

- Fatality rate doubled if bottle-fed

- Fatality rate trebled if both parents smoke

- Incidence of ARI constant world-wide but mortality high in developing countries if bacterial infection

- Most common pathogen is *Streptococcus pneumoniae*, next is *Haemophius influenzae*

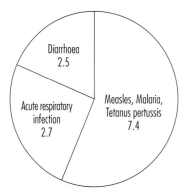

Fig. 22.1. Child deaths (0–4 years) worldwide by main cause (millions). (From Murray, C.L. and Lopez, A.D. 1990, *Bulletin WHO*, 72, 447.)

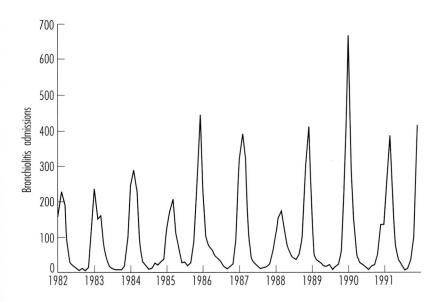

Fig. 22.2. Admissions for acute bronchiolitis in Scotland (1985–91). (From Registrar General's Office.)

Fig. 22.3. Changes in population prevalence of asthma and atopic disease in an urban Scottish setting. (From Ninan, T.K. and Russell, G. 1992. *British Medical Journal*, 304, 873–5.)

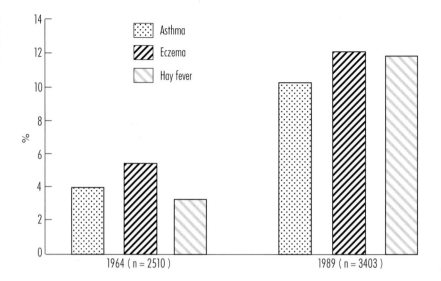

development commences at approximately 18–20 weeks and it is this factor that is largely responsible for the limits of viability at approximately 22–24 weeks gestation age when just enough peripheral lung tissue is available for gas exchange. Any disturbance to the ordered sequence of lung development or failure to establish the correct forms can result in a range of congenital anatomical abnormalities including pulmonary hypoplasia and a range of lung cysts and malformations.

The fetus is well adapted for the intra-uterine environment and rhythmical respiratory movements not only provide a useful rehearsal for subsequent terrestrial life but also help to stimulate airway and alveolar development by a repetitive stretching of the underlying lung. The transition from liquid to air breathing is assisted by rapid lung liquid clearance via the lymphatic system and the vaginal thoracic squeeze during delivery. The fact that the chest wall is a remarkably pliable structure allowing the lung to recoil to a low volume *in utero* reduces the alveolar and interstitial fluid that has to be cleared soon after birth. In some infants, lung liquid clearance is delayed and presents as transient tachypnoea of the newborn (wet lung) (see Chapter 8). The pliable chest wall supports the lung relatively poorly in the first one to two years of life and results in poor peripheral airway function, increased risk of airway obstruction particularly at the lung base (Fig. 22.4).

In the premature, infant this feature is further complicated by poor local production of surfactant (see Chapter 8, p. 168). In later childhood boys have been found to have functionally narrower airways than girls up to about the age of puberty and this may, in part, explain the observed male to female preponderance for lower respiratory symptoms, including asthma in prepubertal boys (Fig. 22.5).

The pattern of upper respiratory tract problems including croup also reflects the growth and development of the nasopharynx. Infants and preschool-age children

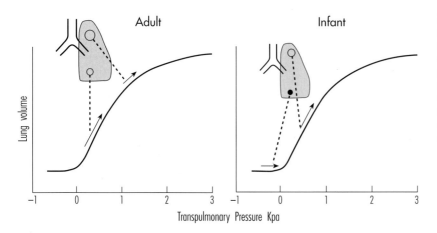

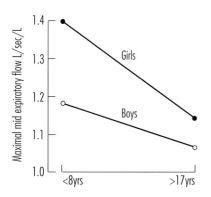

Fig. 22.4. Idealized lung–pressure relationships in a young healthy adult and a normal infant. Intrapleural pressure is relative to atmosphere, and the typical vertical lung–pressure gradient is shown. Note that in the adult the lung apex is relatively overextended, whereas the base is on the optimal portion of the lung volume–pressure curve. In the infant, the lung base is compressed, gas exchange is impaired, and airways likely to close.

Fig. 22.5. Maximal expiratory flow rates at 50% of expired vital capacity in Caucasian boys and girls pre- and postpuberty. Note the higher relative higher flow rates in girls, with a convergence postpuberty. (From Rosenthal, M. et al. 1993. Thorax, 48, 794–802)

have a crowded posterior pharynx which is further compromised by the normal hypertrophy of tonsillar-adenoidal tissue that occurs in early childhood. This can be exacerbated by congenital anomalies such as mandibular hypoplasia and relative macroglossia as occurs in the rare Pierre Robin syndrome or more commonly as slow development of tracheal cartilage in tracheomalacia (Table 22.1).

Important developmental changes take place in infancy and early childhood that affect the upper airway. The infant and young child has a large cranial vault in proportion to other body dimensions but has a small facial skeleton (frontal sinuses do not appear until the age of 5–6 years). Normal growth and development of the face has the effect of lifting the anterior pharyngeal wall away from the posterior pharynx. This is one of the reasons, together with the absolute small diameter of the trachea, why croup (acute laryngotracheobronchitis) is so common in young children. In addition to the important developmental considerations outlined above, significant infections and insults in early childhood may have life-long consequences for adult respiratory health. These influences may even go back to the intra-uterine period, for example, smoking during pregnancy is associated with increased respiratory morbidity in early infancy and the effects are more marked than passive smoke exposure postnatally.

Table 22.1.
Upper airway obstruction

Acute	Chronic
Croup (acute laryngotracheobronchitis)*	Laryngomalacia*
Epiglottitis	Subglottic stenosis
Retropharyngeal abscess	Congenital web
Diphtheria	Recurrent laryngeal nerve palsy
Foreign body*	
Angioneurotic oedema	Haemangioma
	Vascular ring/sling
	Glossoptosis
	Micrognathia (Pierre Robin Syndrome)

*Common causes of obstruction.

Clinical assessment

The developmental features and pattern of disease described above need to be taken into account in clinical assessment. Indeed, the soft and flexible chest wall in the infant and very young child provides a useful marker of disease severity which often alarms parents. When exposed to increased respiratory loads (mainly associated with airway obstruction) diaphragmatic contraction draws the pliable ribcage inwards. If the obstruction is mild it produces an inspiratory groove adjacent to the lower ribcage and anterolateral abdominal boundary. This is referred to as subcostal recession. As in the adult, suprasternal and intercostal recession may be seen but in severe cases and infants and young children, certainly less than 2–3 years of age, sternal recession may also be seen. If the respiratory problem is prolonged a more permanent deformation of the ribcage is seen with a groove or flattening of the lower chest wall so-called 'Harrison's sulcus', named after its first observer. If the problem is associated with air-trapping and hyperinflation the forceful contraction of the diaphragm and associated increased anteroposterior thoracic dimension results in a so-called 'pigeon chest' or pectus carinatum (Fig. 22.6).

Hyperinflation, as seen for example in acute asthma or bronchiolitis, also results in diaphragmatic depression and downward displacement of the liver. This, together with increased anteroposterior dimension of the thorax, results in the 'barrel chest' deformity, lack of cardiac dullness on percussion of the anterior chest and an easily palpable liver edge. All these features can be used to identify chronic lung hyperinflation. Care must be taken not to overdiagnose cardiac failure in the presence of respiratory disease. For the inexperienced, the fine crackles often found in acute bronchiolitis together with the apparently enlarged liver are a frequent pitfall. Crackles with or without wheeze are not infrequently found in prepubertal children with asthma. As in the adult, cyanosis and nasal flaring are signs of significant lower respiratory tract disease.

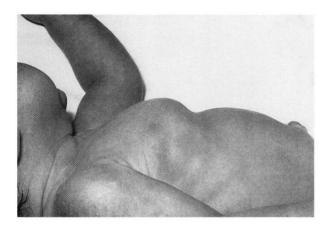

Fig. 22.6. 'Pigeon chest' (pectus carinatum) in an infant with severe lung disease of prematurity.

In addition to the observation of chest wall movement, careful timing of respiratory noises can be used to identify the site of obstruction. In acute upper airway obstruction associated with croup, stridor may be audible during inspiratory efforts from the end of the bed or even on approach to casualty or the ward. In lower respiratory tract obstruction, in addition to fine crepitations, expiratory wheeze will often be audible without the use of a stethoscope. In difficult cases, auscultation of the chest and neck can help in localizing the major site of obstruction. Croup is common and is rarely life-threatening, whereas impacted foreign body or epiglottitis are potentially lethal—a few cardinal features can help to distinguish them (Table 22.2) although first episodes can be difficult for parents and their GP to assess and many affected children are referred to hospital for overnight observation.

Case studies
Croup (laryngotracheobronchitis)

John (aged 18 months), was born at term and was breast-fed for the first four months. He has had an unremarkable development so far. He has had his triple (diphtheria, tetanus, and polio) and Hib (*Haemophilus influenzae* type B) immunizations at 2, 3, and 4 months and has recently received his measles, mumps, and rubella immunization, he has been in excellent health. Over the past 3–4 days he has developed upper respiratory signs of cough and runny nose and over the past 12 hours his parents have noted a rasping cough described as a 'bark like a sea-lion'. They have also noticed increased respiratory distress and that his chest is now seesawing up and down quite alarmingly. His parents have had a dreadful night and are extremely anxious, it is now 6 a.m. The family GP was called out in the early hours of the

Summary
Croup: indications for admission

- Possible foreign body
- Epiglottitis
- Severe chest wall recession
- Parents unable to cope

morning and in view of the increasing respiratory distress and obvious anxiety of both parents has sought hospital admission.

This is a common acute paediatric problem and the history is very suggestive of *croup* or acute laryngotracheobronchitis. The differential diagnosis includes inhaled foreign body and epiglottitis (Table 22.2). Upper respiratory tract infections may also complicate laryngomalacia or more rarely vascular rings, laryngeal webs, or subglottic stenosis (the latter may present as a complication of prolonged intubation in the neonatal period). Underlying problems that are exacerbated by infection are usually characterized by mild but continuous symptoms. The relatively long prodromal period, laryngeal barking cough and gradual worsening over several hours is strongly suggestive of croup. Epiglottitis, which has virtually been eradicated by immunization (see Chapter 4), is *usually* associated with a more rapid onset, quiet respiration, and a sick, pale-looking child with or without drooling of saliva (a consequence of the acutely inflamed and swollen epiglottis). Although John is the appropriate age group for an inhaled foreign body this is unlikely as the onset was not sudden.

On observation, John is tired and irritable but can be diverted into play for at least a short period. He has marked subcostal and sternal recession and an easily audible high-pitched inspiratory stridor. A working diagnosis of croup is made. It is possible to manage mild croup at home but as the parents are exhausted after their broken night, the symptoms are worsening, and John has features of respiratory distress (subcostal and costal recession), he is admitted.

In any acute upper airway obstruction it is dangerous to examine the throat as acute obstruction can rapidly ensue. X-ray imaging is also contraindicated as the

Table 22.2.
Croup: differential diagnosis

Croup	Epiglottitis	Foreign body
Long (2–3 day) prodrome	Short prodrome (< 24 hours)	No prodrome
Loud inspiratory stridor	Often quiet breath sounds	Sudden onset—often during
Well-looking child	Ill-looking child: pale	unsupervised play
Well perfused	Poor perfusion	Coughing
Low-grade fever	Toxic	Age: unusual below 6 months
Age: 6 months–7 years	Drooling saliva	
	Age: 6 months–5 years	

manipulation required may precipitate obstruction. John needs to be kept as calm as possible and symptomatic therapy instituted immediately. Croup often worsens in the late afternoon and evening and in severe cases, such as John's, it is wise to observe overnight. Traditionally, treatment is with humidification of the inspired air using a bedside device (mist tents are no longer used as observation is difficult and water intoxication can ensue). If obstruction is very severe and epiglottitis or inhaled foreign body is a possibility an ear, nose, and throat surgeon and consultant anaesthetist should be alerted prior to the child's arrival. Children with severe obstruction should be taken immediately to theatre and examined under anaesthetic and intubated. Antibiotic therapy is mandatory in epiglottitis and broad-spectrum cephalosporins are now commonly used as many *H. influenzae* organisms, the usual bacteria responsible, are resistant to ampicillin. As a temporizing measure, nebulized adrenaline can be used although there is a rebound phenomenon within 2–3 hours of its use. Steroids (systemic or inhaled) may have a role in shortening the course of the disease.

Bronchiolitis

Calum (age 4 months), is the second child of young parents. His mother is aged 19 and is at home full-time. Calum's father is aged 22 and is a storeman. Calum's older sister Morag, aged 2, has had recurrent coughs and colds and has recently been diagnosed by the family GP as having asthma. She is currently taking salbutamol elixir and has had frequent courses of antibiotics. Both parents smoke, approximately 20 cigarettes per day. The family live in a two-bedroomed council flat which is damp, and which they consider is affecting Morag's health. Calum was born at term, birthweight 2.5 kg (just below the 3rd centile). He has had his first immunizations and is developmentally normal for his age. It is now mid January and Calum has been suffering from a cold for the last 3 days but over the last 24 hours has become increasingly breathless and is now unable to take from his bottle. His parents have also noticed that his chest is see-sawing up and down and that he has developed a wheezy cough.

Risks of symptomatic respiratory disease vary inversely with disposable income, including factors such as overcrowding, damp housing, and passive smoking. Bacterial pneumonia is extremely common in developing countries being less common in developed countries. The range of pathogens is remarkably similar across the globe with *Streptococcus pneumoniae* being the commonest bacterial and respiratory syncitial virus (RSV) the commonest viral pathogen (Table 22.3). A chronic cough without lower respiratory signs, particularly if associated with vomiting, should bring *pertussis* (whooping cough) to mind—particularly if a child has not been immunised.

**Summary
Management of croup**

- Calm reassurance
- Humidification of inspired air
- In severe cases consider nebulized adrenaline, systemic steroids

**Additional Information
Pertussis (Whooping cough)**

- highly infectious
- caused by Bordetella pertussis
- spasms of cough with whoop (this is often absent in infants)
- followed by vomiting
- infant goes red or blue in the face
- symptoms may persist for 3/12
- characteristic lymphocytosis
- may develop pneumonia
- Erythromycin reduces spread but does not cure illness

Table 22.3.
Pneumonia/lower repiratory tract infections

Type	Occurrence
Bacterial	
Strep. pneumoniae	All ages
H. influenzae	Toddlers: becoming less common
Staph. aureus	Uncommon: more in older children
Mycoplasma	In school-age children
Tuberculosis	Rare: risk increased in AIDS
Viral	
Respiratory Syncitial virus (RSV)	Mainly in infants
Parainfluenza virus, 1 & 3	Common in all ages
Influenza, A & B	Common in all ages

Summary
Bronchiolitis

- Winter epidemics
- Peak incidence at 2–4 months
- Respiratory syncitial virus (RSV) in 70%
- 90% of all children acquire RSV before the age of 2 years
- High morbidity/low mortality (in developed countries)
- Bronchodilators usually ineffective

Antibiotics are life-saving in affected children from poorer countries and although widely used in the UK and other developed countries are rarely required. In addition to the classical features of cyanosis, chest and intercostal recession, and nasal flaring, respiratory rate is a useful feature. Indeed, in developing countries village health workers use a respiratory rate of 50 or more as an indication for antibiotic use. Calum's presentation is typical of bronchiolitis, approximately 70% of which are associated with respiratory syncitial virus (RSV). It occurs in epidemics every winter season (See Fig. 22.2) and affects each newborn cohort with approximately 90% of all children acquiring at least one RSV infection by the age of 2 years. Of these approximately 10–20% will have significant lower respiratory signs and symptoms, and 1% will require hospital admission. Peak age of hospital admission is 2 months. Adults and older children are affected but not as severely and this is in part due to improved host defences and to a decreasing tendency to airflow obstruction (see introductory section and Fig. 22.4). Early attempts at vaccination with viral antigen produced worse disease because of a more profound airway inflammatory response with oedema and mucous production. As in many respiratory diseases in prepubertal children, male to female ratios are approximately 2 : 1 which may in part reflect differences in airway growth and development between the sexes (see Fig. 22.5). Other risk factors include passive cigarette smoke exposure (as in Calum's case) and particularly maternal smoking during pregnancy. It has been estimated that approximately 10–15% of all acute paediatric beds could be closed if smoking prevalence in the community could be eliminated or reduced to negligible levels.

On examination, Calum is tachypnoeic at 60 breaths per minute with subcostal recession and absence of cardiac dullness on percussion of the anterior chest wall. He also has a soft easily palpable liver edge at 5 cm below the costal margin. He has a dusky appearance and pulse oximetry confirms hypoxaemia with an oxygen saturation in air of 85%, correcting to 95% in 40% oxygen. Chest auscultation reveals high-pitched expiratory wheeze and showers of fine crackles over all areas. Weight at 5 kilograms places him on the 3rd centile. Both parents smell strongly of tobacco smoke. Calum is admitted, a chest X-ray reveals hyperinflation with areas of collapse/consolidation at the right upper zone, and left base and immunofluorescence of nasopharyngeal aspirate confirms the presence of RSV in mucosal cells. Oxygen therapy is required for the next 2 days in order to maintain an oxygen saturation of more than 90% and a nasogastric tube is passed for enteral feeding. On day 3 he begins to feed on his own, his respiratory rate returns to 30 breaths per minute, and he is discharged on day 5 having made a full recovery, although still wheezing. His parents ask again whether he is going to have asthma like his sister.

The natural history of acute bronchiolitis is often as described above with a gradual worsening over several days—a fact that parents need to appreciate as deterioration often continues after hospital admission. Once a viral cause is established (RSV immunofluorescence is usually available within a few hours of pharyngeal aspirate being obtained) antibiotics are not required as secondary bacterial infection, particularly in developed countries, is extremely rare. Other viral pathogens include parainfluenza viruses 1 and 3 and influenza A and B. A rare form of severe bronchiolitis may be associated with adenovirus which can cause severe illness going on to chronic changes of obliterative bronchiolitis.

Apart from smoking, other risk factors include premature birth, immunodeficiency, and disrupted host defence. As the symptoms are similar to acute asthma, bronchodilators are often given usually by the nebulized route with a variable response to beta agonists and ipratropium bromide. In most infants less than 6 months of age, bronchodilators are ineffective as most of the airflow obstruction is due to mucous plugging and mucosal oedema. Systemic steroids are of no benefit but high-dose inhaled steroids may confer some benefit and may reduce long-term sequelae (definitive results from clinical trials are awaited). Despite the severe obstruction very few infants develop respiratory failure and require ventilation. Severe disease and proven secondary bacterial infection suggest the possibility of other pathologies including immunodeficiency and cystic fibrosis. For a small number of high-risk infants the antiviral agent, ribavirin, may have a role in modifying the natural history of the disease by inhibiting viral replication.

Approximately 50% of affected infants continue with wheezing episodes over the next 6–12 months, probably in response to further viral infections. In some of these infants this will be the first presentation of asthma and this becomes more likely in the presence of a strong family history of atopic disease and parental smoking. Bronchiolitis may therefore be both a cause of subsequent wheezing illness or a consequence of an underlying tendency to airway obstruction and atopy. However, parents should be given an optimistic long term prognosis as the majority of wheezing illness in infants and preschool children resolves spontaneously with growth.

Summary
Signs of acute asthma

- *Acute severe asthma:*
  - too breathless to talk
  - too breathless to feed
  - respiration ≥ 50 breaths/minute
  - pulse ≥ 140 beats/minute
  - peak flow ≤ 50% of predicted or best
- *Life-threatening features:*
  - peak flow < 33% of predicted or best
  - cyanosis, silent chest, or poor respiratory effort
  - fatigue or exhaustion
  - agitation or reduced levels of consciousness
  - pulsus paradoxus

From *British National Formulary* guidelines.

Summary
Cystic fibrosis

- 1 in 2000 births (caucasian population)
- Autosomal recessive: mutation on chromosome 7
- Carriers (1 in 25) are normal
- Multisystem effects particularly on lungs and gastrointestinal tract
- Most present before 2 years with recurrent respiratory infections or failure to thrive
- High index of suspicion required
- Diagnose by high chloride concentration in sweat test (pilocarpine iontophoresis)

Asthma and Wheezing; cystic fibrosis

Mhairi (age 11 years), a child with a 5-year history of recurrent coughs and colds with breathlessness presents to the accident and emergency department distressed and unable to speak. She was first diagnosed as having asthma at the age of 8 and is currently using a dry powder beta agonist inhaler for symptomatic relief. She has never been to hospital before but both parents describe a gradual increase in the number of episodes so that she has been using her inhaler up to 2–3 times per week over the past 3 months. Mhairi's maternal uncle has asthma and her father has hay fever (he also had asthma as a child). Mhairi is a keen hockey player and her parents describe her distress at having to stop playing on occasions because of her episodes of wheezing.

Mhairi is likely to have an acute exacerbation of asthma against a background of poor control and increasing severity of disease. She is a little old for an inhaled foreign body to be likely but this should always be excluded. Other possibilities include a tension pneumothorax and acute bacterial pneumonia, although it needs to be remembered that increased work of breathing associated with asthma often causes a modest elevation of core temperature but not more than 38 °C. Mhairi has severe acute disease or status asthmaticus. Although acute hypoxaemia and death are unusual in children, approximately 2000 adults and children die every year from asthma in the UK. Although asthma prevalence has been rising in the community as a whole (see Fig. 22.3), thankfully very severe asthmatic episodes have not increased in parallel.

Cystic fibrosis (see also Ch. 15, p. 267)

In any child with a chronic history of respiratory symptoms a diagnosis of cystic fibrosis should be entertained but increasing awareness of this condition means that most new cases are diagnosed before the age of 2 years including the 15% of all those diagnosed in childhood who present in the neonatal period with meconium ileus (intestinal obstruction). Features that make cystic fibrosis more likely include failure to thrive and gastrointestinal symptoms, and in older children nasal polyps. (Table 22.4).

On arrival Mhairi, is given nebulized salbutamol driven by oxygen rather than air because her oxygen saturation is 75% in air. An intravenous infusion is established and intravenous hydrocortisone is administered. A calm reassuring approach is adopted and as soon as the nebulized salbutamol has finished, physical examination reveals a central trachea, chest hyperinflation, with an increased AP diameter (approximately equal to the transverse diameter), and a slightly peaked sternum is

Table 22.4.
Features associated with cystic fibrosis

Respiratory system
Bronchiectasis (progressive)
Haemoptysis
Pneumothorax
Nasal polyposis
Chronic sinusitis

Gastrointestinal system
Neonatal meconium Ileus
Rectal prolapse
Meconium ileus equivalent (distal intestinal obstruction syndrome)
Pancreatic insufficiency
Focal biliary cirrhosis/Portal hypertension

Endocrine/Reproductive systems
Diabetes mellitus
Male infertilty (aspermia)

noted. She has no nasal flaring but obvious intercostal recession and tracheal tug. On palpation of the radial pulse it appears to disappear on inspiration. Her oxygen saturation has risen to 90% in 100% oxygen and she is now able to speak in short sentences. She is admitted to the ward.

The priority of acute asthma management remains bronchodilator therapy, sometimes required every 10–15 minutes until symptoms improve together with short courses (2–5 days) of systemic steroids (Table 22.5). Regular and frequent reassessment is required in order to identify impending respiratory failure and the need for ventilatory support. The presence of hypoxaemia, inability to speak, and pulsus paradoxus (defined as a fall in systolic blood pressure of >10 mmHg during inspiration) are all important signs of severe disease and impending respiratory failure and can be used in assessment of severity.

Hypoxaemia must be dealt with as soon as possible and nebulized drugs should always be driven with oxygen rather than compressed air. In the presence of severe disease, rapid administration of appropriate drugs is required.

In addition to clinical criteria, objective assessment of severity can be made using arterial blood gas estimation but this is reserved for children with severe disease in whom ventilation is being considered. In children above $3\frac{1}{2}$–4 years of age regular peak expiratory flow rates (PEFR) are usually possible and should be used as a monitoring tool. Many established asthmatic children will know their best recent PEFR reading and this can be used as a target figure or failing that reference standards can be used to gauge response to therapy and identify the appropriate

Table 22.5.
Treatment of acute asthma

At home	In hospital
• Short acting beta-2 stimulant from metered dose inhaler using large-volume spacer device may be as effective as use of nebulizer; dose is one puff every few seconds until improvement occurs (max 20 puffs), using facemask in very young child	• Short-acting beta agonist by nebulizer
	• Reassess every 10 min plus oral steroids.
	• If too breathless commence IV hydrocortisone
	• Slow (10 min) bolus of IV aminophylline but halve dose if already taking oral xanthines.
• Terbutaline may be given subcutaneously in severe episodes	• Infusion of beta agonist may be used in addition for very severe cases.
• Oxygen is of benefit	• Oxygen is mandatory.
• A child requiring high-dose inhaled bronchodilators should also receive soluble tablets prednisolone 1–2 mg/kg (max. 40 mg) once daily for up to 5 days if necessary; child needs immediate referral to hospital if failure to respond	
• Aminophylline should no longer be used in children at home	

From *British National Formulary* guidelines.

Fig. 22.7. Mhairi's peak flow chart from admission to discharge. Note the target (50th centile PEFR predicted from Mhairi's height of 140 cm). Morning dips can be seen on days 3, 4, and 5, together with measurements before and after bronchodilator.

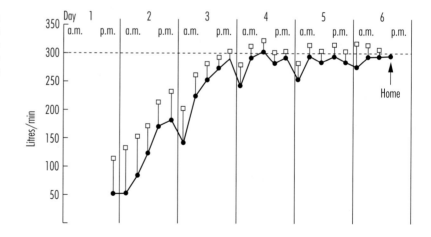

time for discharge (Fig. 22.7). As the asthma improves it is common to find so called early morning dipping as seen in Mhairi's case. Early morning dips are also seen in poorly controlled asthma and may require more intensive therapy particularly in the evenings. Long-acting beta agonist or theophylline preparations may be helpful.

Long-term asthma management needs to be reconsidered in any child with increasing symptoms and with an acute life-threatening episode. Whereas intermittent bronchodilator therapy is quite adequate for very mild intermittent symptoms once bronchodilator therapy is being used more than twice a week, prophylactic therapy is indicated.

It is important to review children on prophylaxis at regular intervals so that attempts can be made to step down the treatment as the disease comes under control.

Most children, whatever their age, can be established on inhaled relievers (beta agonists) and preventers (cromoglycate or steroids) by employing an appropriate delivery system (Fig. 22.8). Infants and very young children can usually be established on aerosols and spacers and older children on dry powder devices or aerosols although there is evidence that dry powder devices are more efficient at delivering active drug to the airways.

Most asthma is managed in the community and specially trained 'asthma nurses' can significantly improve quality of life for affected individuals and their families. Community asthma clinics provide opportunities for patient education including training and monitoring of inhaler technique. Hospital clinics should be reserved for the 1–2% of severe or 'brittle' asthmatics some of whom will require open access to hospital facilities.

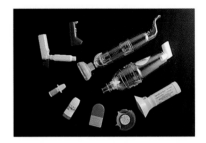

Fig. 22.8. Two spring loaded breath activated aerosols shown top left, one with a spacer attached. Four day powder devices and three spacers designed for use with aerosol devices.

Early use of high-dose inhaled steroids, and in severe cases short courses of oral steroids, can help to reduce the need for hospital admission although clear guidelines need to be given to the child and his/her parents. Exercise-induced symptoms are common and can often be reduced by pre-treatment with beta agonist and the education and support of school staff can be of great benefit to the child. Children must have immediate access to their reliever inhalers at all times, *including at school*. Concerns about possible side- effects of inhaled steroids on growth and bone metabolism need to be addressed and parents can be reassured that the doses used to control the vast majority of children using this form of treatment have no long-term effects.

It is often stated that asthma tends to improve during puberty and adolescence and this is more likely to be so if the asthma is precipitated only by viral infections and is not part of an atopic tendency as in this case. The presence of atopic disease (asthma, hay fever, and/or eczema) in first degree relatives and the precipitation of asthma by a variety of circumstances and agents other than viruses increases the likelihood of continuing symptoms into adult life. Although allergen avoidance, including the two which cause most problems (house dust mites and cats), is difficult to achieve measures to reduce household dust, particularly in the bedroom, may be of benefit.

Pneumonia

John (age 10), came home from school early feeling hot and generally unwell. During the course of the evening he developed an irritable dry cough and pain in his

Summary
Pneumonia

- fever, cough, chest pain, tachypnoea

- common organisms pneumococcus and mycoplasma

- Pseudomonas suggests CF

- if recurrent, consider immune deficiency, CF, congenital lesion (eg cyst), ciliary dyskinesia

- opportunistic organisms: CMV, Pneumocystis in immuno-suppressed

left lower chest, more noticeable on breathing in. His mother telephoned the GP who arrived at the house to find John much the same with a central temperature of 39 °C and with decreased breath sounds with a few crackles at the left base. John was not cyanosed, did not appear to be tachypnoeic, and had no signs of respiratory distress apart from the discomfort in his left chest. Percussion note was slightly reduced at the left base but with no vocal resonance, vocal fremitus, or bronchial breathing. John was started on oral erythromycin with paracetamol 6-hourly as required and his mother was instructed to re-contact the surgery if John failed to improve.

This is a typical presentation of pneumonia with pleurisy and at John's age the most likely organism is *Strep. pneumoniae* (see Table 22.3). *Mycoplasma pneumoniae* is another common organism at this age although it is usually associated with more influenza-type symptoms with muscle aches and pains, and is rarely associated with clinical signs suggestive of collapse/consolidation. The choice of erythromycin is a good one with this differential diagnosis although the gastric irritation associated with its use occasionally makes it unpopular with patients.

Next morning John is no better and his mother has noted that he is more breathless and that his left-sided chest pain is worse. She takes him to the surgery where his doctor finds slight but definite tracheal descent and intercostal recession with quite definite dullness to percussion over the left posterior chest with bronchial breath sounds and increased vocal resonance. He diagnoses lobar pneumonia and refers John to hospital where a chest X-ray shows increased radiodensity in the left, mid, and lower zone consistent with the diagnosis. Blood cultures and full blood count are taken and an intravenous line is set up and penicillin commenced. Oxygen saturation is normal at 95% breathing room air. Over the next 3 days John's temperature slowly settles and his chest signs resolve although at discharge on day 4 he still has dullness to percussion at the left base. Oral penicillin is continued for a further 10 days with an outpatient follow-up appointment.

John is unlikely to have a recurrence of his pneumonia. Any child who has recurrent episodes or continuing symptoms, such as cough and sputum production, needs to have cystic fibrosis (CF) excluded. Evidence of bronchiectasis with chronic radiological changes, particularly in the mid to upper zones is also suggestive of CF which at 1 per 2000 live births remains the single commonest recessively inherited disorder in Caucasian populations. The finding of *Pseudomonas aeruginosa* in infected sputum is virtually pathognomonic of CF.

Recurrent proven pneumonias are also a feature of ciliary dyskinesia and immune deficiencies (see Ch. 18, p. 315), the more severe forms of the latter being associated with failure to thrive and gastrointestinal symptoms. Recurrent pneumonias in the same lobe or lung segment suggests local bronchiectasis or an underlying congenital lung abnormality such as a lung cyst or sequestered segment. Tuberculosis and rare forms of interstitial pneumonia with opportunistic organisms, such as *Pneumocystis carinii* or cytomegalovirus (CMV) are seen in immunosuppressed patients during treatment for cancer, post-organ transplantation or in acquired immune deficiency syndrome (AIDS). Community-acquired pneumonias, such as in John's case, usually respond rapidly to first-line antibiotic therapy and recover without long-term sequelae.

Tuberculosis

In a world context tuberculosis is a major repiratory health problem accounting for 20–30 million deaths per year. With the emergence of AIDS and multiply-resistant organisms there is a distinct possibility of re-emergence of this disease in developed countries. Primary infection is usually pulmonary and its early diagnosis requires a high level of suspicion.

It should be considered in high-risk settings, children at risk of AIDS (e.g. drug-abusing mother), children taking immuno-suppressive drugs, and children from mobile communities with a high risk of endemic disease. Populations at increased risk include those from the Indian subcontinent, Central and East Africa, and South East Asia. Immunization in the first week of life in high-risk communities affords up to 70% protection from the most damaging presentation of tuberculous meningitis. This devastating condition is most common in infants and young children and accounts for the majority of deaths due to primary infection. Routine immunization of school-age children between 12 and 14 years of age remains the policy in the UK but may have to be revised if early disease becomes more widespread and more common. Primary infection is by droplet spread and may occur in the lung, skin, or gut. This infection spreads to the local lymph nodes and forms the primary complex. Most of these complexes heal by fibrosis over the next one to two years and some may calcify. Secondary spread occurs, usually in the lungs, and results in a range of problems from local gas-trapping or collapse secondary to enlarged lymph nodes to local bronchopneumonia, caseation and/or emphysema. The last two problems are most commonly seen in malnourished children.

If the primary focus forms in the upper respiratory tract this presents as cervical lymphadenopathy and may develop into a 'cold' abscess. Although this is usually due to human tuberculosis it may follow the drinking of unpasteurized milk from dairy herds infected with bovine tuberculosis. If the primary infection is in the small bowel this can lead to malabsorption, the formation of strictures, and eventually peritonitis. Atypical mycobacteria can also cause cervical adenitis.

Summary
Tuberculosis

- a major problem worldwide

- established high risk factors e.g. immunocompromised host, endemic areas

- immunization at birth or 12–14 years (70% protection)

- many and varied presentations in many different parts of the body

Summary
Risk factors in Sudden Infant Death

- Passive smoking (particularly maternal)
- Low birth weight
- Male sex
- Winter season
- Prone sleeping

Additional Information
Sources of support

Foundation for the Study of Infant Deaths, Belgrave Square, London SW1X 8QB

The Scottish Cot Death Trust, RHSC, Yorkhill, Glasgow G3 8SJ

Sudden infant death syndrome

Sudden infant death (SIDS), sometimes referred to as cot death, is a diagnosis of exclusion. Thankfully it has reduced significantly in recent years from more than 2 to less than 1 per 1000 live births. The peak age of incidence is between 2–3 months and it has a number of well-established risk factors including maternal smoking (the strongest established single factor), prone (face down) sleeping position, preterm delivery, socio-economic deprivation, male sex and winter season. Whereas careful post mortem examination is required to exclude organic disease it has been estimated that up to 5% may have an underlying metabolic cause and up to 10% may be due to infanticide by what has been termed 'gentle smothering'. The latter suggestion is based on case histories and in part on the pathological findings which almost invariably indicate an asphyxial end stage event.

Although the recent 'back to sleep' campaign has been associated with a welcome fall in incidence, the cause or causes remain unknown. It is likely that they are related to an interplay between the pace of development of cardio-respiratory and/or temperature control and environmental factors such as passive smoking and intercurrent viral infections. The sudden and unexpected death of an apparently normal and much loved infant is devastating for any family particularly as detailed enquiry is required by healthcare professionals and the police in order to exclude other possible causes of death. Families may find support offered by voluntary organizations and charities which support affected families and fund research into the causes of sudden and unexpected infant death.

Further reading

Chernick, V. (ed.) (1990). *Kendig's disorders of the respiratory tract in children*, (5th edn). W.B. Saunders, London.

Dinwiddie, R. (1990). *The diagnosis and management of paediatric respiratory disease.* Churchill Livingstone, Edinburgh.

Loughlin, G.M. and Eigen, H. (ed.) (1994). *Respiratory disease in children.* Williams & Wilkins, Baltimore, MD.

Phelan, P.P., Olinsky, A., and Robertson, C.F. (ed.) (1994). *Respiratory illness in children.* Blackwell, Oxford.

23

General surgical conditions

The appropriate surgical environment

The general surgical care of children falls into two categories:

1. *Specialist care*, which consists of neonatal surgery, correction of congenital anomalies and emergency surgery in young children. This type of care is provided by paediatric surgeons in regional referral centres.
2. *Non-specialist care*, which consists of surgery for common conditions in older children (e.g. appendicitis), inguino-scrotal conditions (e.g. hernia, undescended testis), and common problems related to trauma (e.g. head injury and limb fractures). This care is frequently performed by general surgeons in district general hospitals serving their local community. Trauma is covered elsewhere in this book (Chapter 9).

Irrespective of the setting it is important that surgery in children be carried out in an appropriate environment with facilities for parental attendance and accommodation, and the dedicated support services of anaesthesia, paediatric nursing, physiotherapy, and play therapy. Sometimes, surgery can be performed on a day-care basis. Such treatment requires careful preparation and high-quality anaesthetic care, since maximizing the use of regional anaesthesia as well as general anaesthesia gives optimal pain control.

Emergency surgery in children under five years of age is taxing, both diagnostically and therapeutically. Difficulties include communication, venous access, fluid management and pain relief. These challenges are best met by specialist paediatric surgical units. The significance of bile vomiting in early life and the many manifestations of intussusception in children of any age require familiarity with the conditions and confidence in handling young children in order to give high-quality care.

Familiarity with some of the commoner surgical conditions will avoid unnecessary intervention in many of these conditions which will often resolve with growth. The summary box illustrates the 10 commonest operations carried out in children in order of decreasing frequency.

Case studies
Phimosis and circumcision

Darren (age 2 years), was taken to the accident and emergency department by his parents with the abrupt onset of swelling and discoloration of his penis. He was mildly distressed when passing urine. In addition, there was a white discharge from the tip of his foreskin. A diagnosis of balanitis was made. Antibiotics were withheld and recommendations on perineal toilet with saline baths and high fluid intake given to the parents. Paracetamol was prescribed for the relief of symptoms.

Additional Information
The 10 commonest paediatric operations[*]

- Appendicectomy
- Orchidopexy
- Circumcision
- Herniotomy
- Fracture reduction
- Suture laceration
- Abscess drainage
- Gastrointestinal endoscopy
- Cystoscopy
- Hydrocele repair

[*] Excluding ear, nose, and throat, and ophthalmology

The extrusion of smegma (the white discharge) had allowed natural separation of the previously fused layers of the preputial epithelium and glans.

At subsequent follow-up appointment in the surgical out patient clinic 6 weeks later the swelling had gone. Forward traction on the foreskin showed a good lumen to the prepuce and although retraction still resulted in a tightened and narrow foreskin (phimosis) the lack of any scarring meant circumcision could be avoided.

A clear understanding of the natural history of the prepuce is essential if unnecessary circumcision is to be avoided. The prepuce is often conical and fused to the glans at birth. Complete retraction of the prepuce may not be accomplished until puberty. Retraction of the prepuce is not helpful in assessing the need for surgery and failure of retraction in itself does not point to a need for operation. Indeed, inappropriate and forcible retraction may cause scarring at the opening of the prepuce. Evaluation should therefore be by gentle forward traction of the prepuce which will demonstrate the opening of the epithelial lumen of the prepuce in most boys.

Ballooning of the prepuce during voiding represents redundancy of penile skin and is not in itself a reason for circumcision. Scarring at the tip of the prepuce, with failure to dilate on forward traction, indicates the presence of the rare condition of balanitis xeroderma obliterans, which is a clear indication for surgery. There is also a good case for pre-emptive circumcision to reduce urinary tract infection in boys with dilatation of the upper urinary tract.

While several religions advocate male circumcision as part of their belief, there is no evidence for the value of routine male circumcision in promoting health.

The summary box shows a list of other conditions where natural resolution can be expected with time and unnecessary operation be avoided. However, some other conditions require carefully timed intervention if an optimal outcome is to be obtained (Table 23.1). In particular, children with inguinal hernias require prompt operation and should not be left on a waiting list.

Summary
Conditions that naturally resolve

- Labial fusion
- Tongue tie
- Hydrocele in infants
- Umbilical hernia
- Divarication of recti muscles

Table 23.1.
Conditions requiring prompt elective surgery

Condition	Age at operation
Inguinal hernia	On diagnosis
Undescended testis	Before age 3
Supra-umbilical hernia	After age 1
Epigastric hernia	After age 1
Communicating hydrocele	After age 3

Inguinal hernia and hydrocele

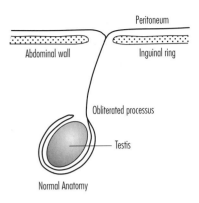

Two days after his first immunization Andrew (age 8 weeks), born at just 34 weeks' gestation, developed a hard lump in his right groin. He vomited his next feed of breast milk and refused to settle following the feed. An emergency visit from his general practitioner resulted in admission to the regional paediatric surgical unit.

The processus vaginalis, which allows descent of the testis *in utero*, normally disappears following birth. If patency of the processus provides a substantial channel then intra-abdominal structures (omentum, intestine, ovary) may herniate down it. If these structures become incarcerated, they may become ischaemic or 'strangulate' (Fig. 23.1).

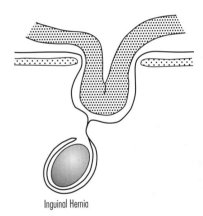

On admission, his incarcerated inguinal hernia was reduced by the surgeon by manipulation. Andrew was observed for a further 24 hours, during which some swelling of his right testis developed. This was considered to be due to transient ischaemia consequent to spermatic cord compression by the hernia.

The elevated pressure within the inguinal canal may compromise blood flow to the testis which can result in testicular atrophy.

The following day Andrew underwent herniotomy on the elective operating list and after an uneventful period of 24 hours, when he was monitored for post operative apnoea, he was discharged.

Repair of the hernia by obliteration of the processus with a suture at the inguinal canal (herniotomy) is best done as an elective procedure soon after the diagnosis is made.

If the processus vaginalis remains only narrowly patent, peritoneal fluid can escape through it and cause scrotal swelling. This congenital hydrocele can be diagnosed by transillumination and the parents can be told that it will probably disappear as the processus closes in infancy or early childhood, since, unlike inguinal hernias, hydroceles often close spontaneously. Persistence of the hydrocele beyond 3 years of age, particularly if associated with diurnal change in size (communicating hydrocele), requires repair by division of the processus vaginalis in the inguinal canal.

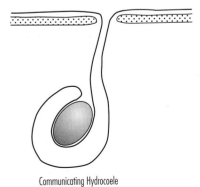

Fig. 23.1. Normal anatomy of the obliterated processus vaginalis (a) with the anatomy of inguinal hernia (b), and communicating hydrocele (c).

Undescended testis

Matthew was found at his neonatal examination to have an underdeveloped right hemiscrotum. Although the left testicle was easily palpable no testis was palpable on the right side.

His parents need to know that some undescended testes come down during the first 3 months, but that if the testis is not down by then, it will need surgery to place it in the scrotum. There are good arguments for screening all boys for undescended testicles either at birth, as here, or at the 6–8 week check, (see Ch. 4, p. 76).

Arrangements were made for review at the surgical clinic at 6 months of age and on examination at this time a right testis was palpable in the inguinal canal. Surgery was carried out in a day-care setting when he was 14 months old. At operation the right testis was found to be within a right inguinal hernia. The testis appeared otherwise normal and following herniotomy and mobilization of the spermatic vessels the right testis was taken down into a dartos pouch (see below). At follow-up 6 weeks later the testis was viable, firmly located in the right hemiscrotum.

Summary
Undescended testis

- If found at birth, check at 3 months (90% descent)
- if persistent, should be placed in scrotum by 3 years
- late risks include infertility, malignancy, and torsion

The testis, although intra-abdominal for most of the fetal life, reaches the groin by the 8th week of pregnancy and will enter the scrotum in the following month. About 1 boy in 14 will have an undescended testis at birth, but 90% of these testes will have descended by 3 months. Routine neonatal examination should find most cases of failure of testicular descent, but some children are not diagnosed until much later. Persistent failure of testicular descent beyond 3 years of age has been associated with infertility, torsion of the testis, and subsequent malignancy.

Surgical correction is achieved by the operation of orchidopexy which is best performed in the second postnatal year. The testis is placed in a retaining cavity created in the scrotal wall (the dartos pouch).

Descent can also be complicated by an anomalous final location, most commonly in a superficial inguinal pouch. This is known as an ectopic testis, and the main complication is torsion.

When the testis is impalpable it must be found, usually by laparoscopy, so that appropriate surgery can be planned. If no testes can be palpated in an apparent male, careful assessment of the genitalia, involvement of a paediatric endocrinologist, and determination of the karyotype for the sex chromosomes is required, (see Ch. 16, p. 286).

Scrotal pain

When a boy develops acute onset of pain in the scrotum, there are several possible causes. Epididymo-orchitis is usually secondary to either urinary tract infection or mumps. However it is concern about testicular torsion which prompts surgical exploration of the scrotum in most boys with this symptom.

Diagnosis is helped by ultrasound of the testis with doppler measurement of testicular blood flow, urine analysis, and sometimes isotope imaging. When clinical doubt remains it is safest to explore the scrotum surgically.

Testicular torsion arises from complete investment of the testis by the tunica vaginalis (see Fig. 23.1). Since this anatomical anomaly is common to both testes, the procedure involves untwisting the affected testis and fixing both testes within the scrotum, using non-absorbable sutures, to prevent recurrence.

A confident diagnosis of torsion of hydatid of Morgagni (a structure just above and behind the testis) can allow conservative management with analgesics, while a diagnosis of epididymo-orchitis implies the need for future imaging of the urinary tract a few weeks after the infection has settled, in order to exclude a structural problem predisposing to infection.

Acute abdominal pain (see also Ch. 14, p. 254)

John (age 10 years) woke his parents at midnight with severe abdominal pain which he could not localize and which came and went over a period of several minutes. It did not get better when he went to the lavatory. After 40 minutes during which it seemed if anything to be getting worse, his parents called out their GP. She examined him and because of tenderness in the renal angle diagnosed urinary tract infection with possible pyelonephritis. Simple analgesia (paracetamol) and an increased fluid intake was advised, and she arranged for him to have a urine sample tested in the morning with subsequent antibiotic treatment if necessary.

Abdominal pain is a symptom commonly reported by children. It is most often managed at home with or without the aid of a doctor. Whilst it may indicate intra-abdominal pathology it is also likely, particularly in the younger child, to be a symptom of a systemic upset. A variety of conditions may present as abdominal pain.

John's urine proved to be negative for blood and was sterile. An abdominal X-ray showed no abnormality other than that the whole large bowel was full of faeces, strongly suggesting that constipation was his problem. Treatment was with a stool softener (lactulose) and a motility enhancer (senna), and he remained symptom-free thereafter.

Summary
The differential diagnosis of acute scrotal pain

- Torsion of the testicular appendages (hydatid of Morgagni)
- Epididymo-orchitis
- Idiopathic scrotal oedema
- Testicular torsion

Summary
Conditions that may present with acute abdominal pain*

- Constipation
- Urinary tract infection
- Gastroenteritis
- Intussusception
- Renal stone
- Pneumonia
- Otitis media
- Diabetes mellitus

* Excluding appendicitis

Summary
Serious symptoms and signs

- Unremitting pain for 6 hours
- Bile vomiting
- Dehydration
- Abdominal mass
- Rectal bleeding
- Signs of peritonitis

Only 2 children in 500 will develop abdominal pain sufficiently severe to warrant hospital admission in any one-year period. Of these, one will have an operation for the condition; the other will not. There are a number of *serious symptoms and signs* which point to the need for evaluation in hospital.

Whilst acute appendicitis is the commonest single condition requiring surgery, more often than not the pain will subside spontaneously without a diagnosis being made. Such children may be said to be suffering from non-specific abdominal pain (NSAP), and they are usually better within 24 hours. In infancy, the differential diagnosis of abdominal pain should include gastro-oesophageal reflux and oesophagitis. The cardinal signs of peritonitis are often difficult to elicit in very young children. (For recurrent abdominal pain see Ch. 14, p. 254).

Appendicitis

At 10 a.m. Colin (age 8), told his primary schoolteacher that he had a tummy ache, and following a vomit 1 hour later, he was sent home. His mother took him to the GP who noted that he was tender on palpation all over (though Colin said it was worst around his umbilicus) and had him admitted to hospital. By this time, his pain had moved to the right lower quadrant and the admitting house officer recognized tenderness and guarding in the lower abdomen. He was reviewed by the surgical registrar who prescribed intravenous fluids because of the vomiting, but being uncertain of the diagnosis decided to review him 2 hours later, according to the unit policy of active observation.

Although appendicitis is less common than it was, suspicion of this diagnosis is still the commonest cause for proceeding to surgery in a child with abdominal pain. Those children in whom the decision is most difficult are those under the age of 3 years, because their less well-developed omentum reduces their ability to confine the spread of inflammation and pus within the abdomen, and their ability to localize their pain is very poorly developed.

Clinical diagnosis is by active observation which can sometimes be assisted by a white cell count, and ultrasound of the abdomen to examine for other possible pathologies. Dehydration requires correction before surgery.

At review the clinical signs of appendicitis were evident. Appendicectomy was carried out together with lavage of the peritoneal cavity using a solution of cephalosporin antibiotic. Prophylactic antibiotics were also given systemically to provide protection against wound infection by bowel organisms including *E. coli*, *Strep. milleri*, and *Bacteroides* species. Pain relief was given by intra-operative injection of bupivacaine into the regional nerves, and postoperatively he received diclofenac suppositories. Colin was well enough to start drinking fluids the following day and was discharged home on the third day after his admission.

It is usual to give perioperative antibiotics at the time of the appendicectomy, and important to give adequate attention to pain relief. Children recover much more quickly than adults from this operation.

Rectal bleeding

Amy (age 2 years), was brought up to the accident and emergency department by her worried 17-year- old single mother. She had seen 'lots of fresh blood on Amy's nappy when she was last changed. Going further into the story it was clear that Amy often passed a rather hard stool, at which she strained, and that her stool frequency could be as little as twice a week.

This history fits very well for an anal fissure which is one of the most common causes of lower gastrointestinal bleeding in childhood. Examination is directed at visualizing the fissure if possible, but it is also important to look for evidence of physical or sexual abuse. Initial treatment is with a stool softener.

Rectal bleeding may be associated with diarrhoea, suggesting an infectious cause such as *Campylobacter* or an inflammatory condition such as Crohn's disease or ulcerative colitis. There may be no other symptoms apart from blood in the toilet bowl if the cause is a juvenile polyp. Bleeding may be associated with acute abdominal symptoms in volvulus or intussusception, or be an unexpected finding on the nappy or in the toilet bowl. In infancy, cow's milk intolerance can cause a haemorrhagic colitis (see also p. 252), and in the newborn necrotizing enterocolitis must be considered.

Other abdominal emergencies in children
Intussusception

This typically occurs between 2 months and 2 years of age as a consequence of lymphoid hyperplasia following the introduction of solid foods or recent gut infection. Although lead points (e.g. a Meckel's diverticulum) may precipitate intussusception, hyperplasia of Peyer's patches can cause the ileocaecal region to prolapse through the colonic lumen (ileocolic intussusception, Fig. 23.2).

Summary
Some common or important causes of lower gastrointestinal bleeding

- *Neonate*
  Necrotizing enterocolitis
  Volvulus

- *Infant*
  Anal fissure
  Cow's milk intolerance
  Intussusception

- *Child*
  Juvenile polyp
  Anal fissure
  Campylobacter, Shigella
  Crohn's disease .
  Ulcerative colitis

Summary
Intussusception

- A disease of infants and toddlers

- Gives severe intermittent abdominal pain

- The child may pass blood and mucus (a 'red currant jelly' stool)

- There may be a sausage-shaped abdominal mass

- May be reduced under fluoroscopic control with a contrast enema

- Carries an appreciable mortality if the diagnosis is delayed

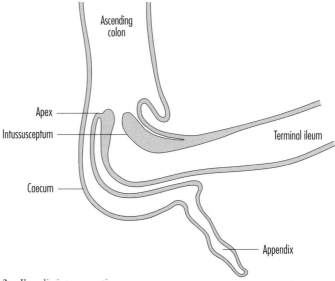

Fig. 23.2. Ileocolic intussusception.

Malrotation

This occurs when there is a lack of fixity of the right colon due to disordered return of the hindgut from the yolk sac into the abdomen in the first trimester of pregnancy. This 'malrotation' of the colon and ileum may allow the midgut to form a volvulus around the superior mesenteric vessels. Bile vomiting is an early presentation of this condition, which can occur in the neonate or later in childhood. If uncorrected, it will progress to ischaemic necrosis of the gut, and death. Diagnosis is made by contrast study of the gut or ultrasound evaluation of the superior mesenteric axis. Surgical correction places the bowel in a state of nonrotation (i.e. colon in the left side of the abdomen and small bowel on the right side). This position is stable and prevents further episodes of volvulus.

Further reading

Nixon, A., O'Donnell, B. (1992). *Essentials of paediatric surgery*, 4th Ed. Butterworth, Oxford.

Morton, N.S., Raine, P.A.M. (1994). *Paediatric day case surgery*. Oxford University Press.

24

Looking forward to adult life

- Definitions

- Legal aspects

- Hazards of adolescence

- Sexual health

- Case study
  coping with chronic illness
  health service needs

- Transition to 'adult' care

Definitions

Transition from childhood to adult life (adolescence) has a broad range of definitions. In North America it has developed as a medical specialty in its own right within an age band spanning 15–25 years. It encompasses a period of life when physical, psychological and sociological maturity, and independence are intermingled. The North American definition is one of convenience as it recognizes the transition from paediatricians who care for families and children to the physicians and surgeons of adulthood. Physical maturation becomes established in girls at about the age of 12 with the onset of menarche or first menstrual period at approximately 13 years of age, although this has a wide variation from 9 to 18 years. There is no such clear-cut identifier in boys but peak height velocity, which is associated with sexual maturation, occurs at around the age of 12 in girls and approximately 2 years later in boys. This gender difference has important implications as girls not only mature physically in advance of boys but they also tend to achieve adult attributes ahead of boys. This is reflected in the fact that young women leave the family home and establish their independence several years ahead of their male counterparts.

Legal aspects

Ratification of the *UN Convention on the Rights of the Child* by the UK government in 1991 has been reflected in national legislation and has significant consequences for parents, young people, and their carers. The Children Act (England and Wales 1989, Scotland 1995) enshrines some of the rights of young people and responsibility of their carers. This legislation supports previous legislation on the age of majority. Although 18 years is commonly regarded as the age of achieving full adult rights and responsibilities, children of any age able to understand what is being proposed for them, whether it be medical, surgical treatment, or custody arrangements, must be involved in the decision-making process. The age at which a child must be included in these important decisions does not have a lower age limit but certainly by the age of 12 years, children and young people need to be involved in consent for any medical or surgical interventions proposed for them. Confidentiality must also be maintained unless there are overriding reasons why this should, not be so. Young people have the right to have confidential information about them withheld from their parents/guardians if they so desire. This has been tested in the English courts for contraceptive advice in young people under 16 years of age (see Chapter 1, p. 22, Ch. 4, p. 70).

Hazards of adolescence

Growing up involves the adolescent in a variety of learning experiences, experimentation, 'testing the boundaries' of rules and accepted social practices. A desire to take part in 'risk-taking' is normal in the transition to adulthood. Some of the most common reasons for admission to hospital at this age are attributable to behavioural factors rather than any disease process. In males, the most

Summary
Adolescence

- Age range 12–25 years
- Physical and psychological maturation occurs earlier in girls
- A period of physical change and emotional turbulence

Fig. 24.1.   Teenagers in a group.

Summary
Characteristics of adolescence

- Changing physiology
- Establishing indepedence
- Experimentation and risk-taking
- Setting of lifetime patterns (e.g. smoking, sexual behaviour)
- Developing self-esteem and value system

Additional Information
Themes from the UN Convention
on the Rights of the Child

- Best interests of the child to be considered particularly in family disputes
- Standards of services and facilities
- Respect and support for parents/carers
- Seek the child's view
- Maintain privacy and confidentiality
- Provide access to relevant information
- Protect from violence
- A right to education
- Rights of disabled
- Rights of minorities
- (see also Ch. 1, p. 22)

common reason for hospital admission is head injuries and open wounds; deliberate self-poisoning increases with age. A third of all legal abortions are in females under the age of 20; and young people between 16 and 24 years of age account for almost 60% of driver fatalities while they represent only 20% of licensed drivers.

In addition to these demands on health services by young people the development of various lifestyles can have important implications for emerging adult patterns of living across the life span as the evidence about smoking, drinking, diet, and physical inactivity reveals. Young people with a chronic health problem such as asthma or diabetes may be particularly 'at risk' in adolescence by taking on embryonic forms of certain aspects of lifestyle such as smoking or drug use in association with their peers or in order to imitate an admired adult role model or to protest against parental values. Attempting to prevent the development of unhealthy lifestyles in adolescence may be the best way of reducing a number of health risks in adult life, particularly for those with a chronic illness.

A number of models of adolescent behaviour have been proposed including the 'focal' theory of Coleman (1990) which suggests that concern about gender role peaks around the age of 13 years, concerns about acceptance or rejection by peers become more important at around 15, while issues regarding the gaining of independence from parents climbs steadily to peak beyond 16 and then tail off towards the early to mid 20s (Fig. 24.2). The fact that adolescents do not usually cope with these crises at the same time but meet them sequentially may provide some resolution of the paradoxes of the huge amount of disruption and crisis implicit in adolescence and yet the relatively successful adaptation to adulthood and maturity in the majority.

In young people with the additional burden of chronic illness attention needs to be given to their and their parents'/guardians' coping skills. A recurrent theme is the need to establish independence while helping parents come to terms with 'the loss' of their dependent sick child for whom they have hitherto taken full responsibility. Compliance with treatment is also a major concern as failure to maintain a regular therapy can have serious immediate and long-term consequences for the health of the affected individual.

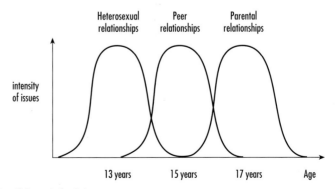

Fig. 24.2. Coleman's focal theory.

Sexual health

There is no doubt that this area of human activity is a major challenge for young people, their parents, and carers. The World Health Organization has defined sexual health as 'the integration of somatic, intellectual, emotional and social aspects of sexual being in ways that are positively enriching and that enhance personality, communication and love'. In making the transition to full adult maturity many pitfalls await the adolescent. Social pressures on young people are enormous with messages from the media portraying sex as glamorous and abstinence and contraception as dull and boring. Partner pressure may present the adolescent with challenges he or she is unable to deal with. Recent surveys from Exeter in the South West of England have revealed that approximately half of all young people are sexually experienced by the age of 17 with 40% having engaged in sexual intercourse before the legal age of 16. Young people are not only at risk of their own experimentation and risk-taking but they are also at risk of exploitation. This further emphasizes their need for principles of accessibility and confidentiality enshrined in the UN convention. Sexual abusers can take advantage of the insecurity and need for acceptance of young people (see Chapter 5, p. 94).

Case study
Coping with chronic illness

Anita (age 13), was first diagnosed with cystic fibrosis after she presented with rectal prolapse and failure to thrive at 11 months of age. She has a brother John who is two years older than her, and her parents made the decision not to have other children after the diagnosis was established. Anita has maintained normal growth and development on her pancreatic supplements and has had three hospital admissions with respiratory exacerbations, one aged 7, one aged 11, and one 6 months previously. She acquired chronic pseudomonas carriage in her sputum between the first and second admission. She has moderately severe disease with an FEV1 of 70% predicted and no chest deformity. She is a regular sputum producer. She is a tall slim girl, height on the 75th centile, weight on the 25th. She is well established in puberty, breast stage 3, but she has not started menstruating yet. Her parents have both been very involved in her treatment with most of the twice-daily physiotherapy by postural drainage and percussion being performed by her mother. Her father is a police sergeant and her mother, who was in the police force when Anita was born, opted to work full-time at home in order to ensure that Anita has the best possible care. Both parents are active members of the local Cystic Fibrosis Group.

Anita has a life-threatening disease and her parents are fully aware of the published data suggesting that Anita might be expected to survive into her thirties or forties. They have a keen interest in recent genetic advances and are anxiously awaiting the introduction of gene therapy or gene replacement. They hope, along with other cystic fibrosis parents, that these advances may extend Anita's life well beyond her expected survival. In the meantime their objective is to keep her as well as possible with currently available treatments for which they have hitherto been largely responsible. All responsible parents invest huge amounts of time and emotional energy into child rearing but these efforts are redoubled by the presence of chronic disease and illness, particularly a life-threatening one such as cystic fibrosis. For most families living in industrialized countries the threat of chronic illness and risk of premature death is not an issue thus making the burden of serious chronic illness harder to bear and accept. Most parents experience feelings of loss at the passing of childhood and the changes in their children as they make the transition to adult life. A common challenge is to achieve a balance between acceptance of the illness by the child and their parents and at the other extreme the denial of the severity of the disease and resultant harm by failing to comply with best possible medical treatments.

Health service needs

Summary
Health service needs of adolescents

- Collaborative not directive
- Confidentiality
- Managed transfer to adult services

At Anita's next clinic visit both parents come with her as is their usual practice. Anita appears uncharacteristically remote and does not take part in the interview. There appears to be some irritation by her mother at Anita's persistent rudeness. Dr Jones suggests to Anita and her parents that perhaps Anita is reaching an age when she should be seen for al least part of the visit on her own. Anita's mother seems somewhat hurt and perplexed at this suggestion but she agrees. Anita becomes more forthcoming and communicative when seen for 5 minutes on her own and it is agreed that in future she will come into the consulting room at the beginning of the interview and then invite her parents into the room to discuss the current situation and plans for the future. Anita is also taught how to perform chest physiotherapy herself using a forced expiration technique reserving the traditional postural drainage and percussion (requiring another person to perform the treatment) for significant respiratory exacerbations

Transition to 'adult' care

The above pattern of transitional care is commonly adopted for children in their early teens but is the first step towards full independence and regular follow-up in an adult-oriented service. Transfer to the adult service would traditionally take place somewhere between 16 and 18 years of age. Parents often find it difficult to

Fig. 24.3. Adolescent with doctor, parent looks on.

separate from their children particularly when they have hitherto been largely responsible for their medical management. Unless responsibility for day-to-day management is transferred to young people there is a danger that the medical management itself may become part of the natural testing of boundaries and need to establish independence. In a life-threatening illness, such as cystic fibrosis, this compliance with therapy can quite literally mean the difference between early death and extended survival in relatively good health. Parents are aware of this fact but as with many young people present behaviour is not often linked to long-term consequences and the young person him or herself therefore has a rather different view. Peer pressure and the need to be accepted as 'one of the crowd' are other influences that are particularly strong in girls and young women. Having to take up to 10–12 pancreatic enzyme capsules with meals draws attention to yourself and constant coughing and sputum production (a feature of the disease) are reasons often cited by young people for poor compliance with regular physiotherapy. Establishing trust between health care professionals and young people and transferring responsibility for disease management and a healthy lifestyle to young people themselves are prerequisites for long-term future health, not only in the presence of chronic illness but also in those without health problems.

Additional Information
Confidentiality and privacy

We (medical practitioners) will

- See you on your own, in private, if that is what you want

- Always keep what you want to tell us confidential (unless there is a very good reason why we should not). If we do have to tell someone else we will tell you who and why

- Give you all the support you need if we are going to tell someone else

- Make sure all our staff know about these rules on confidentiality

From *Child health rights* (1995)

Further reading

Brook, C.G.D. (ed.) (1993). *The practice of medicine in adolescence*. Edward Arnold, London.

Coleman, J.C. and Hendry, L.B. (1990). *The nature of adolescence*. Routledge, London.

McFarlane, A. and McPherson, A. (1996). *The new diary of a teenage health freak*. Oxford University Press.

McFarlane, A. (ed.) (1996). *Adolescent medicine*. Royal College of Physicians, London.

Child health rights. A practitioners guide (1995). British Association of Community Child Health c/o Royal College of Paediatrics and Child Health, London.

Further reading in paediatrics

Textbooks
Large general books for general paediatrics

Nelson's *Textbook of paediatrics*. Philadelphia, Saunders (1992).
This is the main American reference book.

Forfar and Arneil's *Texbook of paediatrics*. Edited by Campbell, A.G.M., McIntosh, N. Churchill Livingstone (1992).

For neonatology

Roberton, N.R.C. *Textbook of neonatology.* Churchill Livingstone (1992).

For community child health:

Polnay, L. and Hall, D. *Community paediatrics.* Churchill Livingstone (1993).

Journals
General

British Medical Journal; *Lancet*; *New England Journal of Medicine*

All of this publish important papers relating to child health issues which may be of wider interest than a specialist paediatric readership.

Paediatrics/Child Health

Archives of Disease in Childhood
This is the main British journal, published by the Royal College of Paediatrics and Child Health.
It has a fetal and neonatal supplement every two months.

Journal of Pediatrics
Pediatrics
These are the two main American Journals. Others are: *American Journal of Diseases in Children*, *Annals of the American Academy of Pediatrics*
Acta Paediatrica
Formerly *Acta Paediatrica Scandanavica*. This is the premier European journal.

Biology of the Neonate
Early Human Development
These two concentrate on the neonatal period.

Pediatric Research
This is the journal of the Society for Pediatric Research, and has tended to concentrate on basic rather than clinical research.

Ambulatory Child Health
A new quarterly journal which is published in both the UK and USA and covers primary care paediatrics, social paediatrics and prevention.
Highly innovative approach.

Developmental Medicine and Child Neurology.
Excellent monthly journal which covers general topics as well as neurology and disability, basic science, and clinical practice.

Review Journals

Examples are:
Pediatric Clinics of North America
Clinics in Perinatology
Current Opinion in Paediatrics
There are several other such review journals.

Series

Ballière's *Clinical Paediatrics*
Recent Advances in Paediatrics
Year Book of Pediatrics

Index